Advances in LIMNOLOGY

Advances in LIMNOLOGY

Smriti Ratna Mishra

2025

Daya Publishing House®

A Division of

Astral International Pvt. Ltd.

New Delhi - 110 002

First Published, 2005
Reprinted, 2025

ISBN: 978-81-7035-255-6 (Hardbound)

Published by : **Daya Publishing House**®
A Division of
Astral International Pvt. Ltd.
– ISO 9001:2015 Certified Company –
4736/23, Ansari Road, Darya Ganj
New Delhi-110 002
Ph. 011-43549197, 23278134
E-mail: info@astralint.com
Website: www.astralint.com

Laser Typesetting : **Classic Computer Services**
Delhi - 110 035

Printed at : **Replika Press Pvt. Ltd.**

PRINTED IN INDIA

Dedicated to my

Respected Mother

Smt. Yaduraj Kumari Mishra

– **Smriti**

Preface

The study of organisms in relation to fresh water habitat constitutes of freshwater ecology whereas chemical, geological, physical and biological aspects come under the term Limnology. The subject matter of limnology derives knowledge from varied disciplines like physics, chemistry, biology, geology, geography etc., and involves a great deal of detailed field and laboratory studies to understand the structural and functional aspects of the fresh water environment from a holistic point of view. Although, the term "Limnology" was originally applied to the study of lakes only, in its current use it also refers to the study of streams. The term stream is used to indicate any mass of water with unidirectional flow, mountain brooks, spring brooks, creeks and rivers.

Limnology now also includes the study of running water as well as that of standing water and has attracted a lot of attention during the recent years. This volume is the outcome of a collaborative effort of the researchers and scientists in this discipline. Thus presenting a wider perspective along with up-to-date information on various aspects of Limnological Research in India.

I wish to express my deep sense of gratitude to all those who have contributed to this volume. My respected mother has always been a constant source of inspiration for me. I pay my sincere regards to her. I thank my wife Seema, for her help in various ways.

Dr. Smriti Ratna Mishra

List of Contributors

Archarekar, R.N.

Department of Botany, Institute of Sciences, Mumbai – 400 032 (Mh.)

Banerjee, Samir

Aquaculture Research Unit, Department of Zoology, University of Kolkata, Kolkata – 700 019 (W.B.)

Chakravarty, Satish K.

Government M.V.M., Bhopal (M.P.)

Chitra, K.Y.

Cell and Molecular Biology Laboratory, Department of Zoology, Nizam College, Basheerbagh, Hyderabad – 500 001 (A.P.)

Fokmare, Anil K.

P.G. Department of Microbiology, Shri Shivaji College of Arts, Commerce and Science, Akola – 444 001 (Mh.)

Ghosh, Mrinalkanti

Aquaculture Research Unit, Department of Zoology, University of Kolkata, Kolkata – 700 019 (W.B.)

Hussain Zahir, M.I.

Manonmaniam Sundarnar University, Shri Pramakalyani Centre for Environmental Science, Alwarkurichi – 627 412 (T.N.)

Jain, D.S.

Department of Botany, Arts, Commerce and Science College, Nagaon, District Dhule – 142 4004 (Mh.)

Jain, Pradeep K.

P.G. Department of Geology, Government Maharaja College, Chhatarpur – 471 001 (M.P.)

Khare, Pushpendra K.

Department of Botany, Government Autonomous Maharaja College, Chhatarpur – 471 001 (M.P.)

Kumari, Anitha S.

Department of Zoology, Nizam College, Basheerbagh, Hyderabad – 500 001 (A.P.)

Md. Babar

Department of Geology. Dnyanopasak College, Parbhani – 431 401 (Mh.)

Mishra, Smriti Ratna

93/2, Trimurti Nagar, Dhar District Dhar – 454 001 (M.P.)

Murugesan, A.G.

A.G. Manonmaniam Sundarnar University, Shri Pramakalyani Centre for Environmental Science, Alwarkurichi – 627 412 (T.N.)

Musaddiq, Mohammad

P.G. Department of Microbiology, Shri Shivaji College of Arts, Commerce and Science, Akola – 44400 (Mh.)

Nandan, S.N.

P.G. Department of Botany, S.S.V.P.S.L.K., P.R. Ghogrey Science College, Dhule – 424 005 (Mh.)

Pradhan, B.

Department of Chemistry, N.I.T., Rourkela – 769 008 (Orissa)

Ruby, John

A.G. Manonmaniam Sundarnar University, Shri Pramakalyani Centre for Environmental Science, Alwarkurichi – 627 412 (T.N.)

Sakhare, V.B.

Department of Zoology/Fishery Science, Yashwantrao Chavan College, Tuljapur – 413 601 (Mh.)

Salgare, S.A.

Department of Botany, Institute of Science, Mumbai – 400032 (Mh.)

Samal, S.K.

Department of Chemistry, N.I.T., Rourkela – 769 008 (Orissa)

Shastry, Yogesh

Department of Botany, S.P.H. Mahila Mahavidyalaya, Malegaon Camp, District Nasik – 429 105 (Mh.)

Shrivastava, Anamika

P.G. Department of Geology, Government Maharaja College, Chhatarpur – 471 001 (M.P.)

Sree Ramkumar, N.

Cell and Molecular Biology Laboratory, Department of Zoology, Nizam College, Basheerbagh, Hyderabad – 500 001 (A.P.)

Sukumaran, N.

A.G. Manonmaniam Sundarnar University, Shri Pramakalyani Centre for Environmental Science, Alwarkurichi – 627412 (T.N.)

Tiwari, D.R.

Government M.V.M., Bhopal (M.P.)

Tiwari, T.N.

Department of Physics, N.I.T., Rourkela – 769008 (Orissa)

Verma, Ashok

Department of Zoology, Dr. Shyama Prasad Mukherjee Government Degree College, Phaphamau, Allahabad (U.P.)

Yadav, B.S.

Arts, Science and Commerce College, Nampur Tal Baglan, District-Nasik (Mh.)

Contents

1

Zooplankton and their Seasonal Variations in a Sewage Collecting River at Gwalior, Madhya Pradesh

☆ *Smriti Ratna Mishra*

Introduction

Though considered as transients of the river Bista by many aquatic biologists, the zooplankton not only form an integral part of the lotic community but also contribute significantly to their secondary biological productivity. The zooplankton of running waters have also been termed as rheozooplankton. The rheozooplankton have not been studied so well as compared to limnozooplankton and heleozooplankton. The information regarding zooplankton of the polluted river water is still scanty. The zooplanktonic organisms showed a reduction in their number in septic zones in Suvaon stream (Banerjea and Motwani, 1960). The rotifer zooplanktons of the river Sokoto in North Nigeria have been studied by Green (1960). Chacko and Rajagopal (1962) have studied rotifers, ostracods, and copepods of Ennore river, Madras, Cronin *et al.* (1962) made a study of quantitative seasonal fluctuation in zooplankton of Delaware river estuary. Zooplankton of Kali river were studied in relation to pollution by George *et al.* (1966) while Ray and David (1966) studied zooplankton of river Ganga at Kanpur. Haertel and Osterberg (1967) described the ecology of zooplankton, benthos and fishes in the Columbia river estuary. The rotifers, ciliates and nematodes in river Sabarmati were studied by Venkateswarlu and Jayanti (1968). Smet and Evens (1972) have recorded various species under Rhizopoda, Ciliata, Rotatoria, Cladocera and Copepoda in polluted Lieve river. In a sewage fed pond in West Bengal, Ghosh *et al.* (1974) found *Brachionus* and *Filinia* as dominant forms in their collections. Relatively smaller zooplanktonic population was observed in polluted Kali river (Verma and Dalela, 1975). In Jamuna river at Agra, Prakash *et al.* (1978) recorded 5 species among Protozoa, 12 species among Rotifera and 7 planktonic species of oligochaetes. Similarly, in Jhelum river 23 species of zooplankton belonging to different groups namely Protozoa, Rotifera, Cladocera and Copepoda were identified by Vass *et al.* (1977). *Arcella, Paramaecium, Phacus, Epistylis* and *Nebela* under Protozoa, *Filinia terminalis* and *Brachionus quadridentata* under Rotifera and *Moina brachiata* and *Cyclops* under Entomostraca have been reported from Kadarabad drain (Verma *et al.*, 1978). The zooplankton of river Betwa in Madhya Pradesh were studied by Adholia (1979). Sampath *et al.* (1979) identified 35 genera and species of rotifers from Cauvery river and designated certain rotifers as biological indicator of water quality. The synecological study of Rotifera in the river Murray in South Australia was carried out by Shiel (1979). The rotifer fauna of river Loire in France at the level of the nuclear power plants was observed by Lair (1980), Bilgrami (1984) designated *Moina* and *Brachionus quadridentata* as bioindicator of pollution in river Ganga. The life history and ecological significance of stream dwelling copepods have been considered by O'Doherty (1985). Patil and Rodgi (1985) have recorded, protozoans, rotifers, ostracods and copepods under zooplankton of municipal waste water with predominance of protozoans. The composition of the rotifera fauna in Morar and Swarnarekha rivers at Gwalior have been described by Saksena and Kulkarni (1986). In the river Reh in Doon valley, Shanker *et al.* (1986) identified 8 species of Protozoa, 5 species of Rotifera, 2 species of Copepoda and 3 species of Cladocera. The rotifer population was minimum throughout the period of study in river, location Ib (Kar *et al.*, 1987). Palharya and Malviya (1988), identified 9 species of zooplankton at various stations established in Narmada river. Studies on zooplankton in relation to water quality of river Kshipra in Madhya Pradesh have been carried out by Kulshrestha *et al.* (1989). The zooplankton of industrial waste water and their seasonal variations at Birla Nagar, Gwalior was studied by Mishra and Saksena (1990). Sai Sastry (2002) had studied pre-pollution status of composition, diel and tidal fluctuations of plankton in the Vasishta Godavary estuary.

In the recent years, attention has been paid to the diversity in population of the living organisms and this diversity has been considered as a measure of water quality. This is based on the principle that in clean waters community diversity is high while in the polluted waters diversity is low (Archibald, 1972). Several workers like Wilhm (1967) and Wilhm and Dorris (1968) employed mathematical expressions of diversity developed by Patten (1962) in monitoring of water pollution. Shannon and Weaver (1963) translated the information theory into ecological terms and introduced the concept of diversity index as a potent tool for studying biological communities. Trauben and Olive (1983) assessed the water quality of Cuyahoga river in Ohio and found low diversity values at polluted sites. Verma *et al.* (1984) had carried out the investigation on the pollution and saprobic status of Eastern Kalinadi and found that the values of Shannon-Weaver index had decreased below the waste outfall. Rao and Shrivastava (1989) conducted a study on biological monitoring of water quality in Chambal and Khan rivers in Central India. In the present study, the zooplanktonic fauna and its seasonal variations in sewage collecting river namely, Morar river had been considered. Shannon-Weaver diversity index was also computed for the zooplankton species observed at various stations in the river to monitor the water quality biologically.

The River and the Sampling Stations

Morar river originates from the forest area near Antri village and flows for about 25 km into the Gwalior district. Obstructed by one dam and two stop dams which are built across the river. Ramaua dam is in the upstream and was constructed in 1956 under an irrigational project. Near Mehra village, *i.e.* approximately 5 km away from the reservoir, it enters in the subcity of Morar. At the outskirts of Morar subcity a stop dam namely, Morar (Jaderua) dam makes a small reservoir from which a canal, (*Behta canal*), takes water to crop fields. This stop dam is much older and was completed in 1917. From Morar dam, the river flows down for about 16.20 km to enter in Bhind district near Bahadurpur. Guthina dam is also situated over this river in the Gwalior district which is about 5 km upstream from Bahadurpur. The river is perennial in most of its stretch Gwalior. The Morar river joins Besli river near Gohad in Bhind district. This river was very important from the fisheries viewpoint as (a fingerling collection centre) was located on this river near Gwalior. The collection of fingerlings of *Catla catla, Cirrhinus mrigala, Labeo rohita, L. gonius* and *L. calbasu* was in practice since long at this centre.

The stretch of Morar river from Mehra village to Morar dam is about 10.25 km which has been selected for the present study. Lot of human activity and disturbances take place in this river right from Mehra village where bathing, cloth washing, cattle grazing etc. are very commonly done. As this river flows through Morar subcity, it receives municipal sewage in the form of partly or fully decomposed organic matter and other domestic refuge along its both the banks at various places. The pollution load due to municipal sewage and domestic water increases as it flows downstream upto Chandra Prastha colony. From this point to Morar dam the water quality is not that bad. Taking the advantage of these characteristics of the river, five sampling stations were established on the stretch of water from Mehra village to Morar dam. These sampling stations are almost equidistant to each other. The course of Morar river and the location of sampling stations have been shown in the Figure 1.1. The sampling stations have been recognised as A, B, C, D and E and their brief description is as follows:

Station A

It was established near Mehra village. At this station human activities like bathing, cloth and utensil washing and cattle grazing by villagers are quite common.

Station B

Station B was established near Morar bridge which joins Morar subcity to the Lashkar subcity. Here, human activities including bathing, cloth washing, and discharge of human faecal are done almost all round the year. A few sewage channels are also entering into the river at various points.

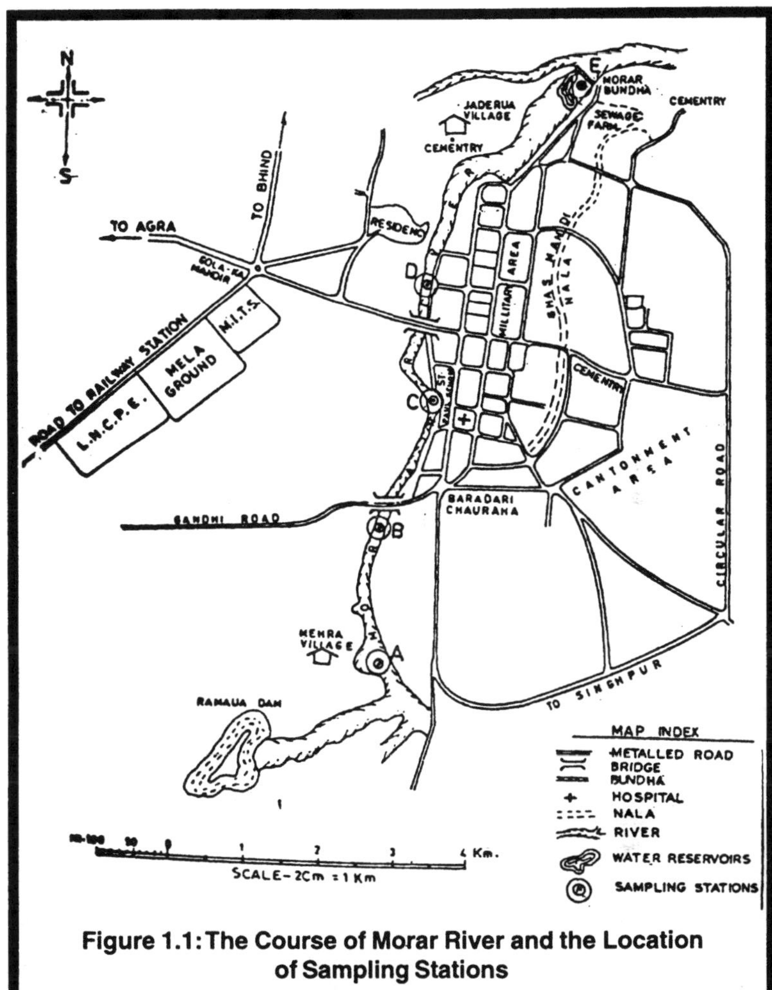

Figure 1.1: The Course of Morar River and the Location of Sampling Stations

Station C

This site was established in the backyard of St. Paul's School, Morar. From station B to this station many sewage channels pour their contents on both sides of the river and increase the pollution load still further making station C as highly polluted.

Station D

It was established near the Kalpi bridge. This bridge joins the Morar to the Gwalior. Here, again all those activities which were observed at station B are also encountered. River continues to receive municipal sewage and garbage on both the sides from adjoining residential colonies. The pollution load is, however, not that great as at station C.

Station E

This is the last sampling station situated at Morar dam. Not much human activity is evident at this station except for human bathing and cattle grazing. The water is comparatively clean at this point.

Materials and Methods

The zooplanktonic samples were also collected from Morar river at five different stations. Samples were collected by filtering a volume of 50 litre surface water through a plankton net made up of bolting silk cloth no. 20. Extreme care was taken in order to keep water undisturbed at the time of sampling and also to avoid spilling of water from the net. The sample concentrates thus obtained, were immediately preserved by adding few drops of 5 per cent formalin.

A volume of 10 ml of concentrate was taken after sedimentation. The samples were thoroughly mixed before their examination. The zooplanktonic organisms were identified upto species wherever possible and upto genus in other cases according to Ward and Wipple (1959), Ruttner-Kolisko (1974), Koste (1978), Victor and Fernando (1979), Sehgal (1983).

For the quantitative analyses of zooplankton, a Sedgwick-Rafter (SR) plankton counting cell was used. A sub-sample of 1 ml was transferred to plankton counting cell for numerical counts. Average of ten counts were taken for the number of zooplanktonic organisms. The number of each species or genus was calculated by the following formula (Welch, 1952):

$$N = \frac{a \times b}{1}$$

where,

N = Number of plankton per litre.

a = The average number of plankton in all counts in a counting cell of 1 ml capacity.

b = The volume of original concentrate in ml.

1 = The volume of original water filtered expressed in litres.

All the zooplankton were represented numerically as organisms per hundred litre of river water.

The Shannon-Weaver's (1963) index was also calculated for zooplanktonic organisms by following formula:

$$H = -\sum_{i=1}^{S} \left(\frac{ni}{N}\right) \log \left(\frac{ni}{N}\right)$$

where,

S = Number of species.

ni = Number of individuals in the ith sample.

N = Total number of individuals.

Observations

32 species of zooplanktonic organisms belonging to 16 families have been identified from the Morar river. *Arcella vulgaris, A. discoides, Diffugia muriformis* and *Centropyxis aculeata* among Protozoa, *Brachionus quadridentatus, B. calyciflorus, B. falcatus, B. bidentatus, B. patulus, B. angularis, Keratella tropica, Platyias quadricornis, Mytilina ventralis, Lecane (L.) luna, Lecane (M.) bulla, Asplanchna brightwelli, Cephalodella auriculata, Polyarthra vulgaris, Lepadella ovalis, Testudinella patina, Filinia longiseta, F. opoliensis* among Rotifera, *Moina brachiata, Simocephalis exspinosus, Ceriodaphnia reticulata, Bosmina longirostris* and *Alona intermedia* among Cladocera, *Cypris* sp., *Cyprinotus qunningi* and *Stenocypris malcolmsonii* among *Ostracoda* and *Mesocyclops thermocyclopoides* and *Thermocyclops crassus* among Copepoda have been identified. The systematic enumeration of zooplankton has been given below:

Phylum: Protozoa

Class: Sarcodina

Order: Testacida

Family: Arcellidae

Arcella (Ehrenberg)

A. vulgaris (Ehrenberg)

A. discoides (Shrenberg)

Family:	Difflugiidae
	Difflugia (Lelerc)
	D. muriformis (Gauthier-Lievre and Thomas)
	Centropyxis (Stein)
	C. aculeata (Ehrenberg)
Phylum:	Rotifera
Class:	Monogononta
Order:	Ploimida
Family:	Brachionidae
	Brachionus (Pallas)
	B. quadridentatus (Hermann)
	B. calyciflorus (Gosse)
	B. falcatus (Zacharias)
	B. bidentatus (Anderson)
	B. patulus (Muller)
	B. angularis (Gosse)
	Keratèlla (Bory de St. Vincent)
	K. tropica (Apstein)
	Platyias (Harring)
	P. quadricornis (Ehrenberg)
Family:	Mytilinidae
	Mytilina (Bory de St. Vincent)
	M. ventralis (Ehrenberg)
Family:	Lecanidae
	Lecane (Nitzsch)
	Lecane (L.) luna (Muller)
	Lecane (M.) bulla (Gosse)
Family:	Asplanchnidae
	Asplanchna (Gosse)
	A. brightwelli (Gosse)

Family:	Notommatidae
	Cephalodella (Bory de St. Vincent)
	C. auriculata (Muller)
Family:	Synchaetidae
	Polyarthra (Ehrenberg)
	P. vulgaris (Carlin)
Family:	Colurellidae
	Lepadella (Bory de St. Vincent)
	L. ovalis (Muller)
Order:	Gnesiotrocha
Family:	Testudinellidae
	Testudinella (Bory de St. Vincent)
	T. patina (Hermann)
Family:	Filiniidae
	Filinia (Bory de St. Vincent)
	F. longiseta (Ehrenberg)
	F. opoliensis (Zacharias)
Phylum:	Arthropoda
Class:	Crustacea
Order:	Cladocera
Family:	Daphnidae
	Moina (Baird)
	M. brachiata (Jurine)
	Simocephalus (Schodler)
	S. exspinosus (Koch)
	Ceriodaphnia (Dana)
	C. reticulata (Jurine)
Family:	Bosminidae
	Bosmina (Baird)
	B. longirostris (O.F. Muller)

Family:	Chydoridae
	Alona (Baird)
	A. intermedia (Sars)
Order:	Ostracoda
Family:	Cyprididae
Sub-Family:	Cyprinae
	Cypris (O.F. Muller)
	Cypris sp.
	Cyprinotus (Brady)
	C. gunningi (Methuen)
Sub-Family:	Stenocyprinae
	Stenocypris (Sars)
	Stenocypris malcolmsonii (Brady)
Order:	Copepoda
Family:	Cyclopidae
	Mesocyclops (Claus)
	M. thermocyclopoides (Harada)
	Thermocyclops crassus (Fischer)

The distribution of zooplankton encountered at various stations in the river has been shown in Table 1.1 and mean values of various physico-chemical characteristics of river water shown in Table 1.2.

Table 1.1: The Distribution of Zooplanktonic Species in Morar River at Various Stations

Zooplanktonic Species	*Stations*				
	A	*B*	*C*	*D*	*E*
Protozoa					
Arcella vulgaris	–	–	+	+	–
A. discoides	+	+	+	+	+
Difflugia muriformis	+	+	+	+	+
Centropyxis aculeata	–	+	–	–	+

Contd...

Table 1.1–Contd...

Zooplanktonic Species	*Stations*				
	A	*B*	*C*	*D*	*E*
Rotifera					
Brachionus quadridentatus	+	+	–	+	+
B. calyciflorus	–	+	+	+	+
B. falcatus	+	–	+	–	+
B. bidentatus	+	+	+	+	+
B. patulus	+	+	+	+	+
B. angularis	+	+	+	+	+
Keratella tropica	–	+	+	+	+
Platyias quadricornis	+	+	+	+	+
Mytilina ventralis	+	+	+	+	+
Lecane (L.) luna	+	+	+	+	+
Lecane (M.) bulla	+	+	+	+	+
Asplanchna brightwelli	+	+	+	+	–
Cephalodella auriculata	–	–	–	–	+
Polyarthra vulgaris	–	–	–	+	+
Lepadella ovalis	–	+	+	+	+
Testudinella patina	+	–	+	+	+
Filinia longiseta	–	+	+	+	+
F. opoliensis	–	+	+	+	+
Cladocera					
Moina brachiata	+	+	+	+	+
Simocephalus exspinosus	+	+	+	–	+
Ceriodaphnia reticulata	–	+	+	–	+
Bosmina longirostris	–	+	+	–	+
Alona intermedia	+	+	–	+	–
Ostracoda					
Cypris sp.	+	+	+	+	–
Cyprinotus gunningi	+	+	+	–	+
Stenocypris malcolmsonii	–	–	+	+	–
Copepoda					
Mesocyclops thermocyclopoides	+	+	+	+	+
Thermocyclops crassus	–	+	–	–	+

+ = Present; – = Absent

Table 1.2: The Mean Values of Various Physico-chemical Characteristics of Morar River Water

Characteristics	Stations				
	A	B	C	D	E
Flow rate, cm.sec^{-1}	5.31	13.60	22.91	23.20	00.00
Water temperature, °C	22.87	23.28	23.85	24.31	23.84
Transparency, cm	33.86	32.50	30.60	37.54	49.06
Depth, cm	53.48	54.95	57.20	66.11	102.91
Turbidity, JTU	23.24	29.20	31.07	26.18	22.23
Total solids, mg.l^{-1}	698.70	862.41	914.16	814.62	737.08
Total dissolved solids, mg.l^{-1}	448.74	523.74	543.41	480.20	481.50
Total suspended solids, mg.l^{-1}	249.95	338.66	370.75	333.91	255.58
pH	7.79	8.13	8.01	8.03	8.04
Conductivity, m S.cm^{-1}	0.43	0.39	0.34	0.38	0.34
Dissolved oxygen, mg.l^{-1}	10.09	7.54	5.08	5.59	7.62
Free carbondioxide, mg.l^{-1}	8.79	14.39	19.15	17.21	10.35
COD, mg.l^{-1}	27.30	68.10	109.33	76.50	43.87
Hardness, mg.l^{-1}	215.03	234.54	278.66	282.08	266.66
Chlorides, mgl.l^{-1}	50.36	62.33	83.82	83.48	75.83
Biocarbonate alkalinity, mg.l^{-1}	265.96	294.08	317.24	312.37	337.62
Sulphides, mg.l^{-1}	1.85	3.87	4.60	3.98	2.66
Residual chlorine, mg.l^{-1}	23.38	24.05	24.42	27.50	27.29
Phosphates, mg.l^{-1}	0.021	0.177	0.176	0.185	0.136
Silicates, mg.l^{-1}	0.244	0.353	0.333	0.342	0.261
Nitrates, mg.l^{-1}	0.024	0.026	0.029	0.025	0.029
Nitrites, mg.l^{-1}	0.003	0.009	0.010	0.009	0.007
Sodium, mg.l^{-1}	11.04	12.33	10.80	14.63	11.12
Potassium, mg.l^{-1}	9.18	10.17	7.60	8.76	8.78
Calcium, mg.l^{-1}	30.86	39.25	50.54	55.55	49.77

Seasonal Variations of Zooplankton During the Year 1987–88 and 1988–89

The number of zooplanktonic organisms under different groups at different stations studied has been given in Tables 1.3–1.7.

At station A

Protozoa

Arcella discoides occurred throughout while *Difflugia muriformis* was present only from January to September during the period of study. *A. discoides* was noted to be minimum in the month of September and August and maximum in the month of June and May during the years 1987–88 and 1988–89 respectively. *D. muniformis* had its maxima in the month of June during two years of the study.

The protozoan increased from October to January except in the month of December when the number decreased. During Feburary, March and April the number of Protozoa was recorded and from May it again increased until June. Later, the number of Protozoa declined upto September during year 1987–88. Almost similar pattern of seasonal variations were noted during year 1988–89 (Table 1.3).

Rotifera

Brachionus angularis, Lecane (L.) luna and *Lecane (M.) bulla* were obtained all round the year at this station. *B. angularis* was found to be maximum in the month of March during first year and in the month of November during second year. *Lecane (L.) luna* occurred in abundance during November and December whereas *Lecane (M.) bulla* was found to be maximum in the month of March and January during first and second year of the study respectively. *Brachionus quadridentatus* and *Platyias quadricornis* were absent in winter months. The former was recorded to be maximum in the months of June and May while the later occurred in high numbers in the month of March during two years of the study. *B. falcatus, B. bidentatus* and *B. patulus* were noticed during summer and winter season only. *B. falcatus* showed its minimum growth in October and November maximum in March during the two year study. *B. bidentatus* was observed to be maximum in November and April while *B. patulus* was noted to be maximum in November and February during the year 1987–88 and 1988–89 respectively. *Mytilina ventralis* was present in the months of November, December, March and April with its highest counts being in the month of December. *Asplanchna brightwelli* was noticed in the months of May, June, March and April with its maxima in the months of June and April during the two year study. *Testudinella patina* appeared during summer and continued upto rainy season. Its highest number was obtained in the month of June and May during first and second year of the study respectively.

Table 1.3: The Numerical Abundance of Zooplankton Belonging to Various Groups in Morar River at Station A*

Months	*Protozoa*	*Rotifera*	*Cladocera*	*Ostracoda*	*Copepoda*	*Eggs, Larvae and Worms*	*Total Zooplankton*
May 1987	252	912	194	174	128	146	1806
June	284	964	166	302	184	98	1998
July	122	322	62	116	48	80	750
August	80	262	74	84	62	72	634
September	76	558	124	108	30	42	938
October	90	122	150	218	108	152	1440
November	100	1070	164	120	56	120	1630
December 1987	80	1104	698	192	166	96	2336
January 1988	182	1102	660	130	94	194	2362
February	152	1052	350	176	82	66	1878
March	142	1418	264	114	46	110	2094
April	118	1154	234	180	104	214	2004
May	264	1222	190	384	112	180	2352
June	206	858	252	288	100	184	1888
July	112	320	106	236	64	46	884
August	96	316	80	104	60	76	732
September	82	282	78	88	72	30	632
October	86	750	144	166	110	56	1312
November	102	1008	210	180	102	204	1806
December 1988	86	1274	358	246	132	200	2296
January 1989	146	1174	378	316	108	94	2216
February	128	1086	332	214	116	172	2048
March	216	1258	476	206	162	194	2512
April 1989	136	1438	306	128	110	114	2232

* The unit of zooplanktonic counts is org.l^{-100}.

It was observed that the total rotifers started increasing from September to December and then decreased upto February and in March second, maxima was observed. The number of rotifers declined once again from April onwards during first year. During second year, almost similar pattern of variation was observed (Table 1.3).

Cladocera

Moina brachiata and *Alona intermedia* were present throughout the period of investigation while *Simocephalus exspinosus* and *Bosmina longirostris* were ephimeral at this station. *M. brachiata* was recorded to be maximum in the months of December and June and *Alona intermedia* was highest in the months of December and November respectively during first and second year of the during year 1987–88 and 1988–89 respectively (Table 1.3).

Copepoda

Only *Mesocyclops thermocyclopoides* was recorded at this station which occurred throughout the year with its lowest number in the months of September and August and maximum in the months of June and March during year 1987–88 and 1988–89 respectively. Without a clear trend of seasonal variation.

Total Zooplankton

The total zooplankton population at this station started increasing from the month of September to January and then decreased upto May. The zooplankton again increased in the month of June, and decreased in July and August during year 1987–88. The abundance of zooplankton followed a sequence as Rotifera > Copepoda > Protozoa > Cladocera > Ostracoda at station A (Figure 1.2).

At Station B

Protozoa

Arcella discoides was present throughout at this station. Its highest counts were obtained in the month of April during 1987–88 and March during 1988–89. *Difflugia muriformis* was noted in the months of November, December, January and April while, *Centropyxis aculeata* occurred in the month of May and June only. Highest number of these species were found in the months of January and December during first year and in June during next year.

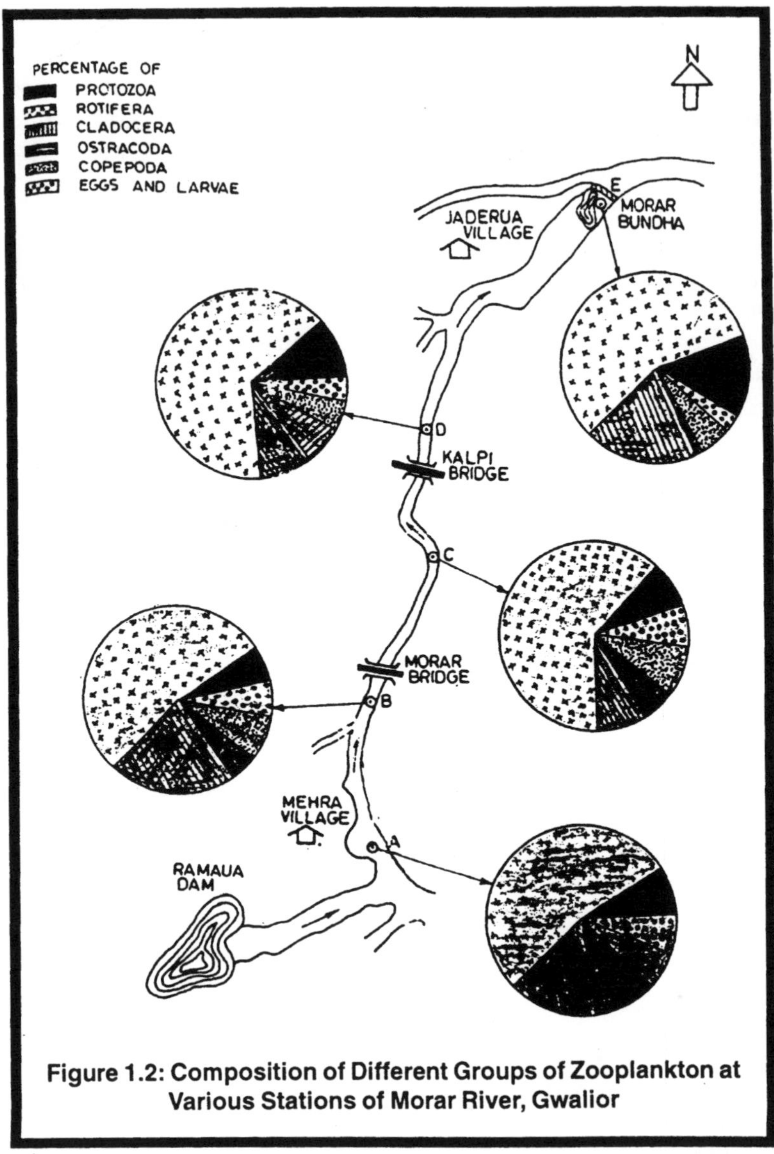

Figure 1.2: Composition of Different Groups of Zooplankton at Various Stations of Morar River, Gwalior

Total number of Protozoa's increased from August to October and then declined in November. In December they again increased and then receded upto March. In April, the number enhanced again and a decline was noted upto July during the year 1987–88. Total number of Protozoa during the year 1988–89 also followed a similar pattern of seasonal variation with a little deviation in the latter half of the year (Table 1.4).

Table 1.4: The Numerical Abundance of Zooplankton Belonging to Various Groups in Morar River at Station B*

Months	*Proto-zoa*	*Roti-fera*	*Clado-cera*	*Ostra-coda*	*Cope-poda*	*Eggs, Larvae and Worms*	*Total Zoo-plankton*
May 1987	122	1168	222	110	128	88	1838
June	188	1352	288	152	232	118	2330
July	36	366	110	28	92	72	704
August	48	418	86	20	82	20	674
September	94	552	164	66	116	20	1012
October	96	962	348	134	108	128	1776
November	88	1170	820	56	118	124	2376
December 1987	350	1086	2096	156	346	68	4102
January 1988	226	1192	564	110	182	188	2462
February	110	1496	330	98	162	104	2300
March	78	1648	322	80	160	210	2498
April	212	1658	314	120	188	156	2648
May	132	1352	326	98	146	188	2242
June	186	1414	270	148	108	130	2256
July	42	510	68	40	86	72	818
August	56	386	128	36	84	20	710
September	96	482	208	66	176	156	1184
October	98	970	202	100	228	146	1744
November	52	1078	490	80	176	172	2048
December 1988	192	1242	2034	230	180	98	3976
January 1989	80	1324	426	98	276	206	2410
February	146	1574	348	82	166	136	2452
March	198	1308	272	110	176	208	2272
April 1989	148	1452	380	128	200	180	2488

* The unit of zooplanktonic counts is org.l^{-100}.

Rotifera

Brachionus calyciflorus, B. angularis and *Lecane (L.) luna* were present all the times at this station. *B. calyciflorus* was noted to be minimum in the months of August and September and maximum in the months of April and June whereas, *B. angularis* was noted to be highest in the months of June and April during year 1987–88 and 1988–89. *Lecane (L.) luna* was recorded to be maximum in the month of June. *B. quadridentatus* was not observed during October, November and December. The maxima of the *B. quadridentatus* was seen in the months of March and February during two years study respectively. *B. bidentatus* and *B. patulus* occurred during winter and summer seasons only. Lowest number of these species were obtained in the months of October and December and highest in the month of March during first year of the study. During second year, minima of these species was noted in the month of November and maxima in the month of February respectively. *Keratella tropica* was found from the months of July to October and January to March. Highest number of this rotifer was recorded in the months of March and February during first and second year of the study. *Platyias quadricornis* was not only absent continuously from September to December but was also absent in April and June. Their highest number was reported in the month of November during both the years of study. *Mytilina ventralis* occurred during October to December and during March and April with its highest population in November during 1987–88 and in December during 1988–89. *Lecane (M.) bulla* was present during summer seasons and rainy seasons only with minimum number in August and March and maximum number in May and June during two year study. *Asplanchna brightwelli* and *Lepadella ovalis* were present during March to June. *A. brightwelli* was maximum in the month of May while *L. ovalis* was recorded to be maximum in the month of April during first and second years respectively. *Filinia longiseta* occurred during winter and early summer seasons while *F. opoliensis* was found during late winter and early summer seasons only. *F. longiseta* was maximum in January and December and *F. opoliensis* in February during two years of study.

The rotifer population increased from August to November and decreased in the month of December and then again increased continuously to show maxima in the month of April. From May onwards the rotifers decreased upto July during year 1987–88. During the next year, the number of rotifers started increasing from

September to February when the trend of variation was unstable. After June, the number decreased continuously to remain minimum in the month of August (Table 1.4).

Cladocera

Moina brachiata occurred throughout the period of study and was maximum in the month of December while *Ceriodaphnia reticulata* was noted during late rainy, winter and summer seasons with its highest number in the month of December. *Bosmina longirostris* was obtained in the months of February and April with its maxima in February during year 1987–88 and 1988–89 respectively. *Alona intermedia* was noticed during winter and summer seasons only. The lowest number of this specie appeared in the month of May and highest in the month of December during two years of study.

The cladoceran population increased from September to December but from January onwards it continued to recede upto August during first year. Same trend of variation was repeated during second year also with minor variation (Table 1.4).

Ostracoda

Cypris sp. was observed during late rainy, winter and mid summer seasons only. Its minimum count was noted in the month of December and September and maximum count in the month of October and December during year 1987–88 and 1988–89 respectively. *Cyprinotus gunningi* was observed first during May to August and then during December to February with highest number in the month of June during the two year study.

Individuals of ostracoda increased from September to December and then decreased from January to March. Again in April the number was increased and in May they started declining. Later in June, the number of ostracods increased and decreased in July and August during year 1987–88. The pattern of seasonal variation in the ostracods during year 1988–89 was not exactly reproducible (Table 1.4).

Copepoda

Mesocyclops thermocyclopoides was present throughout the study with its maxima in the months of December and January during first and second years of the study respectively. *Thermocyclops crassus* was found to be absent in the months of May and June. The number of *T. crassus* was minimum in the months of September and July and maximum in the months of January and December during 1987–88 and 1988–89 respectively.

The number of copepods enhanced from September to December and then declined from January to March. During April, May and June not much variations were observed but from July and August the number once decreased during year 1987–88. During year 1988–89, the number was increased from September to October. From November, the number of copepods declined and then again increaed in December and January. From February to August the number again decreased with an increase in April (Table 1.4).

Total Zooplankton

The total number of zooplankton exhibited an increase from September to December and then decrease upto February. From March to June, the number of zooplankton enhanced excepting May and then they again declined upto August during first year. During second year of the study the number of zooplankton increased from September to December and upto June there were frequent ups and downs. The number was quite low during July and August (Table 1.4). At this station dominant groups had a sequence: Rotifera > Cladocera > Copepoda > Protozoa > Ostracoda (Figure 1.2).

At Station C

Protozoa

Arcella discoides, *Difflugia muriformis* were present in all the months. These species were recorded to be minimum in the months of July and August and maximum in the months of January and March during 1987–88 respectively. During the year 1988-89, these species were minimum in the months of September and July and maximum in the months of April and November respectively. *Arcella vulgaris* was observed during late summer and rainy seasons only with its highest counts in June during first year and in May during second year.

Zooplankton under this group were increased from August to October. In November, their number declined and then again enhanced upto January. In February, they decreased again and then increased upto April, when maxima of this group was obtained. April onwards they decreased once again upto the month of August during the year 1987–88. During the year 1988–89, zooplankton under this group were decreased from June to September and then increased upto November. From December to February the Protozoa receded and from February onwards there was an increase upto May reaching to the highest (Table 1.5).

Table 1.5: The Numerical Abundance of Zooplankton Belonging to Various Groups in Morar River at Station C*

Months	*Proto-zoa*	*Roti-fera*	*Clado-cera*	*Ostra-coda*	*Cope-poda*	*Eggs, Larvae and Worms*	*Total Zoo-plankton*
May 1987	140	1338	216	172	450	228	2544
June	170	1344	74	68	666	236	2558
July	156	638	58	106	140	63	1181
August	74	655	48	40	93	82	993
September	86	458	38	136	28	98	844
October	148	1144	60	108	74	146	1680
November	88	1288	102	100	68	142	1788
December 1987	148	1518	218	80	96	182	2242
January 1988	170	1536	356	78	144	130	2414
February	160	1556	342	84	112	158	2412
March	166	1364	82	98	146	248	2104
April	202	1730	344	132	90	154	2652
May	268	1644	208	188	336	222	2866
June	244	1342	100	80	366	276	2408
July	140	786	70	68	100	174	1338
August	184	776	36	56	86	78	1216
September	110	509	30	170	74	90	983
October	160	988	84	102	106	182	1622
November	194	1310	104	170	100	186	2064
December 1988	176	1910	308	110	142	170	2816
January 1989	174	1534	252	144	104	300	2508
February	146	1542	370	104	162	140	2374
March	170	1896	204	112	152	194	2734
April 1989	254	1830	190	220	292	118	2904

* The unit of zooplanktonic counts is org.l^{-100}.

Rotifera

Brachionus calyciflorus, B. angularis, Lecane (L.) luna, Lecane (M.) bulla and *Filinia longiseta* were recorded throughout the period of study. *B. calyciflorus* was maximum in the months of January and February while *B. angularis* showed its maxima in the months of March and May during two years study respectively. *L. (L.) luna* and *L. (M.) bulla* were noted to be highest in the months of June and December; February and January during year 1987–88 and 1988–89 respectively. *F. longiseta* was observed to be minimum in the month of August and maximum in the months of December and May during two years of the study. *B. falcatus, B. bidentatus* and *B. patulus* were present during winter and summer seasons only. *B. falcatus* showed its highest counts in February and April; *B. bidentatus* in the months of December and March and in the months of January and December during year 1987–88 and 1988–89 respectively. *Keratella tropica* appeared in the month of August and continued upto April with its maximum in January and December during both the years. *Platyias quadricornis* was present during rainy season only at this station with lowest number in September and highest in June and May during two years of the study respectively. *Mytilina ventralis* was recorded first in November and December and then in March and April with its maxima in the first spell *i.e.* in December. *Asplanchna brightwelli* and *Lepadella ovalis* were found to be present from March to May with maxima in summer. *Testudinella patina* was present from August to October but its highest number was seen in the months of September and August during first and second year respectively. *F. opoliensis* was obtained during winter and summer seasons with maximum counts in the months of April and January during the year 1987–88 and 1988–89 respectively.

Total rotifers gradually increased from October to February while in March a decline in number was observed. In April they increased again and from May to September a decrease in population was exhibited during first year. During second year, the total number of rotifers increased from October to December and then declined up to February. In March, the number was again increased and from March onwards and upto September they were on decline (Table 1.5).

Cladocera

Moina brachiata has been reported throughout the period of study. *Simocephalus exspinosus* was observed in the months of December, February, April and May; *Ceriodaphnia reticulatus* occurred in the months of November, December, January and April and *Bosmina longirostris* during January and February. *Moina brachiata* was noted to be minimum the month of September and maximum in the month of January during the year 1987–88. During the year 1988–89, its lowest and highest number was observed in the months of March and February respectively. The maximum counts of *S. exspinosus* was recorded in the months of April and February while *C. reticulata* was maximum in the months of April and January during first and second year of the study respectively. *Bosmina longirostris* had its highest number in February during both the years.

Considering their seasonal trend of variation, the cladocerans were found to increase from September to January. In February and March their number lowered and in April it again enhanced. Their number lowered once again in September when their lowest number was observed during first year. During next year, almost same irregular pattern of variation was noted (Table 1.5).

Ostracoda

Cypris sp. was not recorded in the months of May, July, October and November with lowest number in the month of September and highest in the month of April during the two year study. *Cyprinotus gunningi* was present in the months of September, October, November and in May with its maximum counts in the months of October and May during first and second year of the study respectively. *Stenocypris malcolmsonii* observed in the months of May, July and September only. Its maxima was noted in the month of May during two years of the study.

Low number of the ostracods was observed from June to August and then in September they increased but from October to January the number declined. From February to May ostracods once again increased. During the second year, the trend was quite different with a decrease in number from June to August and then ups and downs in their number was noticed from September to May (Table 1.5).

Copepoda

Mesocyclops thermocyclopoides occurred during all seasons of the year. Highest number of this specie was observed in the month of June during two years of the study.

The number of copepods showed a decreasing trend from July to September and fluctuation was noted until March. From April onwards, the number increased upto June when a maxima was observed during the year 1987–88. Almost same pattern of seasonal variation was recorded during the year 1988–89 (Table 1.5).

Total Zooplankton

The variation in total zooplankton population at this station suggested that it swelled from October to January and then reduced upto March. In April the number of zooplankton increased again and in May it declined. In June, the number was increased a little but from Julay to September, the number declined during the first year. During the second year total zooplankton increased from October to December and then declined upto February. In March and April, the number increased and then again decreased until September (Table 1.5). At this station zooplankton belonging to Rotifera were most abundant followed by Cladocera, Ostracoda, Protozoa and Copepoda (Figure 1.2).

At Station D

Protozoa

At this station, *Arcella discoides* occurred throughout the period of study while *A. vulgaris* occurred during May to September and *Difflugia muriformis* during December to August. Highest population of *A. discoides* was obtained in the month of November and January during two years of the study. The maximum number of *A. vulgaris* was observed in the month of June during both the years. The lowest count of *D. muriformis* was found in the month of August during both the years and maximum in the months of April and March during first and second year of the study respectively.

The protozoan increased gradually from the month of October to January while in February the number declined. The Protozoa fluctuated irregularly from May to September during first year. There was little variation in the seasonal pattern of Protozoa during second year (Table 1.6).

Table 1.6: The Numerical Abundance of Zooplankton Belonging to Various Groups in Morar River at Station D*

Months	*Proto-zoa*	*Roti-fera*	*Clado-cera*	*Ostra-coda*	*Cope-poda*	*Eggs, Larvae and Worms*	*Total Zoo-plankton*
May 1987	234	1340	160	48	144	98	2024
June	256	1096	86	52	108	166	1764
July	118	574	46	38	76	32	884
August	226	462	46	48	82	48	912
September	84	386	32	80	36	52	670
October	94	1660	106	284	86	134	2364
November	148	1306	112	238	76	112	1992
December 1987	190	1800	334	266	158	240	2988
January 1988	294	1394	188	60	86	000	2022
February	234	1430	262	72	88	114	2200
March	244	1766	270	76	88	192	2636
April	278	1390	352	86	84	40	2230
May	230	1476	202	64	80	202	2254
June	254	1402	72	122	102	108	2060
July	104	590	52	56	72	30	904
August	110	566	36	20	86	44	862
September	76	564	46	110	112	82	990
October	156	1516	186	194	112	38	2202
November	94	1374	176	240	72	70	2026
December 1988	206	1464	214	188	102	180	2354
January 1989	306	1310	166	196	96	98	2172
February	144	1308	178	76	198	00	1904
March	260	1670	292	82	92	80	2476
April 1989	148	1598	306	114	76	30	2272

* The unit of zooplanktonic counts is org.l^{-100}

Rotifera

Brachionus quadridentatus, B. calyciflorus, B. angularis, Keratella tropica, Lecane (L.) luna, L. (M.) bulla, Polyathra vulgaris, Filinia longiseta and *F. opoliensis* were present throughout the period of study. *B.*

quadridentatus was noted to be minimum in September and maximum in December and June during first and second year respectively. The maxima of *B. calyciflorus* was observed in the months of December and March while highest number of *B. angularis* was found in the month of March during the year 1987–88 and 1988–89 respectively. *K. tropica* was maximum in the months of May and April whereas *L. (L.) luna* was highest in the months of March and February while *L. (M.) bulla* showed its maxima in the months of March and May during two years of the study. *P. vulgaris* showed its highest count in the month of October during first year and in April during next year. *F. longiseta* was found to be highest in the months of March and December while *F. opoliensis* occurred in maximum number in the months of February and October during the two year study. *B. bidentatus* and *B. patulus* were observed during winter and summer seasons only. The highest number of *B. bidentatus* was recorded in the months of December and November while the *B. patulus* was recorded to maximum in the months of March and February during 1987–88 and 1988–89 respectively. *Platyias quadricornis* was absent during late winter and early summer season. The maxima of this species was observed in the months of June and April during two years of the study. *Mytilina ventralis* was present first in November, December and then in March and April with its highest population being in the month of April during first year and in December during second year. *Asplanchna brightwelli* was noticed in May, June, March and April with its maximum number in the months of April and May. *Lepadella ovalis* occurred during April to June only with its highest counts in the month of April whereas *Testudinella patina* occurred during August, September and October with maximum number in the month of October.

The rotifers increased from October to December and then a little decrease was observed in January. In February and March the numbers increased and then again decreased from April to September during the year 1987–88. Not much variation was noted in the seasonal patterns of rotifers during 1988–89 (Table 1.6).

Cladocera

Moina brachiata was recorded during all round the year while *Alona intermedia* was present from December to May only. The minimum number of *M. brachiata* was observed in the months of September and August and maximum in the months of April and

March during first and second year of the study respectively. Highest counts of *A. intermedia* were observed in the months of February and April during two years of the study.

The total number of Cladocera increased from October to December and then a decline was noticed in January. The number again increased from February to exhibit their maxima in April. From May to September the number of *Cladocera* receded to show their minima in September during the year 1987–88. A similar trend of seasonal variation was observed during the year 1988–89 also (Table 1.6).

Ostracoda

Stenocypris malcolmsonii was found throughout the period of study with its maximum counts in the months of December and January during first and second year of the study respectively. *Cypris* sp. appeared in the month of September and continued upto December with highest number in the month of October and November. It appeared again in the month of April.

The ostracods increased from August to October and from October onwards the number was fluctuated with ups and downs in the seasonal variation during both the years of study (Table 1.6).

Copepoda

Copepoda was represented by *Mesocyclops thermocyclopoides* only which was seen throughout the period of study with its maxima in the months of December and February during two years study. No clear cut seasonal trend of variation for copepods at this station could be observed.

Total Zooplankton

The total zooplankton exhibited an increase in number from October to December and then decrease in January. In February and March they increased again and from April to September they decreased during the first year. During second year, almost same pattern was exhibited by the zooplankton (Table 1.6). The abundance of various zooplanktonic groups followed a pattern: Rotifera > Protozoa > Cladocera > Ostracoda > Copepoda (Fig 1.2).

At station E

Protozoa

All the three species of Protozoa *viz., Arcella discoides, Difflugia muriformis* and *Centropyxis aculeata* were present throughout the period of study at this station. *A. discoides* was recorded to be maximum in the months of April and February while *D. muriformis* was highest in the month of March during two year study. *C. aculeata* was found to be maximum in the month of April and June during first and second year respectively.

The Protozoa increased in number from September to October and then decreased upto December. An increase upto April in population was again noted but from May onwards decrease in number was evident upto August except in June during the year 1987–88. During the year 1988–89, the number of protozoa started increasing from January and continued to be so upto April after that an irregular trend was observed in seasonal variation in their population (Table 1.7).

Rotifera

In addition to the rotifers present at station D, *Cephalodella auriculata* was also found to be present throughout the period of study. The number of *Brachionus quadridentatus* was recorded to be maximum in the month of April while *B. calyciflorus* was noted to be highest in the month of April and March during first and second year of the study respectively. The minima of *B. angularis* was recorded in the months of July and September and the maxima in the months of April and February during two years of the study. *Keratella tropica* was found to be highest in the months of June and January. *Lecane (l.) luna* was maximum in the months of April and January while *Lecnae (M.) bulla* showed its maximum in the months of April and March during 1987–88 and 1988–89. The minimum number of *Cephalodella auriculata* was observed in the months of July and September and maximum number in the months of April and February during first and second year respectively. *Polyarthra vulgaris* was found to be highest in number in November and June; *Filinia longiseta* in the month of December and *F. opoliensis* in the months of April and May. *B. falcatus, B. bidentatus* and *B. patulus* were absent during late summer and rainy season. *B. falcatus* recorded to be

Table 1.7: The Numerical Abundance of Zooplankton Belonging to Various Groups in Morar River at Station E*

Months	*Protozoa*	*Rotifera*	*Cladocera*	*Ostracoda*	*Copepoda*	*Eggs, Larvae and Worms*	*Total Zooplankton*
May 1987	486	1946	388	00	82	104	3006
June	516	1972	346	58	110	124	3126
July	210	642	178	62	164	30	1286
August	184	656	180	48	76	28	1172
September	208	536	134	00	154	46	1078
October	504	1702	290	00	118	68	2682
November	274	1868	592	00	202	68	3004
December 1987	256	1654	728	00	182	120	2940
January 1988	428	2006	1572	52	112	164	4334
February	512	1656	542	68	296	28	3102
March	464	1548	516	00	194	128	2850
April	522	2458	466	00	216	126	3788
May	516	2006	450	00	110	80	3162
June	616	1782	356	66	96	128	3044
July	126	620	148	58	112	28	1092
August	148	610	258	58	98	00	1172
September	146	630	172	00	142	46	1136
October	478	1762	218	00	178	124	2760
November	316	1780	580	00	206	104	2986
December 1988	268	1808	960	00	240	186	3462
January 1989	460	2238	2172	36	160	110	5176
February	490	2002	806	74	170	176	3718
March	522	1848	492	00	186	132	3180
April 1989	562	2202	474	00	212	108	3558

* The unit of zooplanktonic counts is $org.l^{-100}$.

maximum in the months of November and January; *B. bidentatus* in the month of November and *B. patulus* in the month of January during two years of the investigation. *Platyias quadricornis* was observed during late summer and rainy seasons with highest numbers in the months of June and April during the year 1987–88 and 1988–89 respectively. *Mytilina ventralis* was observed in the months of November, December, March and April but *Lepadella ovalis* was found in the months of April, May and June. *M. ventralis* showed their maxima in the months of March and April and *L. ovalis* was recorded at its highest number in the months of May and April during the two year. *Testudinella patina* occurred in the months of August, September, October and then in April with its highest counts in the months of May and April during the two year study.

The rotifers remained low in number from June to September and then an increased was noticed in October and November from December to May irregular curve in seasonal pattern was obtained during first year. During second year, the rotifers were reduced in number from May to August and then the number increased from September to January. The number declined once again in February and March but in April they increased (Table 1.7).

Cladocera

Moina brachiata, Simocephalus exspinosus and *Ceriodaphnia reticulata* were noted throughout the period of study while *Bosmina longirostris* was absent during October. The minimum counts of *M. brachiata* were noted in the months of September and July and maximum in the months of January during two years of the study. *S. exspinosus* was noted to be maximum in the months of January and February while highest number of *C. reticulata* was obtained in the month of January during first and second year respectively. *B. longirostris* showed a maxima in the month of January.

The cladocerans have shown an increase from October to January and then decrease from February upto September during first year. Exactly similar trend of variation was repeated during second year also (Table 1.7).

Ostracoda

Only one specie *i.e. Cyprinotus gunningi* was recorded among Ostracoda at station E with its highest number was in the month of February during both the years.

An irregular trend of seasonal variation occurred in ostracoda's population with its lowest counts in the months of August and January and highest in the month of February during first and second years respectively (Table 1.7).

Copepoda

Mesocyclops thermocyclopoides was present throughout the year except in the months of October and November but *Thermocyclops crassus* was absent in the months of January, May and June. The highest number of *M. thermocyclopoides* were noticed in the months of February and January whereas *T. crassus* showed its maxima in the month of November.

No clear cut trend of seasonal variation could be traced for the copepods at this station.

Total Zooplankton

The total number of zooplankton showed a decrease initially from June to September and then an increase upto January except in the month of December was noted. From February to March zooplanktonic population declined once again during first year. During second year, the zooplankton started increasing from August to January barring September and then a steady decline was observed upto March. In April, they increased again and then decreased until July (Table 1.7). The Rotifera was most dominant group with Protozoa, Copepoda, Cladocera and Ostracoda following the sequence (Figure 1.2).

The coefficient correlation between zooplankton, some physico-chemical parameters and phytoplankton was also calculated and given in Table 1.8. A positive correlation existed between rotifers and water hardness while a negative correlation was observed between rotifers and water temperature. A negative correlation existed between copepodas population and water temperature. The zooplankton population in general was having a high positive correlation with total phytoplankton. At various stations, the rotifers had high positive correlation and the Cladocera has low to high positive correlation while the Protozoa and Copepoda had low to moderate positive correlation with total phytoplankton population.

The Shannon-Weaver's index was computed for the various groups of zooplankton for a period of two years to study diversity in zooplanktonic pattern at various stations. The index for various zooplanktonic groups at different stations has been given in Table 1.9. The values of Shannon-Weaver's index for Protozoa was calculated to be 2.169 at station A, 1.778 at station B, 2.138 at station C, 2.398 at station D, and 2.838 at station E; for Rotifera, 3.488 at

Table 1.8: The Coefficient of Correlation Between Zooplankton, Some Physico-chemical Characteristics and Phytoplankton of Morar River at Various Stations

Characteristics	*Stations*				
	A	*B*	*C*	*D*	*E*
Rotifers v/s Water temperature	– 0.6167	– 0.5466	– 0.4950	– 0.5800	– 0.5961
Rotifers v/s Hardness	0.5842	0.6928	0.5485	0.5862	0.6556
Copepods v/s Water temperature	– 0.2093	– 0.4421	0.3730	– 0.1854	– 0.5793
Protozoa v/s Total phytoplankton	0.4879	0.5305	0.4906	0.5865	0.6979
Rotifers v/s Total phytoplankton	0.8459	0.8219	0.8671	0.8007	0.8704
Cladocera v/s Total phytoplankton	0.4748	0.3762	0.6888	0.7623	0.4912
Copepoda v/s Total phytoplankton	0.5201	0.4825	0.2561	0.2950	0.3662
Total zooplankton v/s Total phytoplankton	0.8508	0.7562	0.8516	0.8071	0.8295

Table 1.9: The Shannon-Weaver's Index ($\bar{H}$) Values for Different Zooplanktonic Groups in Morar River at Various Stations

Zooplanktonic Groups	*Stations*				
	A	*B*	*C*	*D*	*E*
Protozoa	2.169	1.778	2.138	2.398	2.838
Rotifera	3.488	3.327	3.076	2.928	3.315
Cladocera	2.764	3.010	1.874	2.036	3.047
Ostracoda	2.540	1.501	1.701	1.715	0.428
Copepoda	1.749	2.170	1.972	1.650	1.798

station A, 3.27 at station B, 3.076 at station C, 2.928 at station D and 3.315 at station E; for Cladocera, 2.764 at station A, 3.010 at station B, 1.874 at station C, 2.036 at station D and 3.047 at station E; for Ostracoda, 2.540 at station A, 1.501 at station B, 1.701 at station C, 1.715 at station D and 0.428 at station E and for Copepoda, 1.749 at station A, 2.170 at station B, 1.972 at station C, 1.650 at station D, and 1.798 at station E. Thus, the species diversity was decreased from station A to station C and then increased from station D to station E. It is, thus, suggested that the zooplankton diversity was low at polluted stations in comparison to comparatively clean stations.

Discussion

The zooplankton occupy an intermediary position between the autotrophs and the carnivores and form an important link in aquatic food webs. The zooplanktonic animals in fresh waters are dominated by Rotifera, Cladocera and Copepoda. Protozoa also forms a significant part of the fresh water zooplankton, especially in systems exhibiting advanced trophic state. The pollution load affect the presence of zooplankton, their number and behaviour. The abundance of some zooplankton in the aquatic food web is reported to indicate eutrophication (Sprules, 1977; Gannon and Stemberger, 1978 and Halbach *et al.*, 1983). The presence of rotifers throughout the stretch of the river suggested that the nutrient level of the river has gone considerably high in general and the organic pollution has started taking place (Gannon and Stemberger, 1978). The concentration of zooplankton in clear water zone was quite rich as regards to both the number and the species. The crustaceans with nauplius stage of Copepoda formed the dominant group and Rotifera was represented by a large variety of species. But, immediately below the effluent outfall, the zooplanktonic counts showed appreciable fall and further reduction in number was observed in the septic zone of Suvaon stream (Banerjea and Motwani, 1960). In the river Sokoto of North Nigeria, 41 species of rotifers were recorded and it was found that the flood assemblage is dominated by the *Lecane* species in additional to the regularly occurring species such as *Macrochaetus collinsi* and *Trichocerca bicristata* while on all other occasions, the brachionids were the dominant with *Tetramastix opoliensis* as an important auxiliary species (Green, 1960). In Ennore river, Chacko and Rajagopal (1962) found that the rotifers were dominant in the months of May and August while copepods were recorded abundantly from May to January. The highest total number of zooplankton (upto 326 org.l^{-1}) in river Kali, before addition of wastes by Kadarabad drain has been observed by George *et al.* (1966). They found that in heavily polluted zone, the number of zooplankton was low but in downstream their number increased slightly.

In Columbia river estuary, the zooplankton were dominated by a copepod, *Cyclops vernalis* cladocerans, *Daphnia longispina* and *Bosmina* sp. Seasonally abundant species included the cladocerans *Diaphanosoma brachyarum* (summer) and *Ceriodaphnia quadrangula* (summer fall), *Juvenile amphipods, Corophium salmonis* (winter-spring), the rotifers *Brachionus* sp. (spring) and *Asplanchna* sp. (summer) and copepod *Diaptomus ashlandi* (spring and summer) while the cladocerans *Alona costata* and *Chydorus globosus* were present in small numbers throughout the year (Haertel and Osterberg, 1967). Venkateswarlu and Jayanti (1968) have recorded high counts of rotifers at polluted station of Sabarmati river in comparison to clean waters. In the river Godavari, Rajyalakshmi and Premswarup (1975) found that the copepods (*Diaptomus, Pseudodiaptomus* and *Cyclops* being the principal genera) formed the bulk of the zooplankton population, following the order are Cladocera (*Diphanosoma* and *Daphnia, Ceriodaphnia* and *Moina*) and Rotifera (*Brachionus, Keratella* and *Rotaria*). In Kali river, *Bosmina* was designated as sensitive to pollution, *Brachionus* and *Rotaria* moderately sensitive while *Arcella* and *Vorticella* though sensitive only to excessive pollution, were designated as fairly tolerant forms (Verma and Dalela, 1975). Vass *et al.* (1977) recorded 23 species of zooplankton from river Jhelum the water quality of which favoured the good development of zooplankton population. Five species of Protozoa *viz., Carchesium, Amoeba, P. caudatum, Arcella* and *Heteronema* were seen all round the year whereas during summer 17 species (4 flagellates, 7 rhizopods and 6 ciliates), during winter 10 species (3 flagellates, 3 rhizopods and 4 ciliates) and during rainy season 7 species (1 flagellate, 4 rhizopods and 2 ciliates) were observed in river Jamuna. The rotifers were represented by 12 species among which *Brachionus angularis* and *Keratella valga* were present all around the year while *Rotaria scopoli* and *Scaridium longicaudum* were recorded only in monsoon period (Prakash *et al.*, 1978). Verma *et al.* (1978) noticed that the zooplankton were represented by a large number of Protozoa (*Arcella* and *Paramecium*); Rotifera (*Brachionus quadridentatus*); Entomostraca (*Moina brachiata* and *Cyclops*) and several insect larvae in Kadarabad drain but the number of planktonic forms was greatly reduced during rainy season.

In river Ganga, a study of microinvertebrates suggested that the representative Crustacea such as *Cyclops* and *Daphnia* were sensitive to pollution and their concentration was changed with the degree of pollution (Khare *et al.*, 1979). Sampath *et al.* (1979), while studying rotifers in Cauvery river, have identified 35 genera and species of Rotatoria. It was found that *Rotaria rotatoria* was more pronounced at sewage and distillery outfalls, *Anuraeopsis, Monostyla hamata, M. bulla* etc. were found in clean water zone, *Brachionus angularis, B. calyciflorus* and other species of rotifers were found in mildly polluted to highly polluted zones. Shiel (1979) recorded large number of zooplanktonic species (35 species of Rotifera and 31 species of Cladocera and Copepoda) from the river Murray at Mannum, South Australia. He found that the microcrustaceans and rotifers were dominant during winter and summer seasons respectively. The Alknanda river contained low population of zooplankton in general but in those areas where the current was slow, the population was significant and was dominated by *Daphnia* sp. and *Cyclops* sp. (Badola and Singh, 1981). Sharma (1984) has shown that among Crustacea *Cyclops, Cypris, Daphnia* and *Diaptomus* were common, while the rotifers were represented by *Brachionus, Philodina* and *Mytilina* in Bhagirathi river. The maxima of zooplankton were seen in January and then steady decrease was observed with minima occurring in August. In Brahmaputra river system, the zooplankton were dominated by *Filinia, Lecane* and *Brachionus* among Rotifra, *Bosmina* and *Daphnia* among Cladocera and *Cyclops* among Copepoda (Jhingran, 1985). These plankters were more abundant in the upper stretches than in the lower stretches of this river. A full complement of 22 genera of zooplankton was rarely recorded and during April, May and August the zooplankton were altogether absent in Brahmputra river in Gauhati (Singh *et al.*, 1985).

The occurrence of *Arcella* and *Colpidium* (under Protozoa) and *Brachionus* and *Platyias* (under Rotifera) indicated high organic pollution in Reh river (Shanker *et al.*, 1986). The zooplanktonic population was found higher in the upstream of Ib river but rotifers were less dominant in the downstream as compared to other groups (Kar *et al.*, 1987). Among rotifers, *Filinia, Brachionus falcatus, B. forficula* and *Distyla* in Narmada river have been considered as sensitive forms and *Keratella tropica, Polyarthra* sp., *Monostyla* and *Philodina* as the pollution tolerant (Palharya and Malviya, 1988). Sankaran (1988) noted that the protozoan like *Vorticella, Epistylis, Carchesium, Paramecium* and *Sphaerotilus* abundantly found at highly polluted site of Adyar river. In Kshipra river, Kulshrestha *et al.* (1989) have found that *Keratella, Brachionus, Filinia, Daphnia, Cyclops* and

Cypridopsis were tolerant to wide range of physico-chemical conditions. In Morar river, 32 species of zooplankton were identified showing a greater species diversity as compared to other running waters studied. Among which, *Arcella discoides, Brachionus angularis, Lecane (L.) luna, L. (M.) bulla, Moina brachiata, Alona intermediata, Cypris* sp., *Cyprinotus gunningi* and *Mesocyclops thermocyclopoides* at station A, *Arcella discoides, Brachionus calyciflorus, B. angularis, Lecane (L.) luna, Moina brachiata, Mesocyclops thermocyclopoides* at station B, *Arcella discoides, Difflugia muriformis, Brachionus calyciflorus, B. angularis, Lecane (L.) luna, L. (M.) bulla, Filinia longiseta, M. brachiata* at station C, *A. discoides, B. quadridentatus, B. calyciflorus, B. angularis, Keratella tropica, L. (L.) luna, L. (M.) bulla, Polyarthra vulgaris, F. longiseta, F. opoliensis, M. brachiata, Stenocypris malcolmsonii, M. thermocyclopoides* at station D and *A. discoides, Centropyxis aculeata, B. quadridentatus, B. calyciflorus, B. angularis, K. tropica, L. (L.) luna, L. (M.) bulla, Cephalodella auriculata, P. vulgaris, F. longiseta, F. opoliensis, M. brachiata, Simocephalus exspinosus, Ceriodaphnia reticulata, Bosmina longirostris* at station E were found throughout the year.

The species occurred dominantly at all the stations during summer were *Difflugia muriformis, Centropyxis aculeata, Brachionus angularis, B. calyciflorus, B. quadricornis, Keratella tropica, Lecane (M.) bulla, Testidinella patina* and *Filinia longiseta*; during winter season *Brachionus patulus, B. bidentatus, Polyarthra vulgaris, Cypris* sp. and *Stenocypris malcolmsonii* and during rainy season *Arcella discoides, Difflugia muriformis, Platyias quadricornis, Lecane (L.) luna, Asplanchna brightwelli, Moina brachiata, Cyprinotus gunningi, Mesocyclops thermocyclopoides* and *Thermocyclops crassus*. Among rotifers *Asplanchna brightwelli, Cephalodella auriculata, Lepadella ovalis* and *Filinia opoliensis* were added to the rotifer fauna of the river since the study of Saksena and Kulkarni (1986). For Morar river, it may be stated that *Centropyxis aculeata, Cephalodella auriculata* and *Thermocyclops crassus* were sensitive to pollution as they were present at comparatively clean water areas. *Brachionus quadridentatus* and *Alona intermedia* were tolerant to mild pollution while *Arcella vulgaris, Brachionus calyciflorus, Keratella tropica, Lepadella ovalis, Filinia longiseta, F. opoliensis* and *Bosmina longirostris* were tolerant to high pollution. *Arcella discoides, Difflugia muriformis, Brachionus bidentatus, B. patulus, B. angularis, Platyias quadricornis, Mytilina ventralis, Lecane (L.) luna, L. (M.) bulla, Moina brachiata* and *Mesocyclops thermocyclopoides* were able to live in wide range at physico-chemical characteristics as they were observed at all the hypothetical stations on the river.

Haertel and Osterberg (1967) have established that contribution of *Eurytemora hirundoides* varied from 90 per cent to 100 per cent in the total planktonic population in Columbia river estuary. In Kali river, it was noted that the Protozoa were dominant group (90 to 93 per cent) and rotifers were 8 to 9 per cent in composition at polluted site (Verma and Dalela, 1975). The rotifers among total zooplankton at various upstream stations of Cauvery river was from 0.5 to 19.20 per cent while at downstream station they were from 58.19 to 87.9 per cent (Sampath *et al.*, 1979). Shiel (1979) obtained nearly 50 per cent brachionids in the rotifers of river Murray in South Australia. Saksena and Mishra (1988) have observed dominance of rotifers (38.94 per cent) over other groups namely Protozoa (37.81 per cent), Copepoda (8.35 per cent), Ostracoda (7.84 per cent) and Cladocera (7.0 per cent). In the river under study, it was noted that Rotifera (53.03 per cent) was dominant group followed by Cladocera (14.83 per cent), Ostracoda (10.96 per cent), Protozoa (8.18 per cent) and Copepoda (5.57 per cent) at station A, Rotifera (53.04 per cent) was dominant group followed by Cladocera (21.93 per cent), Copepoda (7.93 per cent), Protozoa (6.63 per cent) and Ostracoda (4.75 per cent) at station B, Rotifera (62.06 per cent) followed by Copepoda (8.38 per cent), Protozoa (7.97 per cent), Cladocera (7.90 per cent) and Ostracoda (5.54 per cent) at station C, Rotifera (65.20 per cent) were also dominated over other groups followed by Protozoa (9.93 per cent) Cladocera (8.68 per cent), Ostracoda (6.22 per cent) and Copepoda (5.12 per cent) at station D, Rotifera (56.77 per cent) followed by Cladocera (19.48 per cent), Protozoa (13.78 per cent), Copepoda (5.71 per cent) and Ostracoda (0.86 per cent) at station E. It may also be mentioned here that the rotifer population in general was enhanced with increased load of pollution in the river. A similar phenomenon has also been observed by several workers (Prabhavathy and Sreenivasan, 1977; Gannon and Stemberger, 1978 and Sampath *et al.*, 1979).

Michael (1964), Prabhavathy and Sreenivasan (1977) and Sampath *et al.* (1979) have shown rotifers dominating in hard and alkaline waters. This was also true in the present study. Haertel and Osterberg (1967) have shown that copepods population was greater when temperature was high and turbidity was low. Above observation regarding the turbidity was true in the present study also. Arora (1966) has shown that dissolved oxygen alone can influence the survival of rotifers while Prabhavathy and Sreenivasan (1977) suggested that rotifers are tolerant to low dissolved oxygen. In the present study, rotifers were inversely proportionate to the dissolved oxygen confirming the view of Arora (1966).

In most natural waters either few species having large number of individuals per species or few species with moderate populations existed while in other water bodies only few individuals were represented. But the pollution disturbs all such relationships as the number of species decreases drastically and the individuals of few species increase. This imbalance in zooplanktonic community in the systems could be used in monitoring the effect of pollution. Generally, the species diversity in clean water is high and in polluted waters it is low (Archibald, 1972). Wilhm and Dorris (1968) have computed diversity indices, based on information theory, for the communities above and below the point of introduction of different kind of pollutants into waters and have also concluded that the diversity was high in clean water and low in polluted waters. Trauben and Olive (1983) calculated the Shannon-Weaver diversity index (Shannon and Weaver, 1963) and noted that low values indicated excessive pollution in the river Cuyahoga in Ohio, USA. Verma *et al.* (1984) have also shown decreased community diversity in Eastern Kali river below the point of introduction of wastes into the river and then with time and distance of flow it gradually returned to typical level existed for fresh water. In Kshipra river, value of the diversity index was lowest at highly polluted station and highest at least polluted station (Kulshrestha *et al.*, 1989). In the present study also a similar pattern of species diversity existed among zooplankton. The values of Shannon-Weaver's index were decreased with increased organic pollution. It started decreasing from station B to station C and then with distance of flow it gradually returned at station E to a value which was very near to the value obtained at station A.

Conclusion

Under zooplanktonic fauna, 32 species belonging to 16 families and 24 genera have been identified from the Morar river. *Arcella vulgaris, A. discoides, Difflugia muriformis* and *Centropyxis aculeata* under Protozoa, *Brachionus quadridentatus, B. calyciflorus, B. falcatus, B. bidentatus, B. patulus, B. angularis, Keratella tropica, Platyias quadricornis, Mytilina ventralis, Lecane (L.) luna, Lecane (M.) bulla, Asplanchna brightwelli, Cephelodella auriculata, Polyarthra vulgaris, Lepadella ovalis, Testudinella patina, Filinia longiseta, F. opoliensis* under Rotifera, *Moina brachiata, Simocephalis exspinosus, Ceriodaphnia reticulata, Bosmina longirostris* and *Alona intermedia* under Cladocera, *Cypris* sp., *Cyprinotus gunningi* and *Stenocypris malcolmsonii* under Ostracoda and *Mesocyclops thermocyclopoides* and *Thermocyclops crassus* under Copepoda have been identified. Zooplanktonic species occurring dominantly at all the stations during summer were *Difflugia muriformis, Centropyxis aculeata, Brachionus angularis, B. calyciflorus, B. quadricornis, Keratella tropica, Lecane (M.) bulla, Testudinella patina* and *Filinia longiseta*; during winter season were *Brachionus patulus, B. bidentatus, Polyarthra vulgaris, Cypris* sp. and *Stenocypris malcolmsonii* and during rainy season were *Arcella discoides, Difflugia muriformis, Platyias quadricornis, Lecane (L.) luna, Asplanchna brightwelli, Moina brachiata, Cyprinotus gunningi, Mesocyclops thermocyclopoides* and *Thermocyclops crassus*. It may be stated that in Morar river, *Centropyxis aculeata, Cephallodella auriculata* and *Thermocyclops crassus* were sensitive to pollution, *Brachionus quadridentatus* and *Alona intermedia* were tolerant to mild pollution while *Arcella vulgaris, Brachionus calyciflorus, Keratella tropica, Lepadella ovalis, Filinia longiseta, F. opoliensis* and *Bosmina longirostris* were tolerant to high organic pollution. It is also stated that the population of rotifers was enhanced with increased load of organic pollution in the river.

The co-efficient of correlation was evident that a positive correlation existed between rotifers and hardness of water while a negative correlation occurred between rotifers and water temperature. The zooplankton population, in general, was having a high positive correlation with total phytoplankton population. At various stations, the rotifers had high positive correlation and low to high positive correlation with the Protozoa.

The values of Shannon-Weaver's index (1963) for zooplanktonic community were calculated and they were found in descending sequence from station A to station C and then an increase from station D to station E was evident. This also confirms that the zooplankton diversity was low at polluted sites in comparison to clean stations.

References

Adholia, U.N. (1979). Zooplankton of river Betwa, India. *Geo-Eco. Trop.*, **3**: 267-271.

Archibald, M. (1972). Diversity in some South African diatom associations and its relation to water quality. *Water Res.*, **6**: 1229-1238.

Arora, H.C. (1966). Responses of rotifera to variation in some ecological factors. *Proc. Indian Acad. Sci.*, **63**: 57-66.

Badola, S.P. and Singh, H.R. (1981). Hydrobiology of the river Alaknanda of the Garhwal Himalaya. *Ind. J. Ecol.*, **8**: 269-276.

Banerjea, S. and Motwani, M.P. (1960). Some observations on pollution of the Suvaon stream by the effluents of a sugar factory, Balrampur (U.P.). *Ind. J. Fish.*, **7**: 107-128.

Bilgrami, K.S. (1984). Ecology of the river Ganges: Impact of human activities and conservation of aquatic biota. Proc. IInd–An workshop on MAB projects, New Delhi, 23-25, March, pp. 16-17.

Cairns, J. Jr. (1979). Final summery. In: *Biological indicators and water quality* (A. James and L. Evison eds.), pp. 226-227. Wiley, New York.

Chacko, P.I. and Rajagopal, A. (1962). Hydrobiology and fisheries of the Ennore river near Madras from April 1960 to March 1961. *Madras J. Fish.*, **1**: 102-104.

Cronin, L.E., Joanne, C.D. and Hulbert, E.M. (1962). Quantitative seasonal aspects of zooplankton in the Delaware river estuary. *Chesapeake Science*, **3**: 63-93.

Gannon, J.E. and Stembergar, R.S. (1978). Zooplankton especially crustaceans and rotifers as indicators of the water quality. *Trans. Am. Micros. Soc.*, **77**: 16-35.

George, M.G., Qasim, S.Z. and Siddiqi (1966). A limnological survey of the Kali river with reference to fish mortality. *Environ. Health*, **8**: 262-269.

Ghosh, A., Rao, Z.H. and Banerjee, S.C. (1974). Studies on the hydrobiological conditions of a sewage fed pond with a note on their role in fish culture. *J. Inland Fish. Soc. India*, **6**: 51-61.

Green, J. (1960). Zooplankton of the river Sokoto: The Rotifera. *Proc. Zool. Soc. Lond.*, **135**: 491-523.

Haertel, L. and Osterberg, C. (1967). Ecology of zooplankton, benthos and fishes in the Columbia river estuary. *Ecology*, **48**: 459-472.

Halbach, U., Siebert, M., Wastmayear, M. and Wissel, C. (1983). Population ecology of rotifers as a bioassay tool for ecotoreicological testes in aquatic environment. Ecotoreical. *Environ. Safety*, 7: 484-513.

Jhingran, V.G. (1985). *Fish and fisheries of India*. Hindustan Publishing Corporation (India), New Delhi, India.

Kar, G.K., Mishra, P.C., Dash, M.C. and Das, R.C. (1987). Pollution studies in river Ib III: Plankton population and primary productivity. *Ind. J. Environ. Hlth.*, **29**: 322-329.

Khare, A., Varma, R.L., Hazela, K.P. and Pandey, G.N. (1979). An ecological study of river Ganga at Kanpur-I. Influence of pollutants on Macroorganisms. *Proc. Symp. Environ. Biol.* 383-395. The Academy of Environmental Biology, India.

Koste, W. (1978). Rotatoria. Die Radertiere Mittleuropas. Begrundet Von Mox Voigt. Uberordnung Monogononta. Gebruder Borntraoger, Stuttgart.

Kulshrestha, S.K., Adholia, U.N., Bhatnagar, A., Khan, A.A., Saxena, M. and Baghail, M. (1989). Studies on pollution in river Kshipra: Zooplankton in relation to water quality. *Int. J. Ecol. Environ. Sci.*, **15**: 27-36.

Lair, N. (1980). The rotifer fauna of the river Loire (France) at the level of the nuclear power plants. *Hydrobiologia*, **73**: 153-160.

Michael, R.G. (1964). Limnological investigations on pond plankton, Macrofauna and chemical constituents of water and their bearing on fish production. Ph.D. Thesis, University of Calcutta, Kolkata, India.

Mishra, S.R. and Saksena, D.N. (1990). Seasonal abundance of the zooplankton of waste water from the industrial complex at Birla Nagar (Gwalior) India. *Acta Hydrochim. Hydrobiol.*, **18**: 215-220.

O'Doherty, E.C. (1985). Stream-dwelling copepods: Their life history and ecological significance. *Limnol. Oceanogr.*, **30**: 554-564.

Palharya, J.P. and Malviya, S. (1988). Pollution of the Narmada river at Hoshangabad in Madhya Pradesh and suggested measures for control. In: *Ecology and Pollution of Indian Rivers* (R.K. Trivedy ed), pp. 54-85. Ashish Publishing House, New Delhi, India.

Patil, H.S. and Rodgi, S.S. (1985). Studies on self purification of flowing municipal waste water at Dharwad. In: *Current Pollution Researches in India*, (R.K. Trivedy and P.K. Goel eds), pp. 153-161, Environmental Publications, Karad, India.

Patten, B.D. (1962). Species diversity of the net phytoplankton of Raritan Bay. *J. Marine Sci.*, **20**: 57-75.

Prabhavathy, G. and Sreenivasan, A. (1977). Ecology of warm fresh water zooplankton of Tamil Nadu. Proc. Symp. Warm Water Zool., pp. 319-329. N.I.O. Goa Spl. Publication.

Prakash, C., Rawat, D.C. and Graver, P.P. (1978). Ecological study of the river Jamuna. IAWC Tech. Annual, **5**: 32-45.

Rajyalakshmi, T. and Premswarup, T.V. (1975). Primary productivity in river Godavari. *Indian J. Fisheries*, **22**: 205-214.

Rao, K.S. and Shrivastava, S. (1989). Studies on biological monitoring of water quality in Chambal and Khan rivers of Central India. *Geobios.*, **16**: 78-82.

Ray, P. and David, A. (1966). Effect of industrial wastes and sewage upon the chemical and biological composition and fisheries of the river Ganga at Kanpur (U.P.). *Environ. Hlth.*, **8**: 307-339.

Ruttner-Kolisko, A. (1974). Plankton rotifers, biology and taxonomy. E. Schweizerbart'sche Verlagsbuchhandlung, Stuttgart.

Saksena, D.N. and Kulkarni, N. (1986). On the rotifer fauna of two sewage channels of Gwalior (India). *Limnologica*, **17**: 139-148.

Saksena D.N. and Mishra, S.R. (1988). Characterisation of wastes from industrial complex at Birla Nagar, Gwalior. *IAWPC Tech. Ann.*, **15**: 163-167.

Sampath, V., Sreenivasan, A. and Ananthanarayanan, R. (1979). Rotifers as biological indicators of water quality in Cauvery river. *Proc. Symp. Environ. Biol.*, 441-452. The Academy of Environmental Biology, India.

Sankaran, V. (1988). Pollution studies in Caurvery and Adyar rivers in Tamil Nadu. In: *Ecology and Pollution of Indian Rivers* (R.K.

Trivedy ed), pp. 321-335, Ashish Publishing House, New Delhi, India.

Sastry, A.G.R. (2002). Prepollution status of composition, diel and tidal fluctuations at plankton in the Vasishta Godavary Estuary, East Coast of India. In: *Management of Aquatic Habitats* (Smriti Ratna Mishra ed), pp. 105-115, Daya Publishing House, New Delhi, India.

Sehgal, K.L. (1983). *Planktonic copepods of fresh water ecosystems*. Interprint, Mehta House Naraina II, New Delhi, India.

Shanker, V., Sangu, R.P.S. and Joshi, G.C. (1986). Impact of distillery effluent on the water quality and ecosystem of river Reh in Doon valley. *Poll. Res.*, **5**: 137-142.

Shannon, C. E. and Weaver, W. (1963). The mathematical theory of communication. University of Illinois Press, Urbana.

Sharma, R.C. (1984). Potamological studies on lotic environment of the upland river Shagirathi of Garhwal Himalaya. *Environ. Ecol.*, **2**: 239-242.

Shiel, R.J. (1979). Synecology of the rotifera of the river Murray, South Australia. *Aust. J. Mar. Freshwater Res.*, **30**: 255-263.

Singh, R.K., Kolekar, V. and Chandra, R. (1985). Comparative study of certain ecological aspects of river Brahmaputra and Dighali beel in Assam. *Proc. Nat. Acad. Sci. India*, **22**: 21-30.

Smet, W.H.O. and Evens, F.M.J.C. (1972). A hydrobiological study of the polluted river Lieve (Ghent, Belgium). *Hydrobiologia*, **39**: 91-154.

Sprules, W.G. (1977). Crustacean zooplankton communities as indicator of limnological conditions: An approach using principal component analysis. *J. Fish. Res. Bd. Can.*, **34**: 962-975.

Trauben, B.K. and Olive, J.H. (1983). Benthic macroinvertebrate assessment of water quality in the Cuyahoga river, Ohio: An update. *Ohio J. Sci.*, **83**: 209-212.

Vass, K.K., Raina, H.S., Zutshi, D.P. and Khan, M.A. (1977). Hydrobiological studies on river Jhelum. *Geobios*, **4**: 238-242.

Venkateswarlu, T. and Jayanti, T.V. (1968). Hydrobiological studies of the river Sabarmati to evaluation water quality. *Hydrobiology*, **31**: 442-448.

Verma, S.R. and Dalela, R.C. (1975). Studies on the pollution of the Kali river by industrial wastes near Mansurpur, Part II: Biological index of pollution and biological characteristics of the river. *Acta Hydrochim. Hydrobiol.*, **3**: 259-274.

Verma, S.R., Sharma, P., Tyagi, A., Rani, S., Gupta, A.K. and Dalela, R.C. (1984). Pollution and saprobic status of eastern Kali river. *Limnologica* (Berlin), **15**: 69-133.

Verma, S.R., Tyagi, A.K. and Dalela, R.C. (1978). Physico-chemical and biological characteristics of Kadarbad in Uttar Pradesh. *Indian Environ. Hlth.*, **20**: 1-13.

Victor, R. and Fernando, C.H. (1979). The freshwater ostracods (Crustacea: Ustracoda) of India. Records of the Zoological Survey of India. **74**: 147-242.

Ward, H.B. and Wipple, G.C. (1959). *Freshwater biology*, 2nd edition. John Wiley and Sons, New York.

Welch, P.S. (1952). *Limnological methods*. The Blonkiston Company, Philadelphia.

Wilhm, J.L. (1967). Comparison of some diversity indices applied to populations of the benthic macroinvertebrates receiving organic wastes. *J. Water Poll. Cont. Feb.*, **39**: 1673-1683.

Wilhm, J.L. and Dorris, T.C. (1968). Biological parameters for water quality criteria. *Bioscience*, **18**: 477-481.

2

Surface Water Pollution in and Around Akola District of Maharashtra

☆ *Anil K. Fokmare & Mohammad Musaddiq*

Introduction

Water is a great gift of nature. There is a common saying that "No life without water." It is a universal solvent and is a major constituent of all living organisms, is a well known phenomenon (Gupta and Gupta, 1997). Water is distributed in nature as surface and groundwater in different forms, like rain water, river water, spring water etc. The rain water constitutes to one of the most important and largest sources of water. The original river water is very pure but it takes up suspended impurities as it flows through the plains. The snows that melt in the mountain regions flow in the form of rivers.

Water pollution is due to the alteration in physical, chemical and biological characteristics which may lead to harmful effect on human and aquatic biota (Abassi and Abassi, 1989). It is the addition of excess of undesirable substances to water that makes it harmful for man, animal and aquatic life.

Water pollution is caused due to harmful solids, liquids or gases that are non-permissible, undesirable, unpleasant and objectionable (Shrivastava *et al.*, 2001). Thus water pollution disturbs the normal uses of water for irrigation, agriculture, industries, public water supply and aquatic life.

Water is required everywhere, without which neither the life nor any development is possible. The daily demand of drinking water of a man is normally 7 per cent of his body weight. Thus making it is vital for healthy growth. But it may become harmful for life if one uses water contaminated with harmful or toxic substances and pathogenic microorganisms coupled with poor sanitation (Gupta and Gupta, 1997). The sources of water supply depend almost upon the rainfall and the resulting water bodies like springs, river, lakes etc.

The most important source of drinking water for almost 70 per cent of Indian population is groundwater and considered less polluted. But lack of sanitation, improper waste and surface water contamination and 40 per cent or more of the disease outbreaks were attributed to polluted water consumption (Narain Rai and Sharma, 1995).

The pollution loads contributed to the water courses are physical, chemical and biological in nature. Majority of the industries in India have been located in and around the cities. There are 142 major cities in India of which 112 cities are distributed among 14 river basins, out of balance 30 cities, 17 are coastal, and 13 non-coastal. The river Ganga carries pollution load from 48 major cities in the country (Musaddiq, M., 2001).

Groundwater pollution is usually traced back to four main origins *viz.*, Industrial, domestic, agricultural and over exploitation. Studies carried out in India reveal that one of the most important cause of groundwater pollution is unplanned urban development without adequate attention to sewage and waste disposal. The incidence of ground water pollution is highest in urban areas where large volume of wastes are concentrated and discharged into relatively small areas. The groundwater contamination, however, is detected only some time after the sub-surface contamination begin (Bhujal News, 1997).

Surface water is a major source of water for all big cities. Bacteriological and physico-chemical studies had been carried out in recent years at different places, *viz.*, Bhadravati (David, 1956), Madras (Ganapati, 1956), Nagaur (Gupta, 1991), Pondicherry (Abbasi and Vinithan, 1999), Kolleru Lake (Sreenivasa *et al.*, 1999), Ooty (Sivakumar, *et al.*, 2000), Aurangabad (Thorat and Masarrat, 2000), Beed (Himesh, *et al.*, 2000), Akola (Musaddiq, M., 2000; Fokmare and Musaddiq, 2001).

Due to increased industrial and mining activities, a number of metals find their way to water bodies. Many reports have appeared in literature dealing with environmental pollution due to heavy metals in India (Chandra, 1980). The detection of high levels of manganese in the groundwater in certain parts of the Ganga basin, elevated levels of molybdenum, arsenic, selenium and manganese in the soil and water in some regions of India and possibly part of the heavy metal load of the coastal and estuarine zones in this sub-continent arise from geo-chemical activity (Krishna Murti, 1987, 1989).

Advancement of human civilization has put serious questions to the safe use of groundwater for domestic uses. Studies and survey on metal pollution in surface and groundwater had been carried in recent years at different places *viz.*, Lucknow (Seth *et al.*, 1957), Baroda (Agrawal *et al.*, 1978), Mumbai (Krishnamurti, 1987), Kanpur (Singh and Mahaveer, 1997) and Chennai (Ayyadurai *et al.*, 1998).

Due to improper sanitation, industrial waste disposal and faulty water resource management, the percentage of water borne illnesses are increasing day by day. In developing countries, 80 per cent infectious diseases are water borne and 50 per cent of the deaths among the children are due to diarrhoeal diseases (Manja and Kaul, 1992). The most common water borne infections are typhoid, cholera, shigellosis, viral associated diarrhoea and infectious hepatitis (Pelczar *et al.*, 1988). The enumeration of total viable count, aerobic heterotrophs, *Total Coliforms* and *E. coli* generally indicates the extent of water pollution (Wright, 1986 and Pathak *et al.*, 1988).

Most of the heavy metals, either alone or in certain compounds, exert a detrimental effect upon microorganisms. The most effective are mercury, silver, copper, zinc, cadmium, chromium etc. coagulate cytoplasmic proteins resulting in damage or death of the cell. Salts of heavy metals are also precipitants and in high concentrations such salts could cause the death of a cell (Pelczar *et al.*, 1988).

The discharge of heavy metals by industries pose a serious water problem due to the toxic properties of these metals and their adverse effects on aquatic microorganisms. A number of studies have been conducted on the effect of heavy metals toxicity on metabolism (Lyr, 1997; Corbett *et al.*, 1984; Khillare and Davne, 1991; Collins and Stotzky, 1992). Heavy metal pollutants cause direct toxicity both to humans and other living beings due to their presence beyond specified limits (Rai *et al.*, 1998).

Akola is a district place of central Maharashtra (Figure 2.1). Morna river and Purna river are one of the scared and most important river of the Akola and it provide 60 per cent of water requirement of the population and industries, 40 per cent population depends on groundwater.

According to statistical data of Akola district (Finance and Statistics, Government of Maharashtra, 1998), several number of deaths occurred by water borne diseases in 1998 at Akola. The river Morna of Akola district has been grossly polluted from the city (Musaddiq, 2000). Contamination of Morna and Purna river water, due to untreated domestic waste water from various parts of Akola city is directly discharged into river leading to gross pollution. The Morna river receive millions of litres of sewage, domestic waste, industrial and agricultural effluents containing substances varying in characteristics from simple nutrients to highly toxic substances such as metals.

Kapshi lake is situated about 18 km to the Akola city and Lonar lake (creator) is situated about 1 km to the south-west of village Lonar (Buldhana district) and around 90 km from Akola. The water spread of lake expands and contracts with seasonal precipitation.

Current interest in the state of environment has resulted in increased research to evaluate the global impacts of pollution on the biosphere.

Materials and Methods

Fifty water samples of each surface water source like Morna river, Purna river, Kapshi lake and eight water sample of Lonar lake were collected in glass bottles during October 2002 to November 2002. To achieve uniformity of observations sampling time was uniformly maintained in morning between 10:00 am to 11:00 am.

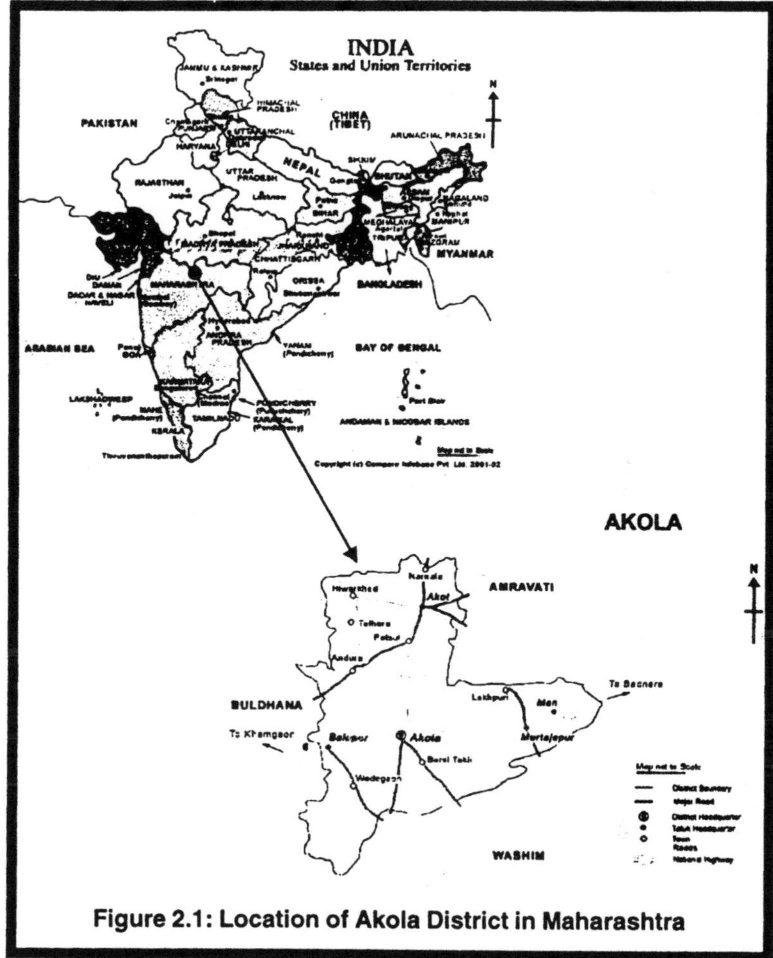

Figure 2.1: Location of Akola District in Maharashtra

The physico-chemical parameters were analysed by employing the standard methods (NEERI Manual, 1988), Kodarkar *et al.*, 1998 and Trivedi and Goel (1986). The results are compared with permissible limits laid down by BIS (IS: 2490, 1981; IS: 10500, 1996) and APHA (1985).

Results and Discussion

The analytical results of the surface water (Morna river, Purna river, Kapshi lake and Lonar lake) for various physico-chemical parameters are presented in the Tables 2.1–2.4 and Figures 2.4–2.8.

Figure 2.2: Sampling Site of Morna River

Figure 2.3: Sewage Mixing with Morna River

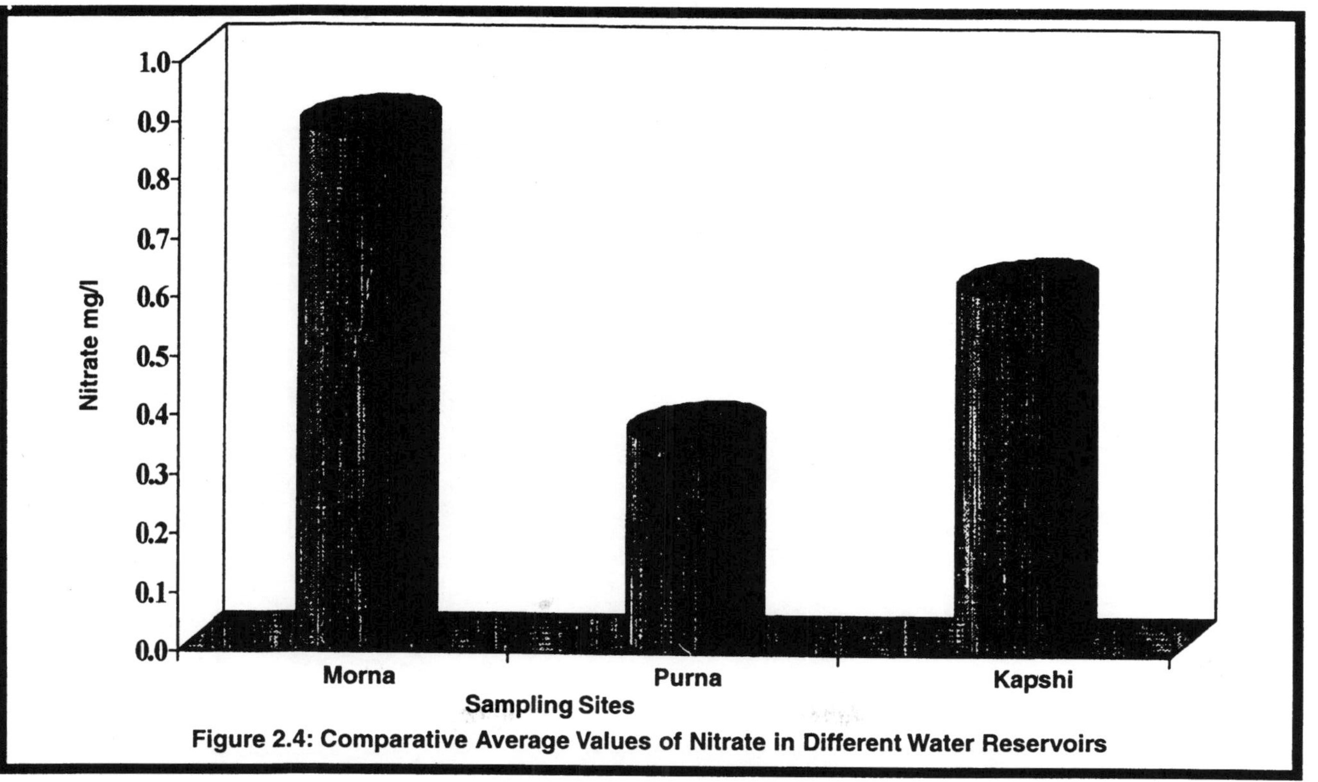

Figure 2.4: Comparative Average Values of Nitrate in Different Water Reservoirs

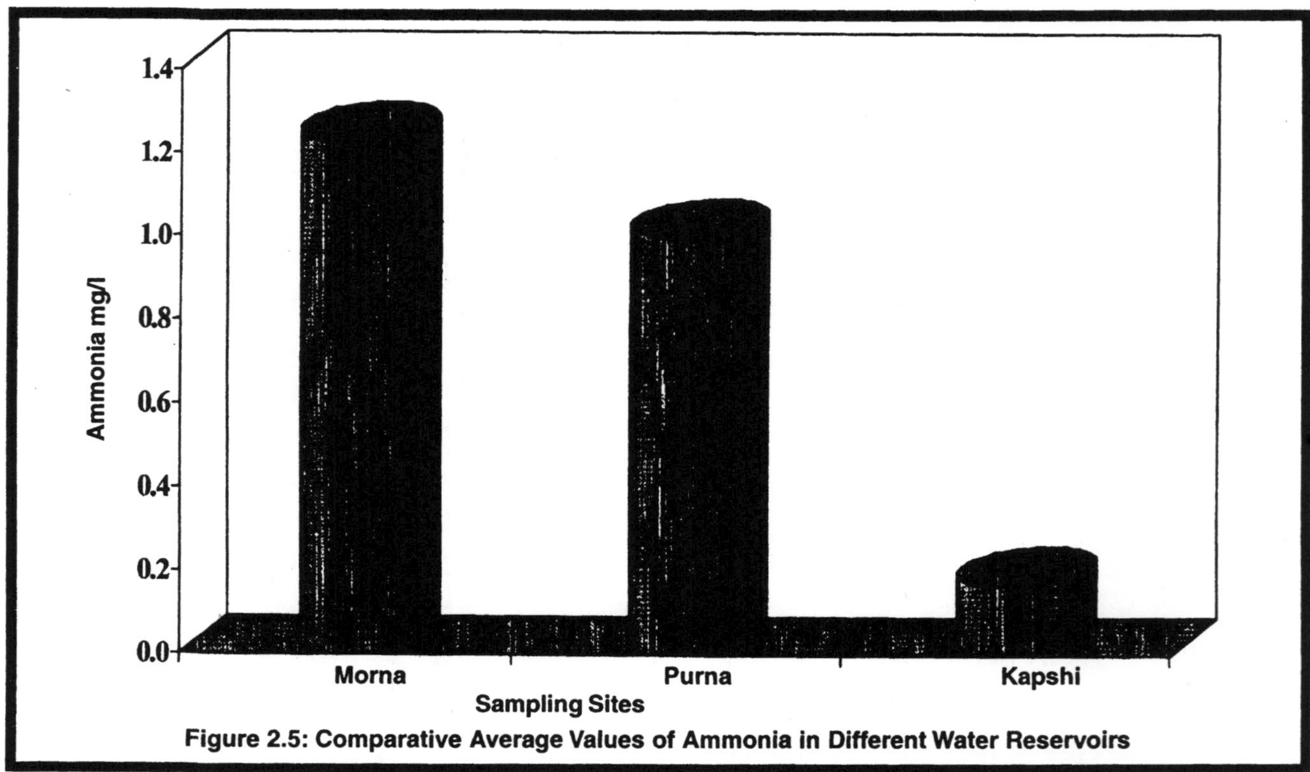

Figure 2.5: Comparative Average Values of Ammonia in Different Water Reservoirs

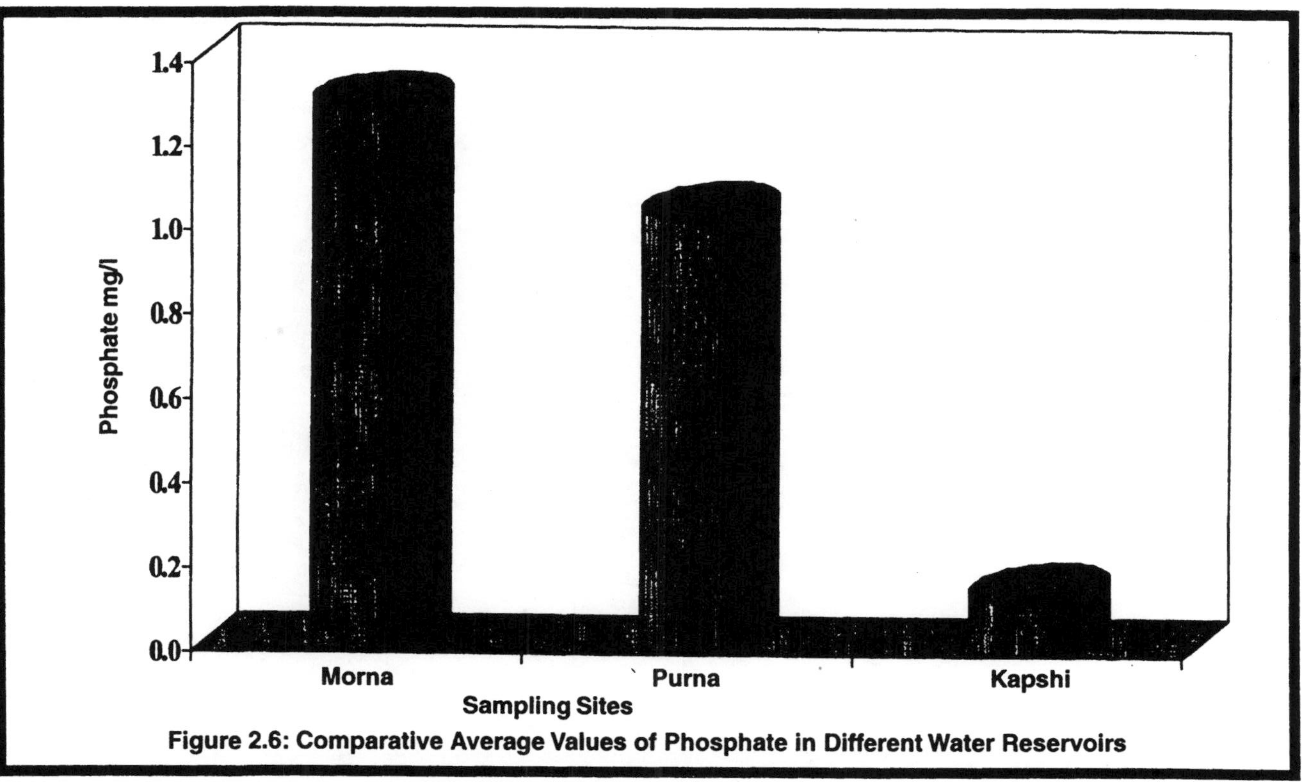

Figure 2.6: Comparative Average Values of Phosphate in Different Water Reservoirs

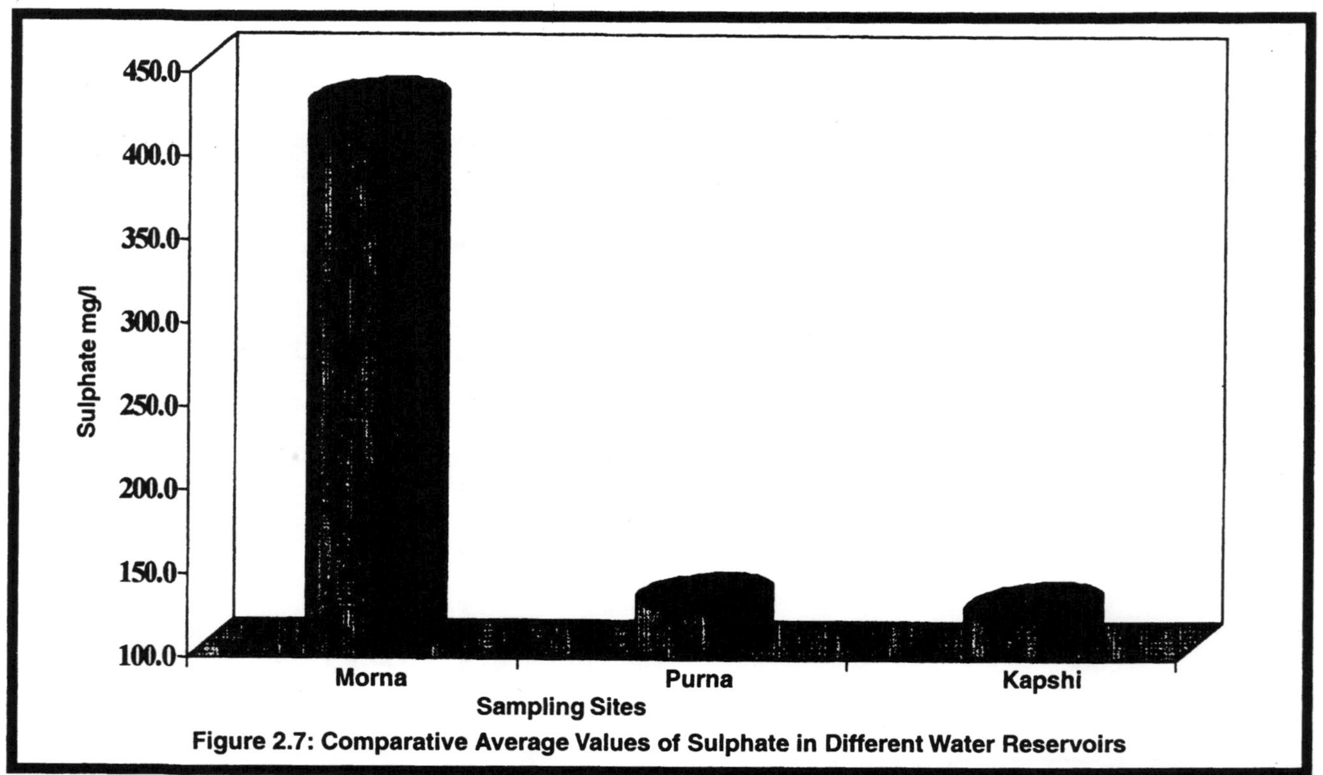

Figure 2.7: Comparative Average Values of Sulphate in Different Water Reservoirs

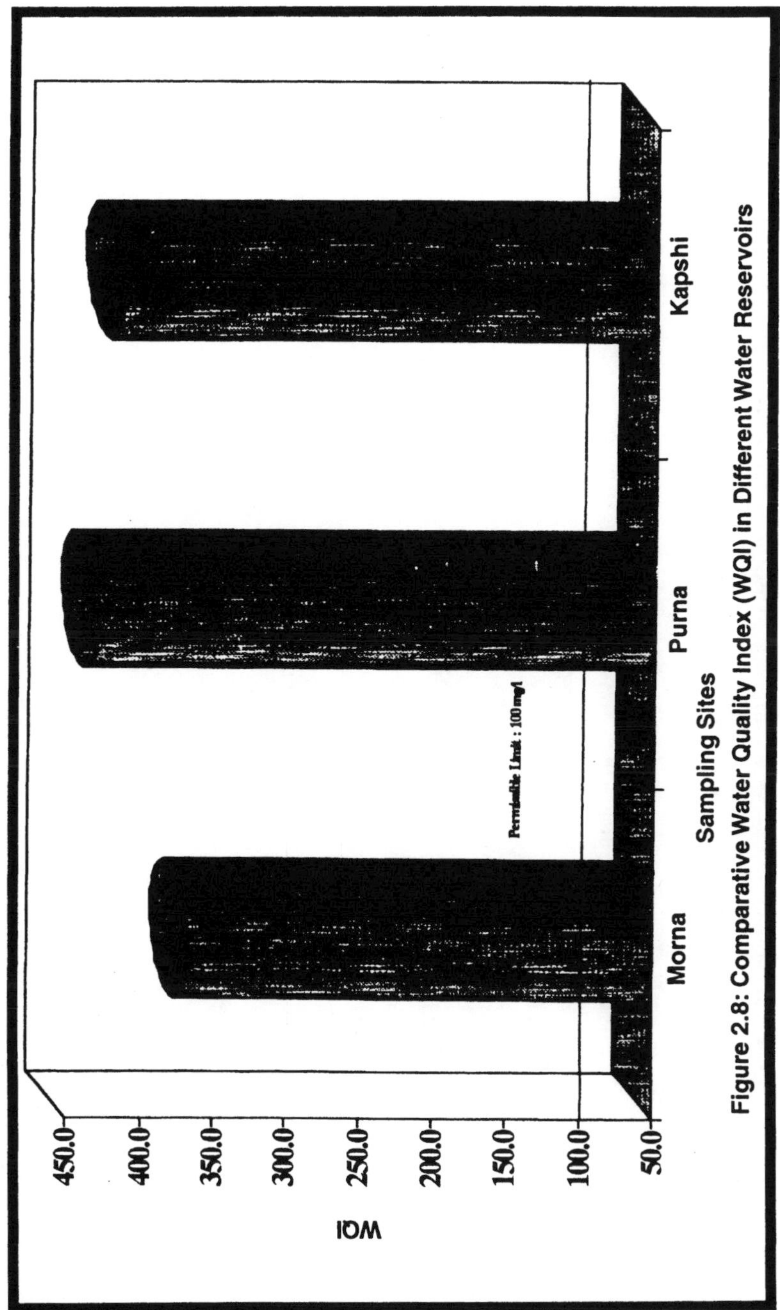

Figure 2.8: Comparative Water Quality Index (WQI) in Different Water Reservoirs

Table 2.1: Physico-chemical Parameters of Morna River at Akola District

Sl. No.	Location	Physico-chemical Parameters in mg/l Except pH, Temperature and EC																	
		pH	Temp. °C	EC µS/cm	TDS	TSS	T.Alk.	TH	Ca	Mg	Cl	Sal.	NO_3	NH_4	SO_4	PO_4	DO	BOD	COD
1.	Chandur	6.2	28	1080	519	43	101	184	68	28.3	175	315.9	0.81	0.4	320	0.8	9.8	18	33.5
2.	Khadki	6.1	28	1079	522	39	100	179	66	27.57	173	312.29	0.8	0.565	317	0.9	9.9	20	33.2
3.	Kaulkhed	6	27	1080	516	38	98	180	63	28.54	175	315.9	0.82	0.795	322	1	10.1	12	34.1
4.	Hingana	6.2	28	1077	520	40	101	176	68	26.35	175	315.9	0.8	0.82	319	1.4	10.1	18	33.3
5.	Khadan	6.2	29	1082	520	37	101	178	68	26.84	174	314.1	0.84	0.82	320	1.65	10.2	17	30.8
6.	Sindhicamp	6.4	28	1083	524	42	104	181	73	26.35	177	319.51	0.84	0.88	321	1.8	8.8	24	28.1
7.	Anikat	6.2	30	1078	521	41	104	180	68	27.32	175	315.9	0.83	0.91	318	1.45	8.5	19	20.5
8.	Kholeshwar	6.3	29	1081	518	40	102	182	70	27.32	176	317.71	0.82	0.91	323	1.87	8.6	16	30.6
9.	Near Fort	8.3	28	1802	853	50	290	500	66	105.9	398	718.42	0.93	2	861.7	0.8	8.8	18.3	34.6
10.	Big Bridge	8.4	28	1800	860	52	290	510	68	107.8	396	714.81	0.95	2.11	862.6	0.9	9.5	18.9	40.1
11	Small Bridge	8.8	28	1800	856	53	306	500	71	104.7	400	722.03	0.94	2.15	861.8	1	9.8	18.4	45.3
12.	Near Industries	8.8	27	1798	859	50	294	505	69	106.4	401	723.83	0.95	1.85	861.9	0.85	8.3	18.7	54.6
13.	Railway Bridge	8.9	28	1801	857	52	306	500	67	105.7	404	729.25	0.93	1.85	861.8	1.2	8.6	18.4	60.3
14.	Balapur	9.2	29	1800	851	48	310	500	69	105.2	402	725.64	0.97	1.89	861.2	1.18	8.1	18.1	45.1
15.	Near Fort Balapur	9.3	29	1800	850	48	310	490	72	102	399	720.22	0.93	1.62	861	1	9.3	17.9	46.8

Contd...

Table 2.1–Contd...

Sl. No. Location	Physico-chemical Parameters in mg/l Except pH, Temperature and EC																	
	pH	Temp. °C	EC μS/cm	TDS	TSS	T.Alk.	TH	Ca	Mg	Cl	Sal.	NO_3	NH_4	SO_4	PO_4	DO	BOD	COD
16. Kholeshwar	8.8	28	1799	854	47	294	495	70	103.7	400	722.03	0.94	1.62	862.4	1.3	10.3	18.5	51.2
17. Kholeshwar	8.3	28	1723	851	48	219	432	73	87.59	270	487.38	0.88	1	472.4	1.12	9.8	15.1	20.6
18. Kholeshwar	8	28	1719	848	52	222	434	75	87.59	269	485.57	0.92	1	472.4	1.1	9.5	14.6	20.1
19. Kholeshwar	8.1	29	1717	853	48	217	437	74	88.57	265	478.35	0.91	1	472	1.4	10.1	15.3	25.6
20. Rajanda 1	8.3	28	1722	847	45	218	436	72	88.81	272	490.99	0.87	1	472.2	1.5	9.8	14.3	26.8
21. Rajanda 2	8.2	28	1721	848	50	223	432	79	86.13	275	496.4	0.9	1	472.4	1.1	7.5	14.5	30.5
22. Anikat	8	28	1718	852	50	220	436	78	87.35	268	483.77	0.88	1	471.6	1.12	8.9	15	66.3
23. Anikat	8.2	28	1720	852	53	221	434	80	86.37	271	489.18	0.9	1.55	472.6	1.3	9.6	15	56.1
24. Sindhi Camp	8.4	27	1720	849	46	219	431	77	86.37	270	487.38	0.86	1.5	472.4	1.5	10.8	14.6	34.8
25. Kholeshwar	8.7	28	1550	880	65	165	488	75	100.8	210	379.08	0.89	1.8	296.6	1.45	11.2	19.6	40.5
26. Rajput Pura	8.9	28	1552	872	70	168	491	78	100.8	218	393.5	0.92	2.2	296.6	1.9	8.8	19.3	45.8
27. Sindhi Camp	8.7	28	1545	862	70	165	497	75	103	230	415.18	0.9	1.11	296.1	1.6	6.8	18.8	40.8
28. Rajput Pura	8.6	27	1554	869	67	164	499	71	104.43	215	388.1	0.93	1.2	296.4	2	7.8	19.1	50.1
29. Akola Fort	8.5	27	1552	877	74	162	494	75	102.23	217	391.71	0.95	1.35	296	2.5	9.5	19.4	50.2
30. Akola Fort	8.8	28	1559	873	72	166	502	79	103.21	225	406.15	0.94	1.8	295.9	1.3	9.8	19.3	48.2

Contd...

Table 2.1–Contd...

Sl. No.	Location	Physico-chemical Parameters in mg/l Except pH, Temperature and EC																	
		pH	Temp. °C	EC µS/cm	TDS	TSS	T.Alk.	TH	Ca	Mg	Cl	Sal.	NO_3	NH_4	SO_4	PO_4	DO	BOD	COD
31.	Big Bridge	8.7	27	1550	865	77	165	496	75	102.72	222	400.74	0.91	1	297	1.4	10.1	19	48.5
32.	Big Bridge	8.7	28	1554	820	79	165	493	72	102.72	223	402.52	0.92	1	297.1	1.6	10.5	19.1	48.2
33.	Big Bridge	7.6	28	1243	883	76	130	203	75	31.23	277	500.01	0.91	0.95	362.5	2.4	11.6	21	48.2
34.	Big Bridge	7.9	27	1240	882	77	145	203	77	30.5	269	485.57	0.98	0.85	363.4	1.1	9.8	23	50.1
35.	Kala Maroti	8	27	1244	880	75	150	197	73	30.25	275	496.4	0.96	1.1	362.9	2	9.9	21	50.2
36.	Kala Maroti	7.8	28	1250	881	78	140	200	78	29.76	272	490.99	0.94	0.95	362.9	1.6	10.1	20	48.5
37.	Small Bridge	7.9	28	1249	886	79	145	198	79	29.03	263	474.74	0.97	0.82	362.4	1.5	10.2	24	48.5
38.	Small Bridge	7.7	29	1246	884	78	135	201	77	30.25	268	483.77	0.99	1.1	362.5	1.1	10.1	24	40.3
39.	Small Bridge	7.8	29	1241	885	81	140	200	76	30.25	271	489.18	0.93	1.4	363	0.88	11.2	23	40.6
40.	Near Industries	7.7	28	1247	883	80	135	199	81	28.79	265	478.35	0.93	1.15	362	0.68	10.8	20	45.1
41.	Near Industries	8	29	1115	710	57	115	150	57	22.69	227	409.76	0.8	2	280	1.1	9.8	18.4	50.1
42.	Near Industries	7.6	28	1108	700	61	111	148	60	21.47	217	391.71	0.83	1	280.2	0.6	10.7	18	50.5
43	Lakadganj	7.8	28	1111	706	63	113	150	61	21.71	223	402.54	0.82	1.95	280.2	1	11.1	18.4	40.5
44.	Lakadganj	7.3	28	1110	702	58	108	150	63	21.22	219	395.32	0.8	0.96	277	0.9	10.9	18.2	45.8
45.	Mahan	7	29	1112	696	62	105	152	58	22.93	213	384.49	0.79	0.88	280	1.8	10.3	17.9	34.1

Contd...

Table 2.1–Contd...

Sl. No.	Location	Physico-chemical Parameters in mg/l Except pH, Temperature and EC																	
		pH	Temp. °C	EC µS/cm	TDS	TSS	T.Alk.	TH	Ca	Mg	Cl	Sal.	NO_3	NH_4	SO_4	PO_4	DO	BOD	COD
46.	Mahan	7.7	28	1109	706	55	112	148	62	20.98	223	402.54	0.77	1.5	280.6	0.8	11.1	18.4	30.2
47.	Mahan	7.2	28	1110	704	59	107	150	60	21.96	221	398.93	0.81	0.82	280.5	1.1	10.9	18.2	28.5
48.	Lakadganj	7.4	27	1105	700	65	109	152	59	22.69	217	391.71	0.78	1	280.2	0.7	10.6	18	55.1
49.	Patur 1	6.6	28	1020	607	48	94	120	60	14.64	180	328.54	0.78	0.6	250.1	1.5	11.6	15	20.1
50.	Patur 2	6.9	28	1030	611	55	99	115	61	13.17	178	321.32	0.87	0.75	250.2	1	11.7	15.4	22.1
	Average	7.82	28.04	1402	772.4	57.86	169.6	318	70.4	60.44	255	462.11	0.88	1.23	424.8	1.29	9.83	18.30	40.47
	S.D.	0.92	0.67	294.5	133.5	13.54	70.1	156	6.64	37.35	72.1	130.08	0.06	0.47	203.3	0.43	1.08	2.63	11.34
	% C.V.	11.76	2.39	21.00	17.28	23.40	41.33	49.50	9.43	61.79	28.20	28.15	6.81	38.21	47.85	33.33	10.98	14.37	28.02
	S.E.	11.597	2.333	0.106	0.095	0.521	0.508	11.290	11.304	46.310	5.858	3.235	69.774	5.181	0.182	3.969	1.907	0.998	@

S.D.: Standard deviation; % C.V.: Percentage coefficient variance; S.E.: Standard error.

Table 2.2: Physico-chemical Parameters of Purna River at Akola District

Sl. No.	Location	pH	Temp. °C	EC µS/cm	TDS	TSS	T.Alk.	TH	Ca	Mg	Cl	Sal.	NO_3	NH_4	SO_4	PO_4	DO	BOD	COD
		Physico-chemical Parameters in mg/l Except pH, Temperature and EC																	
1.	Katepurna	7	27,5	800	300	40	192	90	60	7.324	141.8	255.98	0.3	0.6	110	0.5	8	14.2	80.1
2.	Temple	7	28	900	600	90	192	160	110	12.2	354.5	639.9	0.55	0.4	166	1.1	8	16.2	80.2
3.	Bridge	9.8	28	835	680	112	340	168.5	108	14.76	283.6	511.9	0.56	2	161	1.2	6	18.8	60.3
4.	Mahan	7.3	28	840	750	135	200	160	130	7.32	284	512	0.55	1.5	161	1.1	7	19.2	60.3
5.	Keliveli	7.3	28	840	730	140	205	175	135	9.76	212.7	384	0.6	1.5	161.1	1	8.8	13.2	55.3
6.	Gandhinagar	6.3	27	650	250	140	198	90	60	7.32	212.7	384	0.45	1.11	162	1	8	11.7	55.6
7.	Near Bridge	8.8	29	650	350	150	250	103	75	6.83	283.6	511.92	0.45	1.2	165	2	5	11.6	55.6
8.	Mhaisang	8.1	27	730	630	144	260	150	110	9.76	354.5	639.9	0.4	0.91	166	1.5	6	18.3	40.3
9.	Mhaisang	8.8	28	750	620	150	265	185	125	14.6	141.8	255.98	0.45	0.82	160	1.5	7	20	50.3
10.	Katyar	8.9	28	760	645	148	280	208	140	14.5	212.7	384	0.5	1	170	2	8	15	45.6
11.	Kinkhed	6.81	27	530	610	305	160	60	42	4.392	118.4	213.74	0.2	0.6	105	0.5	6.2	17.2	40.3
12.	Donwada	7.31	27	210	630	120	210	80	31.5	11.83	133.2	240.45	0.2	0.7	110	1.1	6.5	16.5	45
13.	Gopalkhed	7.8	28	210	640	305	200	40	25.2	3.611	118.4	213.74	0.3	1	110	1.1	8.5	12.8	60.3
14.	Nairatwairat	6.11	29	101	650	218	340	60	16.8	10.54	133.2	240.45	0.4	1	115	0.5	6	18.8	60.5
15	Kapileshwar	6.5	27	630	630	205	250	80	23.1	13.88	148	267.17	0.2	1.1	160	1	7	19.5	60.1

Contd...

Table 2.2–Contd...

Sl. No.	Location	Physico-chemical Parameters in mg/l Except pH, Temperature and EC																	
		pH	Temp. °C	EC µS/cm	TDS	TSS	T.Alk.	TH	Ca	Mg	Cl	Sal.	NO_3	NH_4	SO_4	PO_4	DO	BOD	COD
16.	Katepurna	7.61	27	810	320	140	308	70	25.2	10.93	162.8	293.88	0.45	1.1	150	1.1	6.5	16.4	80.1
17.	Katepurna	8.11	27	555	460	105	315	40	16.8	5.66	103.6	187.02	0.21	0.6	111	0.5	6.3	16.2	78.1
18.	Katepurna	8.2	28	301	560	140	320	60	31.5	6.95	118.4	213.74	0.4	0.7	125	1.2	6.4	16.5	60.3
19.	Katepurna	6.61	27	280	481	160	221	70	31.5	9.394	133.2	240.45	0.5	0.8	140	1.1	8	11.3	40.2
20.	Katepurna	7.1	28	290	360	118	310	80	42	9.27	148	267.17	0.4	0.5	120	0.5	7.5	18.3	30.3
21	Big Bridge	6.81	28.5	305	620	280	203	60	42	4.392	133.2	240.45	0.3	0.4	110	0.5	8	11.8	54.5
22.	Big Bridge	5.51	27	630	650	260	108	70	16.8	12.98	103.6	187.02	0.55	1	125	0.5	9.5	12.5	40.1
23.	Big Bridge	6.21	27	610	475	155	210	80	25.2	13.37	162.8	293.88	0.25	1.2	120	1	6.5	15.8	40.3
24.	Temple	6.5	26	388	525	120	230	80	23.1	13.88	118.4	213.74	0.18.	1.1	125	1	8	15	38.2
25.	Temple	7.3	28	555	620	210	350	90	23.1	16.32	103.6	181.02	0.35	1.6	105	1.2	8.5	15.6	40.1
26.	Temple	7.81	29	530	640	208	340	103	42	14.88	118.4	213.74	0.56	0.5	105	1.1	7	20	55.3
27.	Keliveli	7.6	29	460	450	190	320	150	42	26.35	118.4	213.74	0.55	0.6	110	1.1	6	17.8	48.3
28.	Keliveli	8.21	28	530	430	120	410	100	42	14.15	133.2	240.45	0.35	0.7	120	1	8	14.8	45.1
29.	Keliveli	8.1	27	510	425	230	365	90	23.1	16.32	118.4	213.74	0.4	0.8	140	1.2	6	16.5	40.8
30.	Gandhigram	7.91	27.5	505	520	250	305	80	25.2	13.37	162.8	293.88	0.31	0.9	150	1	7	20	60.1

Contd...

Table 2.2–Contd...

Sl. No.	Location	Physico-chemical Parameters in mg/l Except pH, Temperature and EC																	
		pH	Temp. °C	EC µS/cm	TDS	TSS	T.Alk.	TH	Ca	Mg	Cl	Sal.	NO_3	NH_4	SO_4	PO_4	DO	BOD	COD
31.	Gandhigram	7.81	27	488	540	140	398	70	23.1	11.44	103.6	187.02	0.3	1	160	0.5	8	11.9	60.5
32.	Gandhigram	7.92	27	530	550	210	300	60	31.5	6.954	133.2	240.45	0.3	1.1	145	1.1	6	16.9	60.3
33.	Gandhigram	8.01	27	428	630	280	492	80	31.5	11.83	118.4	213.74	0.3	1.1	120	1.1	7	20.1	55.1
34.	Mhaisang	8.1	27	735	400	60	250	90	31.5	14.27	133.2	240.45	0.1	0.6	110	0.5	8	11.6	60.3
35.	Mhaisang	7.88	27	410	350	110	192	90	16.8	17.86	162.8	293.88	0.5	0.8	112	1.1	6	11.8	40.1
36.	Katyar	7.65	27	510	420	120	190	100	25.2	18.28	148	267.17	0.3	1.1	110	1.1	7.8	11.3	30.1
37.	Katyar	7.77	28	620	430	60	191	110	25.2	20.69	118.4	213.74	0.21	1.2	105	1.2	8.1	18.1	33.5
38.	Kinkhed	7.5	28	610	510	120	188	80	42	9.27	133.2	240.45	0.28	1.5	90	0.6	8.1	17.1	66.1
39.	Kinkhed	7.6	27	620	380	80	190	80	42	9.272	103.6	187.02	0.35	0.8	140	0.7	8.5	16.1	60.1
40.	Gopalkhed	8.1	27	710	460	90	220	90	42	11.71	162.8	293.88	0.45	0.7	160	0.8	6.8	16.1	80.1
41.	Gopalkhed	8.2	28	710	510	110	230	90	42	11.71	103.6	187.02	0.35	1	120	0.9	7.3	17.1	50.1
42.	Gopalkhed	8.35	27	401	525	150	240	80	25.2	13.37	133.2	240.45	0.21	1.2	110	0.1	6.1	18.2	60.3
43.	Gopalkhed	8.6	27	510	560	110	250	80	31.5	11.83	162.8	187.02	0.35	0.8	105	1.5	8.3	19.1	40.3
44.	Donwada	8.88	26	680	820	150	288	80	25.2	13.37	118.4	213.74	0.21	1.1	120	1.2	8.5	16.3	45.6
45.	Nairatwanuf	7.61	27	540	425	90	195	90	31.5	14.27	162.8	293.88	0.28	1.1	101	1.1	8.6	17.1	45

Contd...

Table 2.2–Contd...

Sl. No.	Location	pH	Temp. °C	EC μS/cm	TDS	TSS	T.Alk.	TH	Ca	Mg	Cl	Sal.	NO_3	NH_4	SO_4	PO_4	DO	BOD	COD
		Physico-chemical Parameters in mg/l Except pH, Temperature and EC																	
46.	Nairatwanuf	7.33	28	510	460	85	175	90	31.5	14.27	133.2	240.45	0.35	1.5	108	1.1	8.8	18.2	60.1
47.	Wari	6.85	27	310	520	110	168	80	31.5	11.83	133.2	240.45	0.21	1.1	120	1.3	8.9	17.1	60.5
48.	Kapileshwar	7.21	27	310	620	150	180	90	42	11.71	103.6	187.02	0.35	1	90	1.1	9	18.1	60.8
49.	Kapileshwar	7.65	27	460	580	140	178	90	31.5	14.37	118.4	213.74	0.28	1.5	140	1.5	9.1	16.3	78.1
50	Kapileshwar	7.01	28	780	530	120	170	100	42	14.15	148	267.17	0.35	1.8	155	1.8	8.8	15.5	68.3
	Average	8.91	27.49	551.34	529.42	151.46	250.84	95.05	45.76	11.99	154.13	275.96	0.36	1.00	129.18	1.03	7.46	16.11	54.34
	S.D.	6.00	0.71	188.83	122.88	63.18	76.30	36.56	33.60	4.24	60.39	109.81	0.12	0.35	23.69	0.39	1.09	2.65	13.39
	% C.V.	67.04	2.58	34.25	23.21	41.71	30.41	38.46	73.42	35.36	39.18	39.79	33.33	35.00	18.33	37.86	14.61	16.45	24.64
	S.E.	3.218	3.279	0.014	0.022	0.042	0.032	1.800	1.854	7.472	0.241	0.136	21.498	5.914	0.128	6.546	2.364	0.831	—

S.D.: Standard deviation; % C.V.: Percentage coefficient variance; S.E.: Standard error.

Table 2.3: Physico-chemical Parameters of Kapshi Lake at Akola District

Sl. No.	Location	Physico-chemical Parameters in mg/l Except pH, Temperature and EC																	
		pH	Temp. °C	EC μS/cm	TDS	TSS	T.Alk.	TH	Ca	Mg	Cl	Sal.	NO_3	NH_4	SO_4	PO_4	DO	BOD	COD
1.	Kapshi	6.82	27	98	400	10	110	80	60	4.88	85.08	153.6	0.72	0.1	100	0	5.6	13.2	52.1
2.	Kapshi	7.01	27	110	350	12	188	85	60	6.1	92.17	166.4	0.5	0.2	110	0	6.4	11.2	48.4
3.	Kapshi	6.81	29	90	35	30	190	98	70	6.83	92.17	166.4	0.78	0.11	140	0.38	12	12.6	38.8
4.	Kapshi	6.91	27	125	400	40	188	87	60	6.58	92.2	166	0.72	0.2	150	0.42	5.4	11.8	94.4
5.	Kapshi	6.78	28	125	410	40	190	95	70	6.1	92.17	166.4	0.78	0.31	140	0.5	11.2	14.2	75.2
6.	Kapshi	6.79	27	108	400	45	190	80	72	1.95	85.08	153.6	0.79	0.25	148	0.9	10.4	14.2	46.4
7.	Kapshi	6.8	27	98	310	60	168	86	60	6.344	70.09	128	0.82	0.22	140	0.58	16.8	25.6	75.2
8.	Kapshi	7.01	28	98	308	30	178	90	75	3.66	85.08	153.6	0.07	0.2	140	0	20	36	62.4
9.	Kapshi	6.8	27	101	105	10	190	88	60	6.83	92.17	166.4	0.65	0.11	148	0	15	23	41.6
10.	Kapshi	6.8	28	100	120	15	190	88	60	6.832	92.17	166.4	0.88	0.21	150	0	10.4	14.2	72
11.	Kapshi	6.71	27	80	110	40	180	80	31.5	11.83	70.09	128	0.61	0.1	110	0.21	6.1	13.1	40.5
12.	Kapshi	6.8	27	60	210	50	178	90	42	11.71	85.08	153.6	0.55	0.11	120	0.31	5.8	16.3	40.6
13.	Kapshi	6.71	27	70	300	45	160	80	52.5	6.71	99.26	179.19	0.78	0.18	110	0.25	6.8	15.4	45.1
14.	Kapshi	6.81	27	65	115	50	170	90	46.2	10.69	92.17	166.4	0.68	0.1	105	0.26	7.1	16.1	46.1
15.	Kapshi	6.91	28	88	250	42	160	90	48.3	10.17	92.17	166.4	0.78	0.21	90	0.31	7.8	14.2	50.1

Contd...

Table 2.3–Contd...

Sl. No.	Location	Physico-chemical Parameters in mg/l Except pH, Temperature and EC																	
		pH	Temp. °C	EC μS/cm	TDS	TSS	T.Alk.	TH	Ca	Mg	Cl	Sal.	NO_3	NH_4	SO_4	PO_4	DO	BOD	COD
16.	Kapshi	6.8	29	90	260	40	140	90	52.5	9.15	85.08	153.6	0.82	0.18	105	0.21	6.5	14.2	50.3
17.	Kapshi	7.01	28	92	180	38	188	90	58.8	7.613	70.09	128	0.65	0.21	110	0.35	8.1	25.4	60.1
18.	Kapshi	6.85	28	95	210	40	160	80	42	9.272	92.17	166.4	0.88	0.15	120	0.11	7.4	36	40.8
19.	Kapshi	7.1	28	85	280	45	190	80	58.3	5.295	92.17	166.4	0.91	0.16	120	0.05	7.4	23	40.8
20.	Kapshi	6.75	27	82	260	60	150	80	46.2	8.247	92.17	166.4	0.51	0.18	121	0.1	8.2	16.1	30.5
21.	Kapshi	6.8	28	101	250	50	140	90	52.5	9.15	92.17	166.4	0.68	0.2	155	0.1	8.5	14.2	50.6
22.	Kapshi	6.71	28	120	240	40	160	80	42	9.272	92.17	166.4	0.98	0.21	120	0.1	6.5	13.2	50.4
23.	Kapshi	6.55	28	118	190	30	155	85	31.5	13.05	92.17	166.4	0.78	0.18	118	0	7.2	12.6	50.4
24.	Kapshi	7.01	27	88	220	50	185	80	46.2	0.247	92.17	166.4	0.35	0.15	116	0.1	7.1	11.8	60.1
25.	Kapshi	6.81	27	101	225	40	160	80	46.2	8.247	85.08	153.6	0.61	0.21	105	0.1	7.1	16.3	60.1
26.	Kapshi	6.66	26	105	260	80	158	85	63	5.368	70.09	128	0.51	0.11	120	0.1	7	18.4	50.4
27.	Kapshi	6.51	27	98	250	60	160	90	67.2	5.563	92.17	166.4	0.41	0.18	121	0.05	8.3	16.5	55.4
28.	Kapshi	6.75	28	68	240	10	155	90	73.5	4.026	85.08	153.6	0.78	0.15	130	0.05	8.4	16.6	55.1
29.	Kapshi	6.81	28	65	290	20	165	90	73.5	4.026	92.17	166.4	0.55	0.11	140	0.05	8.5	17.8	56.1
30.	Kapshi	6.88	29	105	310	40	165	80	42	9.272	77.99	140.8	0.88	0.16	120	0	8.6	18.6	56.2

Contd...

Table 2.3–Contd...

Sl. No.	Location	Physico-chemical Parameters in mg/l Except pH, Temperature and EC																	
		pH	Temp. °C	EC µS/cm	TDS	TSS	T.Alk.	TH	Ca	Mg	Cl	Sal.	NO_3	NH_4	SO_4	PO_4	DO	BOD	COD
31.	Kapshi	7.1	28	88	280	25	188	85	46.2	9.467	92.17	166.4	0.65	0.15	140	0	8.7	16.3	55.1
32.	Kapshi	6.5	29	95	330	30	168	90	52.5	9.15	85.08	153.6	0.18	0.21	148	0	8.8	15.4	56.1
33.	Kapshi	6.7	28	101	350	40	170	80	58.8	5.173	92.17	166.4	0.72	0.2	130	0.1	8.1	16.4	55.2
34.	Kapshi	6.8	28	88	401	40	160	80	58.3	5.295	92.2	166	0.5	0.1	110	0.1	5.5	14.2	40.3
35.	Kapshi	6.78	27	91	380	50	155	90	67.2	5.563	92.17	166.4	0.72	0.1	140	0.1	5.6	15.6	40.1
36.	Kapshi	6.68	27	93	360	60	160	95	88.2	1.659	70.09	128	0.65	0.2	140	0.15	8.2	14.3	45.6
37.	Kapshi	7.01	28	105	390	30	188	90	73.5	4.026	92.17	166.4	0.4	0.2	130	0.1	10.1	11.8	50.1
38.	Kapshi	7.1	28	100	285	65	192	100	67.2	8.003	92.17	166.4	0.32	0.2	105	0.05	8.8	16.3	56.3
39.	Kapshi	6.12	28	101	410	80	160	100	88.2	2.88	85.08	153.6	0.2	0.21	111	0	7.5	14.8	55.8
40.	Kapshi	6.1	28	95	380	60	155	85	58.8	6.4	92.17	166.4	0.41	0.31	121	0.05	7.4	15.3	58.1
41.	Kapshi	7.2	27	110	290	48	185	90	46.2	10.68	92.17	166.4	0.61	0.18	120	0	8.8	16.2	60.1
42.	Kapshi	6.8	27	65	360	61	165	90	52.5	9.15	70.09	128	0.44	0.11	121	0	9	14.4	55.6
43.	Kapshi	6.55	27	90	350	45	171	85	67.2	4.343	85.08	153.6	0.41	0.12	121	0	10.2	15.3	54.3
44.	Kapshi	7.01	28	85	340	55	189	90	73.5	4.026	77.99	140.8	0.4	0.15	120	0.05	10.5	16.2	55.6
45.	Kapshi	6.68	27	80	405	68	155	80	67.2	3.123	99.26	179.19	0.43	0.16	111	0.1	8.8	17.3	60.3

Contd...

Table 2.3–Contd...

Sl. No.	Location	Physico-chemical Parameters in mg/l Except pH, Temperature and EC																	
		pH	Temp. °C	EC µS/cm	TDS	TSS	T.Alk.	TH	Ca	Mg	Cl	Sal.	NO_3	NH_4	SO_4	PO_4	DO	BOD	COD
46.	Kapshi	6.91	27	68	408	70	160	90	52.5	9.15	92.17	16.4	0.41	0.12	120	0.11	7.8	18.4	65.1
47.	Kapshi	6.82	27	75	390	65	170	95	58.8	8.83	92.2	166	0.38	0.2	121	0.1	6.8	16.3	45.1
48.	Kapshi	6.1	27	78	360	68	160	90	88.2	1.66	92.17	166.4	0.68	0.15	121	0.05	7.5	15.4	50.1
49.	Kapshi	6.3	27	85	370	72	160	80	67.2	3.12	92.17	166.4	0.71	0.15	121	0.05	8.2	14.4	60.2
50.	Kapshi	7.02	27	83	350	62	190	90	73.5	4.02	85.08	153.6	0.65	0.21	120	0	8.8	13.2	40.5
	Average	6.77	27.54	92.1	299.84	45.12	169.24	87.04	59.37	6.61	87.68	158.44	0.61	0.17	123.86	0.13	8.53	16.59	53.12
	S.D.	0.24	0.68	15.5	86.10	17.89	17.06	5.69	13.34	2.95	7.79	13.64	0.20	0.05	15	0.18	2.73	5.05	11.06
	% CV	3.51	2.47	16.83	28.71	39.65	10.08	6.53	22.47	44.63	8.88	8.60	32.78	29.41	12.11	138.46	32.00	30.44	20.82
	S.E.	10.064	3.342	0.138	0.032	0.137	0.132	0.707	0.792	1.719	12.091	6.891	12.327	45.247	0.140	12.044	0.939	0.454	—

S.D.: Standard deviation; % C.V.: Percentage coefficient variance; S.E.: Standard error.

Table 2.4: Physio-chemical Parameters and Heavy Metals Detection from Lonar Lake Water Samples of Different Sampling Stations

Sl. No.	Parameters	Sample No. (1)	Sample No. (2)	Sample No. (3)	Sample No. (4)	Sample No. (5)	Sample No. (6)	Sample No. (7)	Sample No. (8)	Average	S.D.	%C.V.
1.	pH	10.42	10.33	11.12	11.55	10.45	10	11.25	10.33	10.68	0.512	4.8
2.	Temperature °C	270	270	270	270	270	270	270	270	270	270	270
3.	Conductivity (μs/cm)	1876x101	1850x101	1810x101	1860x101	1830x101	1830x101	1819x101	1850x101	16325	5474	33.53
4.	TDS	310	310	320	315	310	310	210	315	300	34.18	11.4
5.	TSS	30	50	40	40	50	60	40	40	43.75	8.57	19.58
6.	T. Hardness	440x101	440x101	440x101	440x101	440x101	370x101	360x101	450x101	4225	334	7.9
7.	T. Alkalinity	500x101	510x101	530x101	510x101	530x101	400x101	420x101	400x101	4750	540	11.36
8.	Chloride	631.01x 101	638.01x 101	638.01x 101	638.01x 101	638.01x 101	545.93	531.75	553.02	6017	454	7.54
9.	Salinity	1157.90x 101	1170.90x 101	1170.91x 101	1001.78x 101	975.76x 101	1014.79x 101	988.77x 101	975.78x 101	10570	857	8.1
10.	Calcium	172.54x 101	176.75x 101	164.12x 101	168.33x 101	172.54x 101	151.50x 101	147.29x 101	147.29x 101	1625	113	6.95
11.	Magnesium	79.90x 101	81.31x 101	89.30x 101	89.36x 101	83.36x 101	87.22x 101	60.63x 101	66.54x 101	796	99.7	12.52
12.	Nitrate	10	10	12.5	12	10	10	10	12.5	10.87	1.13	10.4

Contd...

Table 2.4–Contd...

Sl. No.	*Parameters*	*Sample No. (1)*	*Sample No. (2)*	*Sample No. (3)*	*Sample No. (4)*	*Sample No. (5)*	*Sample No. (6)*	*Sample No. (7)*	*Sample No. (8)*	*Average*	*S.D.*	*%C.V.*
13.	NH_4	80	81	90	100	81	80	80	100	86.5	8.4	9.71
14.	Sulphate	55	70	55	60	60	70	55	55	60	6.12	10.2
15.	Phosphate	140	145	160	140	140	140	150	140	144.3	6.81	4.72
16.	D.O.	0.02	0.06	0.08	0.08	0.08	0.12	0.06	0.06	0.07	0.02	28.57
17.	B.O.D.	0.1	0.4	0.2	0.2	0.2	0.2	0.2	0.6	0.26	0.15	57.7
18.	C.O.D.	0.012	0.014	0.014	0.012	0.011	0.019	0.014	0.014	0.013	0.002	15.38
19.	Copper	1	0.5	0.5	0.5	0.5	1	1	1	0.75	0.25	33.33
20.	Zinc	8	8	10	8	8	10	10	10	9	1	11.11
21.	Cadmium	0.1	0.1	0.1	0.1	0.1	0.1	0.1	0.1	0.1	0	0
22.	Cromium	0.5	0.5	0.5	0.5	0.5	0.5	0.5	0.5	0.5	0	0

* All values are in mg/l except pH, E.C. and Temperature;

S.D.: Standard deviation;

% C. V.: Percentage coefficient variance.

pH

The pH value of Morna river water was in the range 6.0 to 9.3, Purna river water 6.11 to 9.88, Kapshi lake water 6.10 to 7.10, which are not within the limit.

The pH values of extreme environment like Lonar lake water was in the range of 10 to 12, which is very high as compared to all above sources. The higher values of pH can attribute to increased primary production in aquatic ecosystem (Zaffer, 1966) as high rate of photosynthetic activity will raise the pH (Perkins, 1976). Generally very high pH values of river water are due to industrial pollution (Palanivel and Rajaguru, 1999; Fokmare and Musaddiq, 2001a). Most natural water sample are alkaline due to presence of sufficient quantities of carbonates (Trivedy and Goel, 1986).

Temperature (°C)

The temperature of surface water varied between 27°C to 30°C.

Electrical Conductivity (EC µS/CM)

The conductivity value of most of the samples varied from 1000 to 1800 µs/cm in Morna river, 10.1 to 900 µs/cm in Purna river, 60 to 125 µs/cm in Kapshi lake.

In Lonar lake water, the conductivity value ranged between 1810 × 10 to 1876 × 10 µs/cm. This could be due to very high concentrations of ionic constituents present in water bodies (Abbasi *et al.*, 1999; Fokmare and Musaddiq, 2001). It was low in Tap water relatively higher values in Morna river water and Lonar lake water.

Total Dissolved Solids and Total Suspended Solids

The Total Dissolved Solids (TDS) and Total Suspended Solids (TSS) values were 105 to 886 mg/l, 10 to 305 mg/l respectively in Morna river, Purna river and Kapshi lake. High TDS values in the river water might be due to mixing of sewage and industrial effluents with the river (Figure 2.3). Gupta and Singh (2000) also reported high concentrations of TDS in Damodar river due to mixing of sewage and industrial waste. Lonar lake water ranged between 210 to 320 mg/l and 30 to 60 mg/l respectively. According to Todd (1980), when TDS of fresh water is less than 1000 ppm, it can be used safely for drinking purpose. High concentrations of total solids increase water turbidity. This in turn decreases the light penetration and thus affects photosynthesis. Total solids absorb sunlight in greater proportions thus increasing the temperature of water (Sunilkumar and Ravindranath, S., 1998). High TDS in water is mainly due to the presence of bicarbonates, sulfates and chlorides of calcium, magnesium and sodium (Nawlakhe *et al.*, 1995).

Total Alkalinity

The alkalinity values varied 94 to 492 mg/l in surface water, relatively higher values were obtained in Purna river at Gandhigram. The presence of Carbonates, bicarbonates and hydroxide are the major cause of alkalinity in natural water (Jain *et al.*, 1997). Higher values of total alkalinity registered might be due to the presence of excess of free CO_2 product as a result of decomposition process coupled with mixing of sewage and industrial effluents. Similar observation was recorded by Palharya *et al.* (1993), Singh (2000), Mohanta and Patra (2000), Chatterjee and Raziuddin (2002). The values of alkalinity in Lonar lake water were quite high, ranging between 4000 to 5300 mg/l can be ascribed to an interaction between sodium chloride, calcium carbonate and water over a long period of time (Malu, 1999). The evaporation linked concentration of salts over a long period of time is the possible reason from the observed alkalinity of the Lake (Musaddiq *et al.*, 2001).

Total Hardness

The total hardness varied between 115 to 510 mg/l in Morna river, 40 to 208 mg/l in Purna river and 80 to 100 mg/l in Kapshi lake water. It was minimum at Gopalkhed in Purna river and it was beyond the limit in most of the samples of Morna river. In Lonar lake water, total hardness values in the ranged between 3600 to 4500 mg/l. which is beyond the limit. The higher values of total hardness in present study, may be due to evaporation of water and addition of calcium and magnesium salts (Bagde and Varma, 1985; Chatterijee and Raziuddin, 2002).

According to Kannan (1991), water with hardness values more than 180 mg/l is very hard, in this respect, water of various sampling sites were very hard. Hardness below 300 mg/l is considered potable but beyond this limit produces gastrointestinal irrigation (ICMR, 1975).

Calcium and Magnesium Hardness

Calcium and magnesium hardness of surface water varied between 16 to 140 mg/l and 1.0 to 106 mg/l respectively. In Lonar lake water 164 to 176 mg/l and 60 to 80 mg/l respectively. The lake water is characterised by very high concentration of Ca, Mg and total hardness of any water is dependent on them (Jain *et al.*, 1997). Toxicity due to these ions have been studied on biota as earlier reports (Muralikrishna and Sumalatha, 1983).

Chloride and Salinity

The values of chloride and salinity varied between 173 to 404 mg/l and 312 to 722 mg/l respectively in Morna river, 103 to 354 mg/l and 103 to 639 mg/l respectively in Purna river, 70 to 99 mg/l and 70 to 92 mg/l respectively in Kapshi lake water. Olaniya and Saxena (1977) have detected large amounts of chloride near a refuse dump in Jaipur. Septic tank density of a colony largely decides the increase of groundwater. Salinity (Vates, 1986), chloride and conductivity are good indicators for determining plumps of contamination from septic systems (Alhajjar *et al.*, 1990).

The higher values of chloride in water indicates the gross water pollution (Narayana and Suresh, 1989). In Lonar lake water chloride values were very high ranging between 5310 to 6380 mg/l and salinity of water ranged between 3600 to 4500 mg/l making the water brackish (Malu, 1999). The higher concentration of chloride is considered to be an indicator of pollution due to higher organic waste of animal origin (Shanti *et al.*, 2002). There is a direct correlation between chloride concentration and pollution level (Munawar, 1970; Klein, 1957). This could be also due to mixing of sewage, and increased temperature and evapotranspiration by water (Govindan and Sudaresan, 1979; Jana, 1973).

Nitrate (NO_3)

Nitrate of the surface water was noted between 0.77 to 0.99 mg/l in Morna river, 0.10 to 0.60 mg/l in Purna river, 0.07 to 0.98 mg/l in Kapshi lake water. The distribution of nitrates indicates that level of concentration are nearest to desirable limits for drinking purposes as shown in Figure 2.4. The concentration of nitrate above 45 mg/l in drinking water may prove harmful to human health (Jain *et al.*, 1997). The value of Lonar lake water ranged between 10 to 12 mg/l which is within the limit. William and Walker (1969) reported the leachates from nitrogen rich deposits and soils as a source of nitrates in the groundwater. Remo Navone *et al.* (1963) in his studies revealed duration of inhabitation and density of population could significantly affect the nitrate concentrations. Nitrate in water or when it accumulate in soil leads to problem of nitrate pollution in crops like leafy vegetables when grown there which is proved to be carcinogenic (Saxena and Mehra, 1991; Majumdar and Gupta, 2000; Ozha *et al.*, 1993).

Ammonia and Phosphate

The range of ammonia and phosphate in Morna river was 0.04 to 2.15 mg/l and 0.6 to 2.5 mg/l respectively, 0.4 to 2.0 mg/l and 0.5 to 2.0 mg/l respectively in Purna river, 0.1 to 0.21 mg/l and 0.0 to 0.9 mg/l in Kapshi lake water. Ammonia values were beyond the limit in all water reservoirs except Kapshi lake as shown in Figure 2.5.

Lonar lake water value ranged between 80 to 100 mg/l of ammonia and 140 to 160 mg/l of phosphate which is very high. Kofoid (1993) has stated that ammonia a product of decomposition of organic matter tend to be high in the water polluted by sewage which is occurred in present study. Similar results were reported by Bruce (1958); Loster (1975); Shivnikar *et al.* (2000) and Fokmare and Musaddiq (2001). It was reported that ammonia is the most reliable single parameter for measuring the quality of river water (Loster, 1975; Fokmare and Musaddiq, 2001). The presence of large amount of ammonia in surface water indicates pollution (Leonard, 1971). Wetzel (1983) concluded that phosphorus was the most important limiting factor responsible for eutrophication of water all over the world. High levels of ammonia and phosphates in the river water, indicate its gross pollution (Shivnikar *et al.*, 2000).

Sulphate

The sulphate concentration was in the range of 250 to 862 mg/l in Morna river, 90 to 166 mg/l in Purna river, 90 to 150 mg/l in Kapshi lake water, sulphate values of Lonar lake water was 55 to 70 mg/l. Average values were beyond the limit in all sources except Purna river, Kapshi lake (Figure 2.6).

Dissolved Oxygen (DO), BOD and COD

The DO, BOD and COD in surface water were ranged from 7.5 to 11.7 mg/l, 12.0 to 24 mg/l and 20 to 60 mg/l respectively in Morna river water, 6.0 to 9.5 mg/l, 11 to 20 mg/l and 40 to 80 mg/l respectively in Purna river water, 5 to 15 mg/l, 11 to 25 mg/l and 30 to 75 mg/l respectively in Kapshi lake water. Lonar lake water were ranged between 0.02 to 0. 12 mg/l, 0.1 to 0.6 mg/l and 0.011 to 0.019 mg/l respectively.

According to Rana and Palaria (1988), high organic content leads to oxygen depletion. Low oxygen contents in water might be due to higher growth rate of bacteria which utilized oxygen for their metabolic activities (Pandey *et al.*, 2000). Relatively higher values of dissolved oxygen might be due to increased solubility of oxygen at lower temperature (Chatterjee and Raziuddin, 2002). Biochemical oxygen demand (BOD) values increased due to increased biological activities at elevated temperature (Palharya *et al.*, 1993).

Statistical Analysis

The average values ± 95 per cent Confidence limit, Standard Error (SE) of different parameters of water reservoirs were carried out and shown in Table 2.5. The Correlation coefficients (r-value) between each pair of parameters for all the possible correlations are computed and are listed in Tables 2.7–2.9.

The range of r-value of correlations was in the order of – 0.5114 to 1.0 in Morna river, – 0.3626 to 1.0 in Purna river, – 0.4763 to 1.0 in Kapshi lake.

The high correlation coefficient (r) are observed between the pair of chloride and sulphate as well as in salinity and sulphate, the r-value was found to be 0.9401 respectively for each pair. As the value of r is positive, the value is very close to the maximum possible value of 1 and hence the relation of chloride and sulphate as well as salinity and sulphate is very strong in Morna river. There is no strong Correlation coefficient observed in other water reservoirs.

Water Quality Index (WQI)

The concept of indices to represent gradation in water quality was first proposed by Horton (1965). Water Quality Index (WQI) is a measure of overall quality of water which has been statistically arrived at from a number of parameters and is in use more to describe the quality of fresh water systems (Aboo *et al.*, 1968; Handa, 1975; Gupta, 1981; Paliwal, 1983; Anand Walli and Krishnasamy, 1965).

The water quality Index value (WQI) are presented in the Tables 2.10–2.12. The WQI value of Morna river was 366.74, Purna river was found to be 429.52, and 412.70 for Kapshi lake as shown in Figure 2.8. The permissible value for drinking water according to this method is 100 (Sunil Kumar and Ravindranath, 1998) and hence, this implies that the water is not fit for drinking without proper treatment in all above sources except tap water.

Table 2.5: Average Values with + 95 per cent Confidence Limit of Physico-chemical Parameters of Water Samples Collected from Different Water Reservoirs

Parameter	Surface Water				Ground Water			Treated
	Morna River	Purna River	Kapshi Lake	Lonar Lake	Bore Well	Deep Well	Shallow Well	Tap Water
pH	7.82±8.0	8.95±3.27	6.77±4.116	10.68	7.75±40.4	8.08±4.33	7.50±2.51	7.33
Temp. °C	28.04±4.69	27.49±5.85	27.54±2.927	27	27.98±3.19	26.58±0.32	28.92±5.598	27.98
E.C. µs/cm	1402.1±0.0288	551.34±0.021	92.10±0.137	16325	1230.48±0.02	1345.8±0.0023	1212.9±0.0698	279.64
TDS	772.4±0.00693	529.42±0.011	299.84±0.036	300	655.3±0.018	783.2±0.06	624.4±0.0295	269.86
TSS	57.86±0.745	151.46±0.064	45.12±0.070	43.75	46.36±3835	64.02±0.53	39.9±0.7836	21.58
T. Alk.	169.6±0.035	250.84±0.016	169.24±0.0264	4750	283.1±0.1881	231.62±0.28	272.76±0.1064	116.64
TH	318.1±0.270	95.05±0.0	87.04±0.854	4225	350.2±0.022	442.12±5.28	323.14±0.333	105.84
Ca	70.42±0.282	45.76±0.268	59.37±0.622	1625	62.86±0.455	60.48±5.27	54.52±0.5909	29.54
Mg	60.44±4.631	11.99±1.54	6.61±1.897	796	67.82±0.452	93.11±22.6	65.46±1.4412	18.53
Cl	255.9±6.285	154.13±0.33	87.68±2.115	6017	79.96±2.603	179.30±0.03	66.14±9.414	201.48
Sal.	462.11±3.555	275.96±0.184	158.44±1.288	10570	144.6±1.238	319.66±0.0140	120.12±5.499	363.71
NO_3	0.88±101.93	0.36±1.19	0.61±3.661	10.87	0.81±51.46	0.92±6.574	0.80±113.473	0.15
NH_4	1.23±1.461	1.00±01.478	0.17±55.24	86.5	0.51±5.69	1.47±2.50	1.0±1.8574	0
SO_4	424.8±0.0882	129.18±0.242	123.86±0.071	60	2333±0.002	938.46±0.05	2131±0.0015	81.64
PO_4	1.29±2.29	1.03±4.896	0.13±1.890	144.3	0.28±25.21	0.21±75.1	0.29±54.68	0
DO	9.83±2.015	7.46±2.321	8.53±0.0.430	0.07	8.05±0.302	8.28±0.38	6.66±0.622	6.99
BOD	18.30±0.514	16.11±1.455	16.59±0.191	0.26	14.15±0.178	25.31±0.75	10.95±0.0458	1.97
COD	40.47	54.34	53.12	0.013	70.04	46.76	64.6	2.15

Table 2.6: Water Quality with Reference to Standard for Drinking Water

Sl.No.	*Parameters*	*Drinking Water Standard (IS : 10500, 1991)*	*Sources*	*Percentage of Samples Exceeding ISI Limits*
1.	pH	6.5–8.5	Morna river	42%
			Purna river	18%
			Kapshi lake	8%
2.	Total Alkalinity	200 mg/l	Morna river	32%
			Purna river	60%
			Kapshi lake	0
3.	Total Hardness	300 mg/l	Morna river	48%
			Purna river	0
			Kapshi lake	0
4.	Calcium (Ca)	75 mg/l	Morna river	0
			Purna river	14%
			Kapshi lake	6%
5.	Magnesium (Mg)	30 mg/l	Morna river	56%
			Purna river	0
			Kapshi lake	0
6.	Chloride	250 mg/l	Morna river	48%
			Purna river	10%
			Kapshi lake	0
7.	Nitrate (NO_3)	10 mg/l	Morna river	0
			Purna river	0
			Kapshi lake	0
8.	Ammonia (NH_4)	0.5 mg/l	Morna river	98%
			Purna river	92%
			Kapshi lake	0
9.	Sulphate	150 mg/l	Morna river	100%
			Purna river	24%
			Kapshi lake	0
10.	Phosphate	< 0.1 mg/l	Morna river	100%
			Purna river	98%
			Kapshi lake	30%
11.	Copper	0.05 mg/l	Morna river	68%
			Purna river	36%
			Kapshi lake	14%
12.	Zinc	5.0 mg/l	Morna river	38%.
			Purna river	40%
			Kapshi lake	14%
13.	Cromium	0.05 mg/l	Morna river	62%
			Purna river	14%
			Kapshi lake	0
14.	Cadmium	0.01 mg/l	Morna river	80%
			Purna river	40%
			Kapshi lake	4%

Table 2.7: Correlation Coefficient Matrix for Physico-chemical Parameters of Morna River

Sl. No. Parameter	Physico-chemical Parameters in mg/l Except pH, Temperature and EC																	
	pH	Temp. °C	EC μS/cm	TDS	TSS	T.Alk.	TH	Ca	Mg	Cl	Sal.	NO_3	NH_4	SO_4	PO_4	DO	BOD	COD
1. pH	1.000	- 0.01973	0.8226	0.8847	0.3943	0.7552	0.8089	0.4060	0.8072	0.6928	0.6924	0.6824	0.6948	0.5264	0.0368	- 0.2718	0.0772	0.5074
2. Temp.		1.0000	- 0.1344	- 0.2361	- 0.2337	- 0.0445	- 0.1844	- 0.1050	- 0.1833	- 0.0118	- 0.0119	- 0.1315	0.0219	0.0535	- 0.0539	- 0.0525	0.0487	- 0.3327
3. E.C.			1.0000	0.7100	- 0.0530	0.9341	0.9476	0.4698	0.9460	0.7536	0.7533	0.6280	0.5732	0.7418	0.0021	- 0.4552	- 0.2314	0.2730
4. TDS				1.0000	0.6091	0.6160	0.6429	0.5957	0.6296	0.6319	0.6315	0.7924	0.4925	0.3892	0.0969	- 0.1319	0.2106	0.4935
5. TSS					1.0000	- 01588	0.0048	0.3660	- 0.0111	0.0031	0.0027	0.4818	0.0837	- 0.2993	0.2046	#####	0.6016	0.4679
6. T. Alk.						1.0000	0.8252	0.3021	0.8283	0.9152	0.9151	0.6097	0.6300	0.9254	- 0.1485	- 0.4243	- 0.1666	0.2875
7. TH							1.0000	0.4727	0.9993	0.5934	0.5930	0.5977	0.5641	0.6031	0.1287	- 0.5133	- 01781	0.2863
8. Ca								1.0000	0.4385	0.1917	0.1909	0.6972	0.0671	0.1021	0.3018	- 0.2763	0.2331	0.2295
9. Mg									1.0000	0.5967	0.5964	0.5790	0.5724	0.6106	0.1182	- 0.5114	- 0.1920	0.2819
10. Cl										1.0000	1.0000	0.6314	0.6345	0.9401	- 0.2370	- 0.2681	0.0599	0.3809
11. Sal											1.0000	0.6308	0.6341	0.9401	- 0.2368	- 02673	0.0592	0.3801
12. NO_3												1.0000	0.3811	0.4946	0.1659	- 0.3043	0.3793	0.4121
13. NH_4													1.0000	0.5770	- 0.1180	- 0.1996	0.0783	0.3858
14. SO_4														1.0000	- 0.2627	- 0.3562	- 0.0876	0.2299
15. PO_4															1.0000	- 0.1437	0.1432	0.0215
16. DO																1.0000	0.0458	- 0.1340
17.. BOD																	1.0000	0.3210
18. COD																		1.0000

Table 2.8: Correlation Coefficient Matrix for Physico-chemical Parameters of Purna River

Sl. No.	Parameter	Physico-chemical Parameters in mg/l Except pH, Temperature and EC																	
		pH	Temp. °C	EC µS/cm	TDS	TSS	T.Alk.	TH	Ca	Mg	Cl	Sal.	NO_3	NH_4	SO_4	PO_4	DO	BOD	COD
1.	pH	10000	0.0883	0.2584	0.0823	- 0.2060	0.4248	0.3460	0.2916	0.1478	0.1844	0.1589	0.0652	0.8850	0.1618	0.3626	- 0.3090	0.2030	0.0544,
2.	Temp.		1.0000	0.0087	0.0653	0.1269	0.1564	0.3258	0.3268	0.0458	0.1872	0.1987	0.4829	0.0000	0.0147	0.1484	- 0.1823	0.1259	0.0909
3.	E.C.			1.0000	- 0.0038	- 0.3626	- 0.1136	0.5926	0.5913	0.0920	0.5186	0.5190	0.3005	0.2619	0.5316	0.2239	0.0348	0.0162	02690
4.	TDS				1.0000	0.4207	0.0760	0.2542	0.3142	- 0.0823	0.1069	0.1001	0.1025	0.2283	0.0962	0.1295	0.1184	0.3297	- 0.0402
5.	TSS					1.0000	0.2346	- 0.2594	- 0.1712	- 0.2144	- 0.2016	- 0.1881	0.0518	- 0.0867	- 0.0450	- 0.0766	- 0.1724	0.0665	- 0.1755
6.	T. Alk.						1.0000	0.0071	- 0.0650	0.1362	- 0.0887	- 0.8920	0.0685	- 0.1011	0.1012	0.0212	- 0.4089	0.2772	- 0.0309
7.	TH							1.0000	0.8788	0.3734	0.6187	0.6223	0.5623	0.2064	0.5052	0.4762	- 0.0298	0.1716	- 0.0694
8.	Ca								1.0000	- 0.1136	0.7323	0.7358	0.5562	0.1485	0.5997	0.3815	- 0.0554	0.0956	0.0929
9.	Mg									1.0000	- 0.1242	- 0.1236	0.0955	0.1478	- 0.1141	0.2395	0.0404	0.1799	- 0.3184
10.	Cl										1.0000	0.9905	0.4664	0.1483	0.6177	0.3745	- 0.2632	0.0214	0.1099
11.	Sal											1.0000	0.4042	0.1563	0.6344	0.3472	- 0.2776	- 0.0012	0.1311
12.	NO_3												1.0000	0.0451	0.4529	0.2777	- 0.1144	- 0.0290	0.0346
13.	NH_4													1.0000	0.1986	0.3234	0.1860	0.0957	0.0653
14.	SO_4														1.0000	0.3995	- 0.1857	- 0.0617	0.2517
15.	PO_4															1.0000	0.0057	0.0777	- 0.1230
16.	DO																1.0000	- 0.2294	0.0097
17.	BOD																	1.0000	0.0861
18.	COD																		1.0000

Table 2.9: Correlation Coefficient Matrix for Physico-chemical Parameters of Kapshi Lake

Sl. No.	Parameter	Physico-chemical Parameters in mg/l Except pH, Temperature and EC																	
		pH	Temp. °C	EC μS/cm	TDS	TSS	T.Alk.	TH	Ca	Mg	Cl	Sal.	NO_3	NH_4	SO_4	PO_4	DO	BOD	COD
1.	pH	1.0000	0.5200	0.0407	0.1780	- 0.3460	0.4406	- 0.0060	- 0.2607	0.2070	- 0.0243	- 0.0256	0.1055	- 01202	0.0053	0.0903	0.1476	0.1766	0.0140
2.	Temp.		1.0000	0.1679	- 0.0181	- 0.3764	0.0557	0.2065	- 0.0501	0.1902	0.1128	0.1064	0.1173	0.2383	0.1243	- 0.0796	0.1371	0,0841	0.0239
3.	E.C.			1.0000	0.0972	- 0.1895	0.1495	- 0.0540	- 0.0087	- 0.0090	0.0436	0.0417	0.1750	0.5254	0.2486	0.1327	0.1401	- 0.0644	0.3730
4.	TDS				1.0000	0.3180	- 0.1250	0.1324	0.5396	- 0.4763	0.1086	0.1040	- 0.2656	0.2129	0.0847	0.0854	- 0.0316	- 0.2231	0.1898
5.	TSS					1.0000	- 0.1197	0.1550	0.1943	- 0.1459	- 0.1506	- 0.1473	- 0.2856	0.0777	- 0.2330	0.0593	- 0.1463	- 0.0301	- 0.0033
6.	T. Alk.						1.0000	0.2244	0.0994	- 0.0601	- 0.0043	- 0.0043	- 0.0818	0.1790	0.2908	0.2243	0.3385	0.0548	0.1548
7.	TH							1.0000	0.4814	0.0121	- 0.0823	- 0.0834	0.3226	0.1327	0.1772	0.0019	0.2190	- 0.0704	0.1095
8.	Ca								1.0000	- 0.8048	- 0.0489	- 0.0507	- 0.2665	0.1561	0.2676	0.0254	0.3044	- 0.0053	0.1194
9.	Mg									1.0000	- 0.0119	- 0.0107	0.2153	- 0.0896	- 0.1848	- 0.0207	- 0.2071	0.0233	- 0.1178
10.	Cl										1.0000	0.9998	0.0541	0.0919	0.0123	- 0.1468	- 0.1584	- 0.1588	- 0.0477
11.	Sal.											1.0000	0.0569	0.0915	0.0120	- 0.1450	- 0.1581	- 0.1576	- 0.0487
12.	NO_3												1.0000	0.0111	0.0677	0.3428	- 0.1820	- 0.0492	- 0.0608
13.	NH_4													1.0000	0.1680	0.2883	0.1916	- 0.0294	0.3837
14.	SO_4														1.0000	0.1882	0.4626	0.1025	0.2538
15.	PO_4															1.0000	0.1342	- 0.0154	0.1578
16.	DO																1.0000	0.5204	0.2261
17.	BOD																	1.0000	0.0161
18.	COD																		1.0000

Table 2.10: Water Quality Index (WQI) of Morna River

Sl.No.	*Parameters*	*Unit Wt (Wi)*	*ICMR (Si)*	*Observed Value (Vi)*	*Quality Rating (Qi)*	*Qi.W.*
1.	DO	0.2000	5	9.83	49.69	9.94
2.	BOD	0.2000	5	18.30	366.00	73.20
3.	pH	0.0040	7.0-8.5	7.82	54.67	0.22
4.	Cl	0.0040	250	255.90	102.36	0.41
5.	NO_3	0.0500	20	0.88	4.40	0.22
6.	T. Alk.	0.0083	120	169.60	141.33	1.17
7.	TH	0.0033	300	318.10	106.03	0.35
8.	TSS	0.0020	500	57.86	11.57	0.02
9.	COD	0.2000	5	40.47	809.40	161.88
10.	TC	0.1000	1	360	355.63	35.56
	SUM	**0.77**	**1206.00**	**1238.76**	**2001.08**	**282.97**
	AVG	**0.08**	**134.00**	**123.88**	**200.11**	**28.30**
	Water quality index (WQI)			**366.74**		

Table 2.11: Water Quality Index (WQI) of Purna River

Sl.No.	*Parameters*	*Unit Wt (Wi)*	*ICMR (Si)*	*Observed Value (Vi)*	*Quality Rating (Qi)*	*Qi.W.*
1.	DO	0.2000	5	7.46	74.38	14.88
2.	BOD	0.2000	5	16.11	322.20	64.44
3.	pH	0.0040	7.0–8.5	8.95	130	0.52
4.	Cl	0.0040	250	154.13	61.65	0.25
5.	NO_3	0.0500	20	0.36	1.80	0.09
6.	T. Alk.	0.0083	120	250.84	209.03	1.73
7.	TH	0.0033	300	95.05	31.68	0.1
8.	TSS	0.0020	500	151.46	30.29	0.06
9.	COD	0.2000	5	54.34	1086.80	217.36
10.	TC	0.1000	1	158	319.87	31.99
	SUM	**0.77**	**1206.00**	**896.7**	**2267.7**	**331.42**
	AVG	**0.08**	**134.00**	**89.67**	**226.77**	**33.14**
	Water quality index (WQI)			**429.52**		

Table 2.12: Water Quality Index (WQI) of Kapshi River

Sl.No.	*Parameters*	*Unit Wt (Wi)*	*ICMR (Si)*	*Observed Value (Vi)*	*Quality Rating (Qi)*	*Qi.W.*
1.	DO	0.2000	5	8.53	63.23	12.65
2.	BOD	0.2000	5	16.59	331.80	66.36
3.	pH	0.0040	7.0–8.5	6.77	– 15.33	– 0.06
4.	Cl	0.0040	250	87.68	35.07	0.14
5.	NO_3	0.0500	20	0.61	3.05	0.15
6.	T. Alk.	0.0083	120	169.24	141.03	1.17
7.	TH	0.0033	300	87.04	29.01	0.1
8.	TSS	0.0020	500	45.12	9.02	0.02
9.	COD	0.2000	5	53.12	1062.40	212.48
10.	TC	0.1000	1	35	254.41	25.44
	SUM	**0.77**	**1206.00**	**509.7**	**1913.7**	**318.44**
	AVG	**0.08**	**134.00**	**50.97**	**191.37**	**31.84**
	Water quality index (WQI)			**412.7**		

Conclusion

Surface water samples were studied to find their suitability for drinking, domestic and irrigation purposes.

The higher hardness, Ca, Mg in surface water can be sourced to human activities, sewage etc. High levels of ammonia and phosphates in the river water indicate its gross pollution.

It is observed that, values of several parameters exceeded the permissible limits, pointing out the necessity of proper treatment disposal and management of wastes discharged into the rivers and on an open land therefore it is concluded that most of these water samples were found unsuitable for both domestic and drinking purposes.

It is revealed that, the pollution load of surface water is more in Akola district.

Suggestions

During the above investigations some facts are clear which have been discussed earlier. In view of these studies the following suggestions are recommended:

1. The appreciation of the nature for a healthy environment is the most important need for today which will create a value to conserve and protect our atmosphere.
2. Since the pollution and imbalancement of the ecosystem is the gift of urbanisation and civilisation, there is a need of sustainable development, use of modern technologies and proper town planning.
3. Environmental impact assessment and monitoring of environment specially the water and waste water at proper time intervals are important to tress the extent of pollution in our water reservoirs, such as Rivers, Lakes, Wells and Bore Wells.
4. Proper management, treatment and disposal of solid wastes and waste water. Priority must be given to this aspects by the industrialists and pollution control boards, Government of India and State Governments should monitored this process. If small scale industries can't manage such treatment system, Government should increase the facilities of centralised waste management plants.
5. Purification of water should be regular, systematic, scientific and chlorination must be undertaken regularly with sufficient dose, so that it should be at least 0.5 mg/l residual chlorine at distribution point regularly.
6. Deep wells, Shallow wells and Bore wells should be well constructed and well protected, so as to prove as a best source of sate water–not a source or disease transmission.
7. Environmental education programme should be implemented properly at both formal and non-formal educational level by taking the help of educational boards, State and Central Government and Non-Government Organisations.
8. The environmental laws should be implemented more vigorously. The industries need to comply to the effluents standards before discharging in to the surface drain, so that surface and groundwater contamination can be controlled.
9. Proper sanitation including sanitary latrines, underground drainage systems and disposal plants

protect the water from entry of pathogens like coliforms and others. In developing countries like India a special attention should be given to this aspects.

Our healthy atmosphere is a heritage to us, we have to use it safely and protected it from all unwanted nuisances. So that we could handover a healthy protected environment to coming generations.

Acknowledgement

The author express their sincere thanks to Dr. V.B. Wagh, Principal, Shri Shivaji College of Arts, Commerce and Science, Akola and Mr. Vasantrao Dhotre, Chairman, Shri Shivaji Education Society, Amravati for providing necessary facilities.

References

Abassi, S.A. and Abssi Naseema (1989). Studies on the water quality of Kuttadi river at Kerala. In: *Pollution and Biomonitoring of Indian Rivers*, (Ed. R.K. Trivedy), ABD Publishers, Jaipur, pp. 154-158.

Abassi, S.A. and Vinithan, S. (1999). Studies on water quality in and around an industrialized suburb of Pondicherry. *Indian J. Env. Hlth.*, **41(4)**: 253-263.

Abbasi, S.A., Khan, F.I., Sentilvelah, K. and Shabuden, A. (1999). *Indian J. Env. Hlth.*, **41(3)**: 176-183.

Aboo, K.M., Sastry, C.A. and Alex, P.G.A. (1968). Study of well water in Bhopal City. *Indian J. Environ. Hlth.*, **10**: 189-203.

Agrawal, V.K., Raj, K.P.S., Panchal, D.I., and Balaraman, R. (1978). Heavy metal content in the water and sediment of Surasagar Lake of Baroda, (Gujarat). *Inter. J. Env. Stud.*

Akola District Map (1996). Govt. of India.

Alhajjar, Chesters, B.J. and Harking, J.M. (1990). Indicators of chemical pollution from septic systems. *Groundwater*, **28(4)**: 559-568.

Anandawalli, M. and Krishnasamy, S. (1981). A quality profile of river Vaigai (South India). *Indian J. Env. Hlth.*, **23**: 191-202.

Aneja, K.R. (1993). *Water Microbiology: Experiments in Microbiology, Plant Pathology and Tissue Culture*. Wishwa Prakashan, New Delhi, pp. 299-314.

APHA (1992). *Standard Method for the Examination of Water and Waste Water*. APHA, AWWA, WPET, Washington DC, U.S.A., 18th edition.

Ayyadurai, K., Kamalam, N. and Rajgopal, C.K. (1983). Mercury pollution in water at Madras city. *Indian J. Env. Hlth.*, **4**.

Bagde, U.S. and Varma, AK. (1985). Physico-chemical characteristics of water of J.N.U. Lake at New Delhi. *Indian J. Ecol.*, **12**: 251-256.

Bhu-Jal News (1997). *Ground Water Pollution and Quality Hazards Scenario in India*. Central Groundwater Board, Ministry of Water Resources, New Delhi, **12(3-4)**: 55-58.

Bruce, A. (1958). *Report on Biological and Chemical Investigation of the Water.* Pollution Advisory Council, Marine Dept. Wellington.

Chandra, S.V. (1980). *Toxic Metals in Environment: A Status Report of R&D Work Done in India.* Industrial Toxicology Research Centre, Lucknow.

Chatterjee, A.A.K. (1992). Water quality studies on Nandankanan Lake. *Indian J. Env. Hlth.*, **34(4)**: 329-333.

Chatterjee, C. and Raziuddin, M. (2002). Determination of water quality index (WQI) of a degraded river in Asansol industrial area (W.B.). *Nature Environment and Pollution Technology*, **1(2)**: 181-189.

Collins, Y.E. and Stotzky, G. (1992). Heavy metals after the electrokinetic properties of Bacteria, yeast and clay minerals. *Applied and Env. Microbiology*, **58(5)**: 1592-1600.

Corbett, J.RK., Wright and Bacillie (1984). *The Biochemical Mode of Action of Pesticides, 2nd edition*. Academic Press, New York, pp. 24-28.

David, A. (1956). Studies on the pollution of Bhadra river fisheries at Bhadravati Mysore state with industrial effluent. *Proc. Nat. Inst. Sci., India*, **22**: 132-139.

Finance and Statistics of Akola District (1998). Govt. of Maharashtra, Mumbai, p. 133.

Fokmare, A.K. and Kulkarni, N.S. (1999). Bacteriological status of drinking water in and around Buldhana town of Maharashtra. *National Cof. on Wetland Conservation*, p. 107.

Fokmare, A.K. and Musaddiq, M. (2002). A study on physico-chemical characteristic of Kapshi lake and Purna river waters in Akola district of Maharashtra. *Nature Environment and Pollution Technology,* **1(3)**.

Fokmare, A.K. and Musaddiq, M. (2001). Comparative studies of physico-chemical and Bacteriological quality of surface and groundwater at Akola (M.S.), *Pollution Research,* **20(4)**: 651-665.

Fokmare, A.K. and Musaddiq, M. (2001a). Levels of ammonia and phosphates indicates organic pollution in Morna river of Akola (M.S.). *Env. Pollution Control Journal,* **4(4)**: 20-22.

Fokmare, A.K. and Musaddiq, M. (2002a). Studies on metal pollution in surface and groundwater of Akola district. *Environmental Pollution Control Journal,* **5(4)**: 12-15.

Fokmare, A.K., Musaddiq, M. and Purushottam, T. (2001). Physico-chemical and bacteriological analysis of groundwater at Akola, *J. Microb. World,* **3(2)**: 51-58.

Ganapati, S.V. (1956). Studies of the source of Madras quality water supply and on other water of the state. Ph.D. Thesis, Madras University.

Govindan and Sundaresan, B.B. (1979). Seasonal succession of algal flora in polluted region of Adyar river. *Indian J. Env. Hlth.,* **21**: 131-142.

Gupta, B.K. and Singh, G. (2000). Damodar river water quality status along Dugda-Sindri industrial belt of Jharia coalified, In: *Pollution and Biomonitoring of Indian Rivers* (ed. Trivedy, R.K.), ABD Pub., Jaipur, pp. 58-69.

Gupta, S. and Gupta, M. (1997). *Water and Its Management: Water and Basic Environmental Technology,* 1st edition. Anmol Publications Pvt. Ltd., New Delhi, pp. 1-17.

Gupta, S.C. (1991). Chemical character of groundwater in Nagaur district, Rajasthan, *Indian J. Env. Hlth.,* **33(3)**: 341-349.

Gupta, S.C. (1981). Evaluation of well waters in Udaipur district. *Indian J. Env. Hlth.,* **23**: 191-202.

Handa, B.K. (1975). Geochemistry and genesis of fluoride containing groundwater in India. *J. Groundwater Tech. Div. NWA.*

Hashmi, Ilyas Fazil, Shakir, M., Musaddiq, M. (2000). Water quality of Khazana Well at Beed district, (M.S.). *J. Microbial World*, **2(2)**: 29-33.

Horton, R.K. (1965). An index number system for rating water quality. *J. Water Poll. Cont. Fed.*, **3**: 300-305.

Hem, J.D. (1959). Study and interpretation of the chemical characteristics of natural water. *U.S. Geological Survey Water Supply*, p. 1473.

ICMR (1975). *Manual of Standards of Quality for Drinking Water Supplies*. ICMR, New Delhi.

Indian Standards (IS: 10500) (1993). *Drinking Water Specification*. Bureau of Indian Standards, New Delhi.

Indian Standards (IS: 2490) (1991). *Domestic Water Specification*. Bureau of Indian Standards, New Delhi.

Jain, C.K., Bhatia, K.K.S. and Vijay, T. (1997). Groundwater quality in coastal region of Andhra Pradesh in coastal region of Andhra Pradesh. *Indian J. Env. Hlth.*, **39(3)**: 182-192.

Jameel, A. (1998). Physico-chemical studies in Vyyakondan channel water of Cauvery. *Pollution Research*, **17(2)**: 111-114.

Jana, B.B. (1979). Seasonal periodicity of plankton in a fresh water pond, West Bengal, India. *Int. Rev. Ges. Hydrobiol.*, **58**: 127-143.

Kannan, K. (1991). *Fundamentals of Environmental Pollution*. S. Chand and Co. Ltd., New Delhi.

Khillare, Y.K. and Davne, P.M. (1957). Effect of pollutants on pyruvic acid level in fresh water fish, *puntius stigma*. *Env. Poll. Research Land and Water*, pp. 271-277.

Klein, L. (1957). *Aspects of River Pollution*. Butter Worths Scientific Publication, London.

Kodarkar, M.S., Diwan, A.D., Murugan, N., Kulkarni, K.M., Anuradha Ramesh (1998). *Methodology for Water Analysis*. Indian Association of Aquatic Biologist, (IAAB), Publication No. 2.

KoFoid, C.A. (1993). The plankton of Illinois river. *Bull. Lab. Nat. Hist.*, **6**: 395-629.

Krishna Murti, C.R. (1989). Chemicals in tropical and arid regions. In: Phillippe Bourdeau, John, A. Haines, Werner, Werner Klein and C.R. Krishnmurti, (eds.), *Ecotoxicology and Climate. SCOPE* 38, John Wiley and Sons, pp. 97-136.

Krishnamurti, C.R. (1987). The cycling of arsenic, cadmium, lead and mercury in India. *SCOPE-31*. John Wiley and Sons, pp. 315-333.

Leonard, L. (1971). *Water and Water Pollution Handbook, Vol. 1*. Marcel Dekker Inc., New York, p. 449.

Loster, W.S. (1975). *Polluted River*. River Trent, England, Lack Well Scientific Publication.

Lyr, H. (1977). *Antifungal Compounds*, Seigle'kd Sister ed. Decker, New York, **2**: 332.

Majumdar, Deepanjan, and Gupta, N. (2000). Nitrate pollution of groundwater and associated human health disorders. *Indian. J. Env. Hlth.*, **42(1)**: 28-39.

Malu, R.A. (1999). Lonar lake: A case protection and conservation as Ramsar site. *Wetland Conservation*, pp. 13-18.

Manja, S.K. and Kaul, R.K. (1992). Efficacy of a simple test for bacteriological quality of water. *Journal IAEM*, **19**: 18-20.

Mohanta, B.K. and Patra, A.K. (2000). Studies on the water quality index of river Sanamachhakandana at Keonjhar Garh, Orissa. *Poll. Res.*, **19(3)**: 377-385.

Munawar, M. (1970). Limnological studies on fresh water ponds of Hyderabad, India. *The Biotope, Hydrobiology*, **35**: 127-162.

Muralikrishana, K.V.S.G. and Sumalatha, V. (1983). Groundwater quality in Kakinda. *Proc. of the All India Seminar on Ground Water Management in Coastal Area*, Visakhapatham.

Musaddiq, M. (2001). Current Trends in wastewater management in India. In: *Environmental Pollution and Management of Wastewaters by Microbial Techniques*, Pathade, G.R. and Goel, P.K. eds., ABD Publishers, Jaipur.

Musaddiq, M., Fokmare, A.K. and Khan R. (2001). Microbial diversity and ecology of Lonar lake, (M.S.). *J. Aqua. Biol.*, **16(2)**: 1-4.

Musaddiq, M. (1990). *Microorganism Responsible Environmental Pollution and Suggestive Methods to Control*. Ph.D. Thesis, Marthwada University, Aurangabad.

Musaddiq, M. (2000). Surface water quality of Morna river at Akola. *Pollution Research,* **19(4)**: 685-691.

Narain Rai, J.P. and Sharma, H.C. (1995). Bacterial contamination of groundwater in rural areas of North-West (U.P.). *Indian J. Env. Hlth.,* **37(1)**: 37-47.

Narayana, A.C. and Suresh, G.C. (1989). Chemical quality of ground water of Manglore city of Karnataka. *Indian J. Env. Hlth.,* **31(3)**: 228-236.

Nawlakhe, W.G., Lutade, S.L., Patni, P.M. and Deshpande, L.S. (1995). Groundwater quality in Shivuri district in Madhya Pradesh. *Indian J. Environ. Hlth.,* **37(4)**: 278-284.

NEERI Manual (1988). *Water and Wastewater Analysis.* National Environmental Engineering Research Institute, Nagpur.

Olaniya, M.S. and Saxena, K.L. (1977). Groundwater pollution by open refuse dumps at Jaipur. *Indian J. Env. Health,* **19**: 176-188.

Ozha, D.D., Varshney, C.P. and Bohra, J.L. (1993). Nitrate in groundwater of some districts of Rajasthan. *Indian J. Env. Hlth.,* **35(1)**: 15-19.

Palanivel, M. and Rajaguru (1999). *The Present Status of River Noyyal.* Workshop on environmental status of rivers in Tamil Nadu, Coimbatore, pp. 53-57.

Palharya, J.P., Siriah, V.K. and Malviya, S. (1993). *Environmental Impact of Sewage and Effluent Disposal on the River System.* Ashish Publishing House, New Delhi.

Paliwal, K.V. (1983). Pollution of surface and groundwater. In: *Water Pollution and Management,* C.K. Varshney, ed. Wiley Eastern Ltd., New Delhi.

Pandey, B.N., Gupta, A.K., Mishra, A.K., Das, P.K.L. and Jha, A.K. (2000). Ecological studies on river Panar of Araria (Bihar) with emphasis on its biological components, In: *Pollution and Biomonitoring of Indian Rivers,* Trivedy, R.K., ed. ABD Pub., Jaipur, pp. 130-147.

Pathak, S.P., Bhattacharjee, J.W., Kalra, N., and Chandra, S. (1988). Seasonal distribution of *Aeromonas hydrophila* in river water and isolation from river fish. *J. Appl. Bact.,* **65**: 347-352.

Pelczar, M.J., Chan, E.C.S., Krieg, N.R. (1988). *Heavy Metals and Their Compounds: Microbiology, 5th ed.*, McGraw-Hill Book Company, New York, pp. 497–498.

Perkins, E.J. (1967). *Physical and Chemical Features in Estuarine Waters: The Biology of Estuarine and Waters*, Academic Press, London, pp. 25-37.

Rai, AK., Upadhya, S.N., Kumar, S. and Upadhya, P.D. (1998). Heavy metal pollution and its control through a cheaper method: A review. *Journal of IAEM*, **25(1)**: 25-51.

Rana, B.C. and Palaria, S. (1988). Physiological and physico-chemical and biological study of water quality at Lilari open cast project, Udaipur. *Phycos.*, **27**: 211-217.

Ravichandran, S. and Pundarikanthan, N.N. (1991). Studies on groundwater quality of Madras. *Indian J. Environ. Hlth.*, **33(4)**: 481-487.

Remo Navone, Judson, A., Harmone and Charles, F.V. (1963). Nitrogen content of ground water in southern California. *J. Amer. Water Works Association*, **55**: 615-619.

Saxena, S. and Mehra, K. (1991). Groundwater pollution due to Nitrogenous Fertilizers. *Indian J. Evn. Hlth.*, **33(1)**: 91-95.

Seth, T.D., Hasan, M.Z. and Sircar, S. (1975). Lead levels in urban water supply. *J. Ind. Med.*, **21**: 97.

Shanthi, K., Ramasamy, K. and Lakshamanaperumalasamy, P. (2002). Hydrobiological study of Singanallur Lake at Coimbatore, India. *Nature Environment and Pollution Technology*, **1(2)**: 97-101.

Shivnikar, S.V., Vaidya, D.P. Bandella, N.N. and Patil, P.M. (2000). Levels of ammonia and phosphates as indicators of organic pollution of river Godavari at Nanded (M.S.). *J. Aqua. Biol.*, **15(1&2)**: 52-55.

Shrivastava, R.K., Shrivastava Seema and Shukla, A.K. (2001). River pollution in India: A brief Review. *J. Environ. Research*, **2(2)**: 111-115.

Singh, D.N. (2000). Evaluation of physico-chemical parameters in an Ox-bow lake. *Geobios*, **27(2-3)**: 120-124.

Singh, H.P., Mahaveer, L.R. (1997). Preliminary observation on heavy metals in water and sediments in a stretch of river Ganga and some of its tributaries. *J. Env. Biol*, **18(1)**: 49-53.

Sivakumar, R., Mohanraj R. and Areez, A. (2000). Physico-chemical analysis of water sources of Ooty, South India. *Pollution Research*, **19(1)**: 143-146.

Sreenivasa, R., Rammohan, P.R. and Someswara, R.N. (1999). Degradation of water quality of Kolleru lake. *Indian J. Env. Hlth.*, **4(4)**: 300-301.

Sunilkumar, M. and Ravindranath, S. (1998). *Water Studies: Methods for Monitoring Water Quality*. Centre for Environment Education, Bangalore.

Thorat, S.R. and Masarrat Sultana (2000). Pollution status of Salim Ali lake Aurangabad. *Pollution Research*, **19(2)**: 307-309.

Todd, D.K. (1980). *Groundwater Hydrology, 2nd ed.* John Wiley and Sons, New York, p. 535.

Trivedy, R.K. and Goel, P.K. (1986). *Chemical and Biological Methods for Water Pollution Studies*. Environmental Publications, Karad.

Vates, M.V. (1986). Septic tank density and groundwater contamination. *Groundwater*, **23(5)**: 586-590.

Wetzel, R.G. (1983). *Limnology, 9th edition*. Michigan State University, CBS College Publishing House, Philadelphia, New York, Chicago, p. 763.

WHO (1993). *Guideline for Drinking Water Quality, 2nd edition*. **1**: 188.

William, H., Walker, I. (1969). Groundwater pollution. *J. Amer, Water Works Association*, **61**: 31-40.

Wright, R.C. (1986). Seasonality and bacterial quality of water in a tropical, developing country. *J. Hyg. Comb.*, **96**: 75-82.

Zaffer, A.R. (1964). On the ecology of algae in certain fish ponds of Hyderabad: India physico-chemical complexes. *Hydrobiology*, **23**: 179-195.

Zaffer, A.R. (1966). Limnology of Hussain Sagar lake, Hyderabad, India. *Phykos, 5th edition*, pp. 126-155.

Zaheeruddin, Khurshid, Shadab, Usman and Shabeer (1996). Heavy metal pollution in parts of Delhi. *Indian J. Environ. Port.*, **16(11)**: 828-830.

3

Water Quality Status of Hirakud Reservoir and Its Suitability for Irrigation

☆ *S.K. Samal, B. Pradhan & T.N. Tiwari*

Introduction

Mahanadi is the largest river of Orissa. It originates from Amarkantak plateau of the Chhattisgarh state and enters into Orissa through the district of Jharsuguda. The largest artificial lake of Asia was constructed in the year 1957 across the Mahanadi at Hirakud, which is located about 15 km upstream of Sambalpur town in the state of Orissa. This was the first post-independence major multipurpose river valley project in India. The full potential was achieved in 1966. The project provides 1,55,635 ha for *Kharif* and 1,08,385 ha of *Rabi* crops irrigation in the districts of Sambalpur, Baragarh, Bolangir and Subarnpur. The water released through power house irrigates another 4,36,000 ha of *Rabi* crops in the districts of Sambalpur, Baragarh, Bolangir and Subarnpur. It has also 4,36,000 ha of crops in Mahanadi delta region. The power generation capacity is 307.5 MW through its two power houses. One is at Burla, which is on the right bank of the reservoir, and the other is at Chiplima, 22 km downstream of reservoir. The project also provides flood protection to 9500 km^2 of Mahanadi delta area in the district of Cuttack and Puri. The reservoir has a live storage of 5,818 m^3 with gross storage of 8,136 m^3.

The quality of reservoir or impounded water is subjected to major physical, chemical and biological changes due to influx of sediments and dissolved substances from the catchments area with which water comes into contact with rocks, soils and vegetation during run-off. The reactive amounts of these two components are constantly changing due to change in the rate of precipitation and evaporation. The problem of pollution is due to the domestic wastes, untreated industrial effluents; unscientific use of fertilizers and pesticides in agricultural fields, and by algae also in catchments area. Time is perhaps not too far when pure and clean water will be inadequate for maintaining the normal living standards (Sharma, 2001).

The reservoir water is generally used for irrigation purposes, but the water of the reservoir is also used for drinking and bathing purposes by the people of a number of villages surrounding the reservoir. This will have a profound impact on the treatment of different physico-chemical characteristics to assess the water quality for drinking and bathing proposes at different sampling sites of the Hirakud Reservoir.

Materials and Methods

The water samples were collected in precleaned and sterilized polyethylene bottles of 3 litre capacity, about 2 meter away from the bank and at a depth of 1 meter, in the morning between 8 to 10 a.m., at temperature in summer and winter of 40°C and 20°C respectively. The water samples were collected from 11 different places with a distance of about 1 km each from the neighbouring site during the year 2000. The samples were preserved as per APHA, AWWA and WPCF (1985) and Trivedi and Goel (1986), and brought to the laboratory within 12 hours and stored at 4 to 15°C. Analysis was completed within 48 hours to avoid any variation in the result. All the glassware and other containers were thoroughly cleaned by soaking in detergent followed by 10 per cent of HNO_3 for 48 hours and finally rinsed with double-distilled water. All the chemicals used were of analytical reagent (AR) grade. The salts used for the preparation of the reagents and standard solutions were dried at 110°C for 24 hours before use, and double distilled water was used to prepare all the reagents and dilution. The physico-chemical analysis was carried out according to standard methods given by APHA (1985) and Trivedi and Goel (1986) for the parameters like colour, odour, taste, temperature, pH, EC, TDS, DO, turbidity, Na^+, K^+, Ca^{+2}, Mg^{+2}, TH, Cl^-, SO_4^{-2}, CO_3^{-2}, HCO_3^-, PA, TA, NH_3, NH_3-N and NO_3–N. The parameters temperature, pH, EC, TDS and DO and turbidity were measured using a portable water analysis-kit (model–191 E) and with the standard methods supplied by the manufacturer. Sodium and potassium were measured by flame photometer (Toshniwal–model no.–Chemito 1000). CO_3^{-2}, HCO_3^-, OH^-, PA and TA were determined by titrimetric methods and argentrometric method, respectively, while SO_4^{-2}, NH_3 and NO_3^-N, were measured spectrophotometrically. The statistics like max-min, mean, variation, standard deviation and standard errors were evaluated by Trivedi and Goel (1986).

Results and Discussion

The standard WHO, ICMR, IS, ISI and values of different parameters are shown in Table 3.1. Table 3.3 shows the tolerance limits prescribed by the Bureau of Indian Standards for inland surface water to be used for irrigation and average observed values of Reservoir water. The analysis results of different physico-chemical parameters of the reservoir for the year 2000 are summarized in Table 3.2.

Table 3.1: Standards for Drinking Water*

Parameter	*WHO*		*ICMR*		*IS*		*ISI*	
	PL	*EL*	*PL*	*EL*	*PL*	*EL*	*PL*	*EL*
pH	7–8.5	6.5–9.2	7–8.5	6.5–9.2	6.5–8.5	6.5–9.2	7–8.5	6.5–9.2
EC			300					
Turbidity	5	25	5	25	10	25		
DO			3	6				
TA			120				200	600
TH	200	600	300	600	300	600	300	600
TDS	500	1500	500	1500			500	1500
Cl^-	500		250				600	
Ca^{+2}	75	200	75	200	75	200	250	1000
Mg^{+2}	50	150	50	150	30	100	30	100
SO_4^{-2}	200	400	200	400	300	600	200	400

* All parameters are expressed in mg/l except EC and pH.

PL = Permissible limit; EL = Excessive limit.

Table 3.2: Statistical Analysis of Physico-chemical Characteristics of Hirakud Reservoir in the Year 2000*

Parameter	*Min.–Max.*	*Mean*	*Variation*	*S.D.*	*S.E.*
pH	7.64–7.9	7.52	0.18	0.42	0.12
EC	117.5–165.38	147.73	197.12	14.04	4.05
TDS	93.75–177.5	128.69	905.34	30.09	8.69
Turbidity	5.1–11.5	7.63	5.11	2.26	0.65
DO	5.9–7.7	6.87	0.31	0.55	0.16
Na^+	6–17	10.67	11.88	3.45	0.99
K^+	2–9	5.08	4.45	2.11	0.61
Ca^{+2}	11.63–25.9	17.58	19.65	4.43	1.28
Mg^{+2}	4.19–8.56	6.16	1.71	1.31	0.38
TH	51.84–95.5	69.2	181.78	13.48	3.89
Cl^-	63.36–117.86	89.68	191.61	13.84	4
SO_4^{-2}	31.75–44.64	38.09	24.54	4.95	1.43
CO_3^{-2}	6.92–28.75	16.87	52.92	7.27	2.10
HCO^-_3	98.75–143.07	124.6	269.49	16.42	4.74
PA	3.64–21.15	13.19	43.47	6.59	1.9
TA	119.75–163.07	142.53	138.28	11.76	3.39
NH_3	0.131–0.265	0.197	00	0.04	0.01
NH_3-N	0.16–1.41	0.785	0.18	0.42	0.12
NO_3-N	0.10–1.25	0.70	0.16	0.4	0.11

* All parameters are expressed in mg/l except EC and pH.

S.D.: Standard deviation; S.E.: Standard error.

The observed values of parameters, when compared with standard values, indicate that all the parameters are within the prescribed upper limit. The mean, variance, standard deviation and standard error values show typical representative, inferring values, degree of spread and reliability of observed parameters respectively. The colour, odour and taste of the reservoir water vary in different seasons. The water is mostly light green to deep green colour, algae odour and swampy taste during the months of October to middle of the June, and muddy brown to light brown colour, soil odour and muddy taste during the rest of the year. The temperature variations are from 10° to 45°C during the year. The pH values, varying from 7.46 to 7.9, indicate that the reservoir water is slightly alkaline in nature. The EC is maximum during summer and minimum during monsoon. The DO of the reservoir is maximum during the winter season and minimum during monsoon, as expected. The cationic and anionic composition of the reservoir was found in the sequence of $Ca^{+2} > Na^{+} > Mg^{+2} > K^{+}$ and $HCO^{-}_{3} > Cl^{-} > SO_4^{-2} > CO_3^{-2}$.

Table 3.3: Classification of Irrigation Parameters and Average Observed Values of Reservoir Water*

Parameter	*Range*	*Classification*	*Average Value*
SAR	< 10	Excellent	0.557
	10–18	Good.	
	19–26	Fair	
	> 26	Poor	
SSP	< 20	Excellent	23.46
	20–40	Good	
	41–60	Permissible	
	61–80	Doubtful	
	> 80	Unsuitable	
EC	< 250	Excellent	147.73
	250–750	Good	
	751–2000	Permissible	
	2001–3000	Doubtful	
	> 3000	Unsuitable	
TDS	< 200	Excellent	128.69
	200–500	Good	
	501–1500	Permissible	
	1501–3000	Doubtful	
	> 3000	Unsuitable	
PSS	> 6	Unsuitable	2.926
RSC	> 2.5	Unsuitable	1.22
PI	> 120	Unsuitable	102.22
OP	> 21	Unsuitable	0.053

* All parameters are expressed in e.p.m except EC and TDS.

The TDS, turbidity, total alkalinity, hardness, NH_3, NH_3^-N and NO_3^-N were gradually increased from winter onwards, and the maximum values were noted during monsoon. The water quality index value 59.043 indicates that the water quality of the Hirakud reservoir goes under the "poor quality" category for drinking purposes. It is a meaningful, uniform method for assessing the overall quality of water body, for comparing the quality of different waters and for assessing the effectiveness of pollution abatement measures (Dwivedi *et al.*, 2000). Irrigation parameters such as SAR (Sodium adsorption ratio), SSP (Soluble sodium percentage), PSS (Potential soil salinity), RSC (Residual sodium carbonate), PI (Permeability index) and OP (Osmotic pressure) are given in the last column of Table 3.3 and it is found that all the irrigation parameters are within the prescribed limit.

Conclusion

The present communication gives the water quality and its suitability for irrigation of the Hirakud reservoir–Asia's largest artificial lake. Water samples were collected and analysed for 19 physico-chemical parameter during the year 2000. The results of the investigation revealed that the reservoir water comes under the "poor" category for drinking, but suitable for irrigation purposes. Therefore, if the following corrective steps are carried out, then the Hirakud reservoir and also other reservoirs, rivers and lakes of India may become pollution free:

1. The domestic and industrial effluents should be treated before releasing to water bodies.
2. The forest in the catchment area should be preserved and conserved.
3. The various crop cultivation should be done scientifically, so that residual fertilizers and pesticides should not be discharged along with run-off water.
4. Steps should be taken to control the pollution in the industries situated in the upper part of the river Mahanadi.

Acknowledgements

The authors are greatly indebted to Prof. G.K. Roy, Ex -Director, N.I.T., Rourkela, for permitting to carry out the research work. Sincere thanks are also due to Chief Engineer, Burla Main Dam Division, Burla, Chief Research Officer, Hirakud Research Station, Hirakud and Orissa Pollution Control Board, Rourkela, for their encouragement and esteemed help to conduct this study.

References

APHA, AWWA, WPCF (1985). *Standard Methods for the Examination of Water and Waste Water* (16th edition), Washington DC.

Dwivedi, Smriti, Tiwari, I.C. and Bhargava, D.S. (2000). Classification of Water Quality of River Ganga at Varanasi. *Indian J. Env. Prot.*, **20(9)**: 688-697.

Manivaskaram (1986). *Physico-chemical Examination of Water Sewage and Industrial Effluent*. Pragati Prakashan, Meerut.

Sharma, B.K. (2001). *Environmental Chemistry*. Goel Publishing House, Meerut.

Trivedy, R.K. and Goel, P.K. (1986). *Chemical and Biological Methods for Water Pollution Studies*. Env. Publications, Karad.

4

Water Quality and Hydrobiological Profile of the Perennial River Tamirabarani

☆ *A.G. Murugesan, John Ruby, M.I. Zahir Hussain & N. Sukumaran*

Introduction

The freshwater is life, but is the most exploited system. Since man strode the earth knowingly or unknowingly the water resources are misused and this poses great danger for the supply in future. Among all the other natural resources water is one of the most vital for all kinds of life. Of the total amount of 1500 km^3 of water available in the hydrosphere, about 95 per cent of it is in the sea, 4 per cent frozen as snow and glaciers in the mountains and colder regions on earth and 1 per cent is available for human utilization (World Resources, 1986).

India is a country studded with scenic beauty, comprising of various landforms criss-crossed by rivers. There are 14 major rivers in India that share 83 per cent of the total drainage basin and contribute 85 per cent of the total surface flow (Nilay Chaudhuri,

1983). The total annual flow in all the rivers is estimated to be 18,18,100 million m^3 according to the Central Water and Power Commission (Kayastha, 1989). In India estimated rainfall is about 105 cm. India is rich in precipitation and surface water resources. In addition to this there are 44 medium and 55 minor rivers that also contribute to surface flow, have 80 per cent of their discharge during monsoon months (Gautam, 1992). The consumption of freshwater has increased markedly during last few years, due to the rapid population growth and industrial development which will further increase in future. Nag and Kathpalia (1975) made an estimation in 1974, in which they found that at present water consumption is 9.5 per cent of annual precipitation (400 m.ha.m) and will rise to 26 per cent in 2025 AD.

Water resources, in general, in India are facing a growing onslaught of environmental deterioration. Deforestation along the catchment area is leading to soil erosion and flooding. The discharge of industrial effluents and organic domestic sewage is affecting water resources adversely. Coupled to these, few other anthropogenic activities and malpractices such as construction of roads, dams, houses, fishing etc. are also stressing the aquatic environment in terms of deterioration.

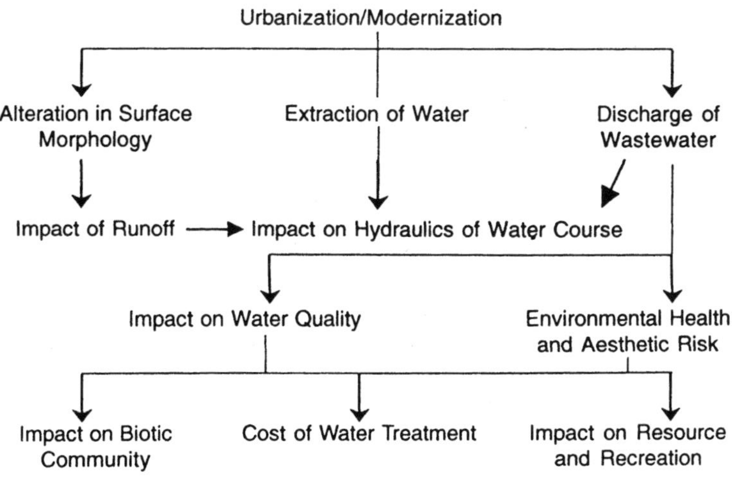

Network of Impacts on Aquatic Environment

Rivers play a significant role because they not only serve the purpose of water supply for domestic, industrial and agricultural and power generation, but also utilized for the disposal of sewage and industrial wastes and therefore it is put under tremendous pressure due to human activities. In the last few decades, there has been increasingly greater emphasis on the deterioration of the quality of Indian rivers. Most of the rivers have been unmindfully used for the disposal of domestic and industrial effluents far beyond their assimilative capacities and have been rendered grossly polluted (Agarwal and Sharma, 1982). According to an estimate, about 80 per cent of total population in India is deprived of pure and safe drinking water. A recent study revealed that there were 1,53,000 problem villages in India, which had infected water supply 90 per cent of total drinking water is severely polluted. Ganga is the most polluted river of the world. Other Indian rivers include Damodhar, Hooghly, Kulu have almost the same story to reveal.

In Southern river Tamirabarani serves as the principal source of water for Ambasamudram, Vickramasingapuram, Cheranmahadevi, Tirunelveli, Sankarankoil, Alankulam, Tenkasi and most of the villages of Thoothukudi district. With increasing number of industries and populations especially in the lower reaches, concern over the water quality of the river began to be strongly felt.

A Review of River Pollution in India

Industrialization coupled with population explosion and urbanization has created an acute problem of disposal of wastes and industrial effluents into the rivers. This has led to excess load of the organic and inorganic materials into the river system, altering the water quality and making the water unpalatable were water is grossly polluted. Several workers have documented the water quality profile of the major rivers in India.

Chacko and Sreenivasan (1955) studied the hydrobiological characteristics of three major rivers in Chennai. Observations on the pollution of Yamuna River, Delhi was taken by Balani and Sarkar (1965). George *et al.* (1966) have pointed out that the wastes of various industries discharged into Kali river contain a heavy load of organic matter. Sreenivasan and Ray (1967) studied the effect of certain wastes on the water quality and fisheries of rivers Cauvery and Bhavani. Rajagopalan *et al.* (1973) studied the pollution status of river Subarnarekha. Vass *et al.* (1977) studied hydrobiological features of river Jhelum in Kashmir. Prakash *et al.* (1978) have carried out ecological study of river Jamuna at Agra. Verma *et al.* (1980) observed the pollution of Hindon river in relation to fish and fisheries.

Badola and Singh (1981) studied the hydrobiological characteristics of the river Alaknanda of Garhwal Himalaya. Agarwal (1982) have studied the nutrient level of Chambal river at Kota. Mitra (1982) studied the chemical characteristics of surface water at selected stations on the rivers Gadavari, Tungabhadra and Krishna. Chandra and Krishna (1983) studied the impact of tannery waste disposal on the quality of the river Ganga at Kanpur. The water quality profile of Vaigai river in Madurai has been determined by Mahadevan and Krishnaswamy (1983). Patil and Patil (1983) have studied the water quality of Ulhas river in Bombay. Sikander and Tripathi (1983) observed the physico-chemical characteristics of Ganga river water at Varanasi. Potamological studies on lotic environment of river Bhagirathi in Garhwal Himalaya have been carried out by Sharma (1984) while Somasekar (1984) studied the water pollution of river Cauvery in Karnataka. Sangu and Sharma (1985) studied pollution in Yamuna river at Agra. Rao and Rao (1986) have observed the pollution in Kshipra river at Ujjain due to pilgrim bather during *kumbha mela*. Distributory pattern of dissolved oxygen at the selected station of Cooum river at Madras was found to affect aquatic fauna (Jebanasan *et al.*, 1987). Ajmal and Razi-ud-Din (1988) studied the pollution of Hindon river and Kalinadi in Uttar Pradesh while the hydrological profile of Hooghly sector of river Ganga was studied by Datta *et al.* (1988). Water pollution in Niva river in Andhra Pradesh has been studied by Nandkumar *et al.* (1988). Chavadi and Nalawadi (1989) have used hydro-environment indicators in the environmental impact assessment of Kali river system in Karnataka. Pollution studies on river Subarnarekha around industrial belt of Ranchi have been carried out by Singh and Singh (1990).

The distribution of algal flora in polluted regions of Karwan river at Agra was studied by Mittal and Sengar (1991). Unni *et al.* (1992) have noted the physico-chemical and biological characteristics of river Narmada from Amarkantak to Jabalpur. The impact of textile mill effluent on the fish in river Tamirabarani, Tamil Nadu has been documented by Murugesan and Haniffa (1992).

Krishnan *et al.* (1994) have studied the contamination of faecal coliforms in river Vaigai. Impact of sewage and industrial effluents in river Betwa at Vidisha, Madhya Pradesh has been observed by Mishra (1996). Physico-chemical characteristics of river Ganga near Patna were observed by Singh (1997). Studies have been conducted by Jain *et al.* (1998) to assess the effect of waste disposals on the water quality of river Kali. Sivasubramani (1999) have studied the physico-chemical, microbiological and hydrological characteristics of river Periyar in Tamil Nadu. An attempt has been made by Aditya Tyagi *et al.* (2003) to study the spatial and temporal water quality trends of one of the pristine river Kshipra (Madhya Pradesh) using water quality index. Physico-chemical and biological parameters have been assessed to study the status of a river Umshyrpi at Shillong (Rajurkar *et al.*, 2003).

Tamirabarani River Basin

Tamirabarani, one of the perennial rivers in Tamil Nadu, originates from Pothigai hills on the eastern sloped of the Western Ghats and drains its water with the Bay of Bengal at Punnakayal of the Gulf of Mannar. The total area of the basin is 5969 sq. km out of which the hilly catchment area comprises of 688 sq. km. It travels about 120 km. through Tirunelveli and Thoothukudi districts, a total area of about 5969 sq. km. The reference of river Tamirabarani in great Indian epics, Ramayana and Mahabharatha and various Tamil literatures reveals its historical importance. An eminent sage and poet Agasthiar is much associated with this river. Etymologically the term "Tamirabarani" is treated in different contexts. In Sanskrit "Tamira" means copper or red, "Parna" means a leaf or a tree and the "Verna" means colour. Hence it may be called as the river of red. Other derivation is the "river flowing among red (sandal) trees". Those days the river basin had sandal trees along the banks of the river and the water had the fragnance of it (Pate, 1916). The tributaries of Tamirabarani river basin in the Western Ghats are Peyar, Vellar, Karaiyar, Pambar and Servalar. Three important reservoirs namely, Pabanasam, Servalar and Manimuthar have been constructed across this river. The following are the important channels of this river, Kannadian, Palayam, Maruthur, Srivaikundam (Murugesan and Sukumaran, 1999). The perennial flow of Tamirabarani river basin is regulated by Pabanasam and Servalar reservoirs that impound and regulate its flow to meet out the irrigation demands and to produce hydroelectric power. The eight anicuts endowed with this river basin are Kodaimelagian, Nadiyunni, Kannadian, Ariyanayagipuram, Palavur, Suthamalli, Maruthur and Srivaikundam.

Rainfall

The percentage component of annual rainfall is very high on the Western Ghats with Sivasailam and Valaiyar recording more than 400 cm in a year. However, the heavy rainfall is confined only to the upper reaches of Tamirabarani and Manimuthar. The annual rainfall of the entire basin is calculated based on its division into 4 zones, zone IA being the upper reaches of Tamirabarani and Manimuthar, zone IB being the Chittar region and zone II and zone III being the Uppodai and middle reaches of Tamirabarani and the lower reaches of Tamirabarani. The dependable rainfall in each of the zone in 50 per cent, 75 per cent and 95 per cent are given in Table 4.1 (WRO–Water Resources Plan, 1996).

Table 4.1: Dependable Rainfall (cm) in Tamirabarani Basin Area

Zones	*50%*	*75%*	*95%*
Zone IA	187.0	160.0	130.0
Zone IB	82.0	67.0	47.0
Zone II	81.5	68.5	50.5
Zone III	68.0	50.0	24.0
Basin	107.2	89.0	67.0

Geomorphology of the River Basin

Tamirabarani river basin comprises of crystalline rocks of Archaean age on the western portion and sedimentary formation of tertiary and quaternary age. Altisol, major soil type found in this basin is well suited for most of the common crops. They are well-developed red or brown soil and they vary from coarse loamy to fine loamy soils. Here vertisol, heavy textured clayey soils are found in scarce amounts. The thickness of weathered zone varies from 1 to 3 bgl. At few locations the sheared and fractured zone extends upto 60 m. The bed rock depth vary from 1 to 60 m. The aquifer thickness in the sedimentary ranges from 5 to 40 m underlain by clay, standstone, shale or weathered rock. The existence of water table at

10–18 m bgl depends on local topographical, geological and structural conditions. Adjacent to Tamirabarani river, the thickness of alluvium ranges from 10–15 m while the teri sand thickness ranges from 20–30 m. The depth of bed rock in tertiary sand stones extend beyond 150 m bgl in the eastern coastal areas. Transmissivity of aquifer varies from 0.5–8 sq. m/day in the Chittar sub basin, whereas in Tamirabarani basin it ranges from 0.5–13.0 sq. m/day and in Uppodai sub basin it is between 0 and 6 sq. m/day. The transmissivity in the hard rock area of Tamirabarani sub-basin ranges from 3–56 sq. m/day whereas in sedimentary formation is 100–800 sq. m/day. The alluvium of the sedimentary formation occurs along Tamirabarani river from Seevalaperi. The width of the alluvium is 0.5–1 km near Seevalaperi and Srivaikundam. It widens out to more than 10–15 km in the delta below Srivaikundam and it exceeds further when it reaches nearer to sea. The wind blown red sand dunes of Sawyerpuram teri and Kudiraimozhi teri occur in the northeast and southern part of the Tamirabarani river respectively in the eastern side of the basin. A narrow strip of wind blown coastal sands occur parallel to the coast (WRO–Water Resources Plan, 1996).

Hydrology of River Tamirabarani

Out of the total area of Tamirabarani basin (5969 km), the area which falls under hilly and mountainous terrain limited scope for the development of ground water is 1262 sq. km. Altogether the four zones comprises to work out to a recharge area about 4707 sq. km. The chemical composition state of groundwater from 77 shallow dug wells were reviewed for the quality assessment of this basin. The salinity value in terms of electrical conductivity falls within the acceptable limits of 3000 microsiemen/cm. High salinity was found in some of the places in the basin. The poor groundwater quality might be due to the return flow of irrigation or the nature of the parent material of the soil. In addition the chloride values exceed the limit of 1000 mg making the water unfit for use. The use of water in Seevalperi, Palayamkottai and Kayalpattinam for irrigation is highly restricted as their sodium concentration is invaluably high. The nitrate and fluoride values fall within the limits in this basin. Of the chemical composition data available for the period of 10 years for Manimuthar reservoir, the quality of water was found to be fit for human consumption and irrigation use.

Pollution Due to Industrial Discharge

Pollution of river courses associated with industrial discharge and refuse from human settlements is a global problem. In India it is reported that about 70 per cent of the available water is polluted (Citizens Report, 1982). The chief source of pollution is identified to be sewage constituting 84–92 per cent of the wastewater. Industrial wastewater comprises 8–16 per cent (Chaudhari, 1982). Tamirabarani river basin on its banks has a number of industrial units including Pulp and Paper, Textile, State Transport Corporation Workshops, Photographic industries and other small-scale industries. It provides industry the water needed for power generation. Every industry draws fresh water and discharges wastewater at some stage or the other.

The public that confronts the public is the liquid waste with inorganic and organic content. The waste liquids from textile mill comprises mainly of dye stuff sulphates, sulphide, copper, zinc, lead, phenolics and wastes from the manufacture of pulp and paper contains sulphides, chlorides, lignocellulosic wastes, mercaptans, mercury etc. The pollutants that are concerned to the ecosystems are those that do not breakdown and those that resist to biodegradation and are found to enter the aquatic and human (Murugesan, 1988).

Tamirabarani, mainstream receives the effluent from the major textile industry, Madura Coats situated on the banks at Vickramasingapuram. It draws water from river for the manufacturing process and emptying about 800 KLD of trade effluent and 400 KLD of sewage with partial treatment through a channel into the river (Murugesan and Haniffa, 1992). About 10,400 KLD of oily waste water from the Tamil Nadu State Transport Corporation Workshop at Pabanasam and much more pollution into river. Another major industry, M/s Sun Paper Mills situated on the banks at Cheranmahadevi also draws water and generates about 24,450 KLD of effluent a part of which ultimately reaches the river. In addition there are about 4961 numbers of small scale industries which include 4203 number in Tirunelveli district and 758 in Thoothukudi district. There are 19 large and medium industries functioning in the river basin. Near the cities in the mainstream where the effluents are discharged from municipal and industrial wastes the river is highly polluted. Systematic observations made at various points of the

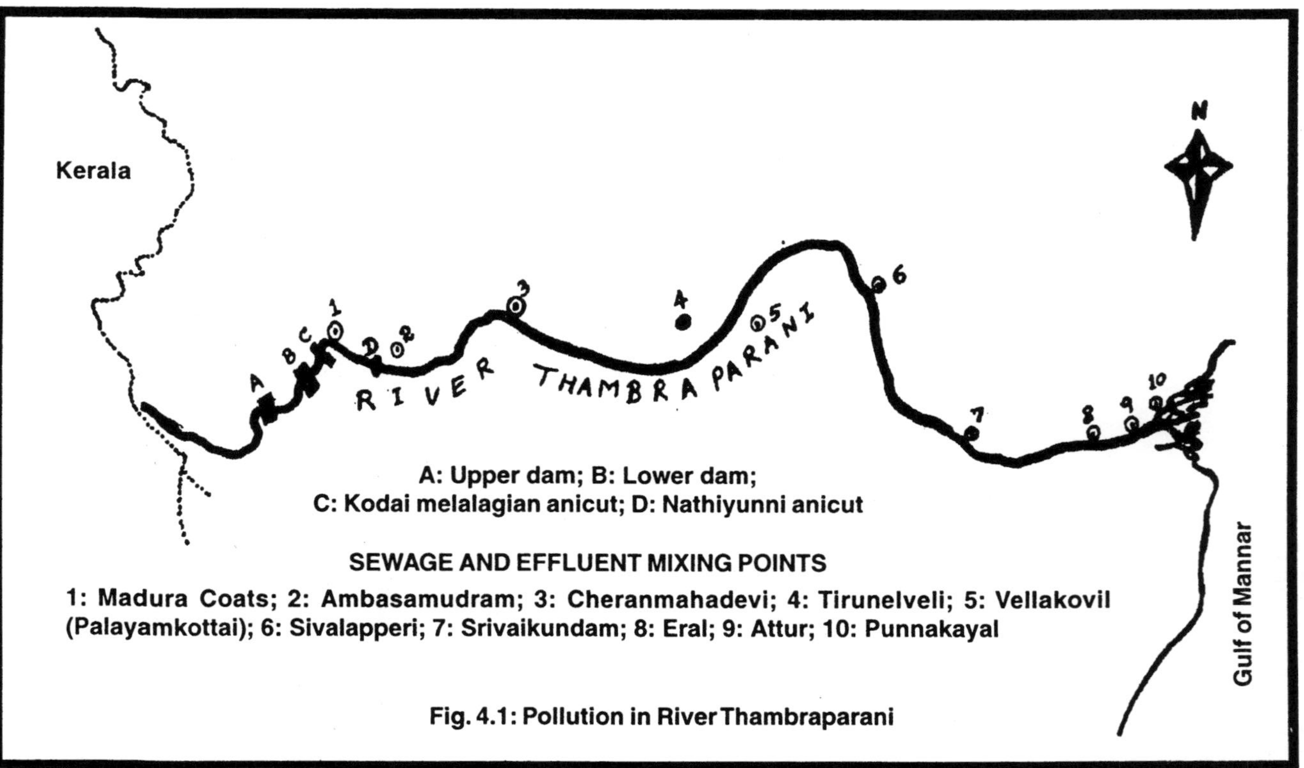

Fig. 4.1: Pollution in River Thambraparani

riverine basin shows that the effluents are being regularly discharged and cause serious pollution problems (Kundra *et al.*, 1997; Murugesan and Sukumaran, 1999).

Sewage Entry Points into River Tamirabarani

The sources of pollutants and sewage and the entry points into the river Thamirabarani starting from Papanasam to Punnakayal were investigated. At Papanasam, the state transport corporation workshop the wastes mix in the river through the Kodaimelagian channel (14 km. away from the origin of the river). The Kodaimelalagian channel receives domestic sewage from Vickramasingapuram, Sivanthipuram and Gnaniar thoppu areas. A part of the domestic sewage from Vickramasingapuram enters directly into the river, near Madura Coats bridge. The agricultural run-off from Sivanthipuram also enters the river directly.

A part of effluent from the Madura Coats and sewage from Kasirthoppu area are discharged into the East Kodaimelalagian Channel that supplies water to Thevarkulam of Keela airmalpuram located 11 km away from Madura Coats. Major part of Madura Coats effluent mixes into the river near the bridge. Large number of polythene bags and solid wastes from Madura Coats Mill are dumped on the roadsides on the way to Chettimedu.

At Ambasamudram, the government hospital wastes and domestic, sewage from the nearby area finds entry into the West Kodaimelalagian Channel. The water of this channel is used for irrigation and further the agricultural run off from the fields mixes with the river water at a site 20 km away from its origin. At Kallidaikurichi domestic sewage mixes with the channel water. The agricultural run off from various sites at Thiruvidaimarthur, Polikarai and South Sankanthiradu area enters into the mainstream of the river.

The Tamirabarani river water is contaminated by domestic sewage coming from the C.N. Village, Mayilvannapuram, Mettutheru and Meenakshipuram areas. At Kailasapuram area the river is severely polluted by sewage, human activities and workshop wastes. Human defecation is common in Kokkivakuḷam area into the river. Kannamman Koil Street, Perumal Koil Street, Periyathattakudi Street and Sangetha sabha areas dump their wastes into the river. The wastes from the railway station area, Getwel hospital region and Sindhupoondurai area is discharged into the river at the northside of the river bridge at Kokkirakulam. Domestic sewage from Udayarpatti and Megalingapuram also finds entry directly into the river.

In Tirunelveli, the western bank of the river Tamirabarani is polluted by domestic sewage and workshop wastes, while, the eastern bank is polluted by the agricultural run off from the fields of Thirumoolavar Ashram and Nattam areas (South of the railway bridge). Domestic sewage and wastes from the TNSTC Workshop, Sakunthala hotel area, Valayapathy street, Vetriveladivinayakar street and Ettuthokai street is discharged directly into the river at Vannarpettai.

Human activities are prevalent at Moolikulam. Sewage mixes into the Melakulam, near Moolikulam. At Vellakoil, the crematory wastes are dumped into the river. At a few kilometers before Aruvankulam the agricultural run off is drained into the mainstream. Domestic sewage entry points were identified in the river at Seevalaperi. In addition to the sewage, human activities in the river also pollute the water. At Vallanadu, the river is highly polluted by the agricultural run off from the fields located in the river banks. Human activities are also contaminating the water considerably in this site. At Srivaikundam, Azhwarthirunagari and Aeral, the river is polluted by domestic sewage and human activities. Abundant growth of *Echhornia crassipes, Ipomea carnia,* and *Typha angusta* was recorded in these sites during our visit. Domestic sewage and agricultural run off from the fields enters into the river at Aathur. Water hyacinth and water lily were recorded as aquatic weeds in this site. At Punnakayal, human activities including defecation were found along the riverside and aquatic weeds were absent in this site.

It is necessary to take preventive measures to control the discharge of sewage into the river system. Local bodies of every area should take steps to reroute the sewage away from the river and employ the sewage for other purposes. The sewage from the village or town should be collected in a particular place and can be employed for fish culture.

Impact of Pollutants on the Biotic Communities of River Tamirabarani

The focus on environment protection and restoration has widened from the stress specific water quality standards to the achievement of broad objectives of restoring self maintained ecosystems and quality of human life maintenance. Biological. data at a particular location in a stream or river not only reflect the conditions resisting at the time of monitoring but also the past conditions as well (Murugesan *et al.*, 2000). Not many studies have been initiated to assess the status of pollution of river Tamirabarani and its tributaries, except for the studies of Murugesan *et al.* (1994), wherein the physico-chemical parameters of the river water at various stations were recorded for a year. The impact of textile mill effluent in the air breathing fish *H. fossilis* has been documented by Murugesan and Haniffa (1992). The histopathological and histochemical changes were quite significant in the oocytes of the effluents treated fish, vacuolation in the nuclear extensions were observed in addition to the reduction in number of unclear extrusions. The sudden abrupt check in the RNA synthesis gave rise to vacuoles in place of par fibrosa. Haniffa *et al.* (1986) showed that level of erythropoiesis was enhanced due to the effect of the distillery and paper mill effluents on *H. fossilis* and *S. mossambicus* of river Tamirabarani. This was revealed by the elevated demands for O_2 or CO_2 transport in effluent exposed fish as a result of increased physical or metabolic activity or by destruction of gill membranes, which resulted in faulty gaseous exchange. Murugesan and Haniffa (1985) also document the increase in RBCs, haemoglobin and haemocrit values of fish exposed to textile mill effluent. Elevated levels of 3 red cell indices, MCV, MCH and MCHC revealed that the fish exhibited macrocytic anemia due to malabsorption of ions. Murugesan (1998) reported that almost all the systems of the fish living in Tamirabarani underwent histopathological changes and the biochemical pathways will also upset by the industrial effluents.

Exuberance of Aquatic Weeds

The head reach of Tamirabarani river basin is almost free of perennial weeds. However, the luxuriance of few exotic weeds like *Clerodendron inflortunnatum, Clerodendron phomidis, Lantana camara* and *Chromolaena odorata* were observed here and there. The exuberance of weeds was highly noticed in the places especially where the solid and liquid wastes confluences with the river. Weeds like *Xanthium indicum, Cyperus corymbosus, Typha angustifolia, Parthenium hysterphorus, Prosopis chilensis, Ipomoea carnea, Ipomoea obsura* and *Eichornia crassipes* were found to be observed invariably in pools, ponds, puddles, channels canals and ditches associates with the river. Moreover, free floating weeds like *Salvinia molesta, Pistia stratiotes, Lemna minor, Azolla pinnata, Spirodela polyrchiza* were found to be seasonal. Also, the algae like *Microcystis aeruginosa*, Sps. of *Anabaena, Oscillatoria, Chlamydomonas,* and *Pandoriana, Eudorina, Volvox* etc. are common. These invasive species produce undesirable odour and taste in the impounded water and makes it unhygienic (Abraham and Samuel, 1999).

The rapid colonisation and distribution of aquatic weed *Ipomoea carnea* have reflected the abundance of sewage pollution in places like Ambasamudram, Veeravanallur, Cheranmahadevi, Melapalayam, Tirunelveli Town and Sindupoondurai. The encroachment of aquatic weeds due to adverse pollution along the course of river has changed the water flow and it's current (Palmer, 1970; Pandey, 1973). The luxuriant growth of invading aquatic weeds in the main stream could however attributed to the domestication, industrialisation and clearance of riparian vegetation along the course of river banks. To the worse, these weeds compete with the wild florae, reduces species bed and rapid consumption of water by aquatic and semi aquatic weeds are the serious problems that the riverine ecosystem has at present and their eradication has to be seriously look out for replace. Moreover, due to the competitive ability, they replace the native vegetation in other habitats. Unless or until if the threatened habitats are identified for immediate management and conservation the problems of the change of vegetation and environmental change in different vegetation zones favourable conditions like the light intensity, stream bed and DO, aquatic weeds like *Cladophora* and *Vancheria* increased well but this in turn put down the algal biomass because the N_2 and P are being used up. On the zone affected by the organic enrichment there was an increase in the total biomass of the benthic fauna due to good supply of waste where the effects are severe, pollution can lead to the elimination of fish population from long reaches of the stream (Murugesan and Sukumaran, 1999).

Table 4.2: List or Some Important Weeds Observed in River Tamirabarani

Botanical Names	*Family*
Ageratum conyzoides	Asteraceae
Arundo donax	Poaceae
Ceratopteris thallictroides	Parkeriaceae
Chromolaena odorata	Asteraceae
Clerodendron infortunatum	Verbenaceae
Clerodendron phlomidis	Verbenaceae
Croton bonblandianum	Euphorbiaceae
Cyperns corymbosus	Cyperaceae
Cyperus exaltatus	Cyperaceae
Eichhornia crassipes	Pontederiaceae
Hydrilla verticillata	Hydrocharitaceae
Hygrophila auriculata	Acanthaceae
Ipomea aquatica	Convolvulaceae
Ipomea obsura	Convolvulaceae
Ipomoea carnea	Convolvulaceae
Jatropha gossypifolia	Euphorbiaceae
Nelumbo mucifera	Nyphaeceae
Nymphaea stellata	Nyphaeceae
Ottelia alismoides	Otteliaceae
Parthenium hysterophoros	Asteraceae
Persicaria glabra	Polygonaceae
Phragmites karka	Poaceae
Pistia startiotes	Araceae
Proposis chilensis	Mimosaceae
Salviniae molesta	Salviniaceae
Spilanthes calra	Asteraceae
Typha angustata	Typhaceae
Vallisneria spiralis	Hydrocharitaceae
Xanthium indicum	Asteraceae

The causes for the growth of aquatic weeds are discharge of domestic sewage and agricultural drainage water containing nutrients *viz.*, nitrates and phosphates. Both nitrogen and phosphorus occur in small amounts in natural waters, but their concentrations get greatly increased by the activities of man. As much as 80 per cent of nitrogen and 75 per cent phosphorus added to surface water originate from the discharge of domestic sewage. The phosphorus content of domestic sewage is becoming a topic of concern. It is estimated that up to 70 per cent of this comes from the use of household detergents. Continuous addition of nitrates and phosphates enrich the water body and accelerate the proliferation of aquatic weeds. Aquatic weeds are found to infest pristine waters but their proliferation rate is enhanced in waters rich in nutrients.

Water Quality Profile of River Tamirabarani

River Tamirabarani, the principal source of water to the Tirunelveli and Thoothukudi districts is also a major site for the disposal of the solid and the liquid wastes in the two districts. Hence this makes it necessary to monitor the quality of the river regularly to protect the river from deterioration. The present study was under to assess the status of the water quality of the river in the years 2001 and 2002 and also to identify the sources of pollution into the river.

Materials and Methods

21 stations were selected along the river Tamirabarani, in the Tirunelveli and Thoothukudi districts. Station 1 is situated at the foot of the Pothigai hills, at Pabanasam and there are no chance of pollution except the bathing and washing activities of the people. Station 2 is the place where the effluent from the Madura Fabrics is mixing with the Tamirabarani through a canal at Vickramasingapuram (VK Puram). Station 3 is situated at a distance from this site. Stations 4, 5, 6 and 7 are Ambasamudram, Ambasamudram, Thirupadai Maruthur and Mukkudal where the only possible source of pollution is human activities. Stations 8 and 9 are at Cheranmahadevi, here the effluent from the Sun Paper Mills is the major source of pollution. Stations 10, 11, 12, 13 and 14 are Tharuvai, Kurkkuthurai, Kokkirakulam, Vellakoil and Pottal that are places in and around Tirunelveli. Sewage is the major problem in Kurkkuthurai and Kokkirakulam, whereas Vellakoil is the place where the crematory wastes are dumped into the river. Stations 15–16 are place in the Thoothukudi districts wherein domestic sewage and agricultural runoffs confluence with the river. Water samples

were collected monthly and the various parameters were estimated using standard methods (APHA, 1995). Water for the dissolved oxygen was fixed at the site itself and preserved for the analyses. The following are the parameters analysed: Temperature, pH, dissolved oxygen (DO), biological oxygen demand (BOD), chemical oxygen demand (COD), total solids (TS), total suspended solids (TSS), total dissolved solids (TDS), free CO_2, acidity, alkalinity, hardness, calcium, magnesium, chloride, sulphate, total coliforms and faecal coliforms.

Results and Discussion

The entry of a pollutant into flowing stream or river set off a progressive series of physical, chemical and biological events in the downstream water. The nature of the polluted body is thus governed by the character and quantity of the polluting substances. River Tamirabarani is not an exception to this. The values of the DO, BOD, COD and nitrate are given in Table 4.3. It is observed that the dissolved oxygen content of the water has been quite similar in all the stations, however the BOD and the COD levels have been elevated slightly in areas where there is entry of effluents or sewage into the river. The BOD level has been greater in Kokkirakulam which receives sewage and other solid matter. This is because the paper mill effluent at Cheranmahadevi is poured into the river. The paper mill effluent is generally colloidal and tends to contain high COD, suspended solids and dissolved solids.

Figure 4.2 shows the physico-chemical characteristics of the river Tamirabarani during the year 2001. It is evident that the DO level has been uniform through the year except for the dry seasons like June and July. Then soon after the rains the level of DO has come up. The BOD level has been slightly elevated during lower DO levels, however during the latter half of the year the BOD level had been increased to cross the permissible limit. This may be due to the high organic content in the water. The COD level has showed a marked fluctuation in the different months. It has been very high during the months of January, April, June, August and September 2001. When the temperature and pH of the water is considered it is clear that the variation in the water temperature and pH is not very significant. The level of bicarbonate is very high during May and September (Figure 4.3). The water is slightly alkaline in all the sites but for Kokkirakulam where it is slightly (Table 4.4). Bicarbonates are in the

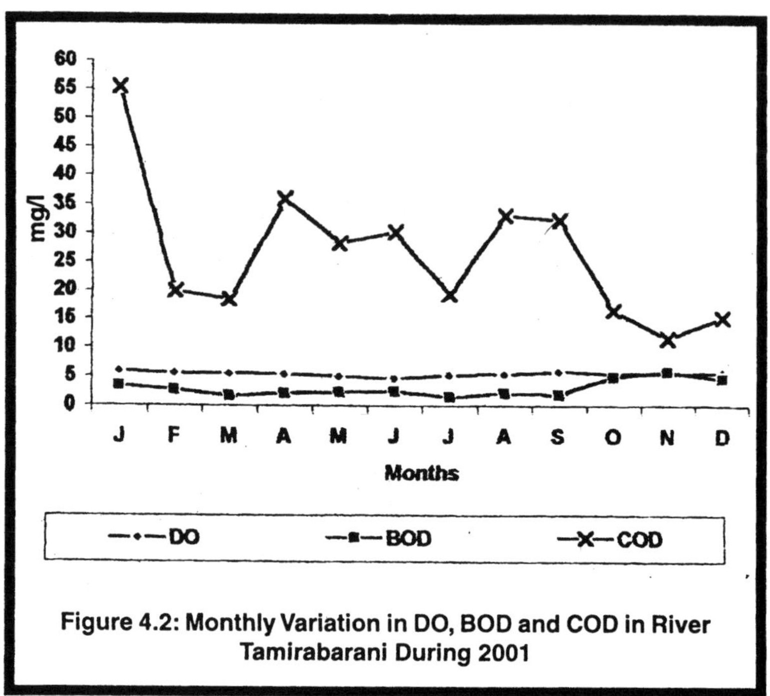

Figure 4.2: Monthly Variation in DO, BOD and COD in River Tamirabarani During 2001

range of 17.83 to 188.83. The bicarbonates are higher in stations such as Kokkirakulam, Vellakoil, Eral and Aathur where there is entry of sewage, hospital wastes and also detergents due to human activities. When the level of Total Solids, TSS and TDS are analysed it is observed that generally the solids are found in ample amounts in all the sites downstream. Aathur is found to contain the highest amount, when compared to the other sites, however they are within the permissible limits of 500 mg/l. The water analysis also reveals higher amounts of hardness of the water. The ions such as calcium and magnesium are found in higher amounts in all the sites after Cheranmahadevi. Pollution due to industrial effluent and domestic sewage is a major factor for this nature of water quality. The total hardness of water in Aathur is 122.75 mg/l which is above the permissible limit of 100 mg/l (Table 4.5). Chloride content of river water shows that they do not exceed the permissible limit, but the downstream stations contain comparatively higher amounts of chlorides. Higher chloride concentration in the downstream water may be an indicator of pollution of industrial, municipal and organic wastes (Thresh *et al.*, 1994; Ownbey and Kea, 1967). Venkateshwarlu and Jayanti (1968) reported chlorides to range between 29 to 396 mg/l as a result of pollution from sewage and industrial waste. The chloride themselves do not possess any hazards as far as water quality is concerned but their concentration works out as a very good indicator of organic pollution. The monthly variation of hardness shows that hardness had been more during May, September and December (Figure 4.5).

Table 4.3: Physico-chemical Parameters of River Tamirabarani at Different Sampling Stations During 2001

Sampling Sites	*Distance from Origin (km)*	*DO*	*BOD*	*COD*	*Nitrate*
Agasthiar Falls	12	5.86±0.93	2.45±1.21	22.17±14.56	1.42±0.90
V.K. Puram	14.9	5.37±0.55	2.37±1.08	21.17±11.96	1.75±1.36
V.K. Puram	15	5.42±0.43	2.58±1.30	26.50±25.85	2.33±1.83
Ambasamudram	19	5.47±0.32	2.99±1.70	29.83±24.32	1.75±1.14
Ambasamudram	21	5.61±0.35	2.96±2.08	28.50±21.29	1.83±1.27
Thirupadi Maruthur	30	5.50±0.37	3.21±2.60	30.50±21.94	2.42±1.68
Mukkudal	35	5.57±0.55	2.29±1.18	21.33±0.07	2.25±1.76
Cheranmahadevi	39	5.80±0.68	2.43±0.92	23.17±13.63	2.00±1.41
Cheranmahadevi	39.2	5.66±0.67	2.25±0.10	30.50±30.79	3.83±5.29
Tharuvai	52	6.26±0.87	2.08±0.68	23.83±11.58	2.33±1.37
Kurukku Thurai	57	5.58±0.88	3.86±3.00	30.83±15.41	2.58±1.38
Kokkirakulam	59	5.51±0.76	4.26±2.40	48.50±30.56	5.42±5.53
Vellakoil	61	5.08±0.64	3.22±1.64	47.50±55.24	2.67±1.67
Pottal	64	4.95±0.92	2.59±1.09	23.67±11.96	2.33±1.30
Manapadai Veedu	67	4.88±0.80	3.67±3.04	24.50±12.00	2.00±1.04
Seevalaperi	71	4.30±0.58	3.30±2.25	22.5±12.03	2.11±1.19
Maruthur Anicut	77	5.25±0.58	2.61±1.11	23.83±15.69	2.08±1.08
Karungulam	88	5.70±0.21	2.66±1.21	19.17±08.46	2.08±0.67
Srivaikundam	96	4.50±0.88	3.08±2.44	18.17±07.46	2.17±0.58
Eral	110	4.55±0.54	2.13±0.67	18.00±07.03	1.91±0.83
Aathur	115	5.67±0.45	3.42±2.74	19.83±11.86	2.75±4.22

All the values are in mg/l. Each value represents the mean of 12 samples ± STD.

Table 4.4: Physico-chemical Parameters of River Tamirabarani at Different Sampling Stations During 2001

Sampling Site	*Water Temperature °C*	*pH*	*Bicarbonate*	*TS*	*TDS*	*TSS*
Agasthiar Fall	24.50±1.15	7.04±0.35	17.83±07.26	63.25±37.73	43.83±27.17	21.92±20.37
V.K. Puram	24.38±1.30	7.02±0.29	28.25±12.61	105.17±70.00	64.67±37.25	44.50±49.64
V.K. Puram	24.50±1.00	6.99±0.42	42.00±28.55	142.25±85.40	103.00±67.98	43.25±45.64
Ambasamudram	25.17±1.13	7.08±0.22	31.67±16.70	94.83±57.92	60.17±33.22	38.67±32.27
Ambasamudram	26.25±1.16	7.09±0.33	22.08±09.91	77.25±66.88	48.42±33.42	33.50±37.67
Thirupadi Maruthur	27.08±1.68	7.04±0.47	47.50±31.09	107.42±51.84	81.50±45.01	30.25±27.91
Mukkudal	27.67±2.02	7.31±0.37	39.83±15.75	92.83±43.02	68.25±32.63	30.25±22.66
Cheranmahadevi	29.04±2.47	7.26±0.28	43.33±23.45	91.50±45.01	66.33±30.30	29.50±21.89
Cheranmahadevi	29.17±1.95	7.24±0.28	38.67±15.59	91.25±51.44	62.42±33.42	33.92±22.14
Tharuvai	28.83±1.19	7.25±0.33	46.83±17.93	100.33±35.88	74.58±30.06	31.75±22.43
Kurukku Thurai	29.67±1.50	7.22±0.35	49.42±24.97	115.33±75.78	72.92±31.88	39.42±39.19
Kokkirakulam	30.25±1.60	6.98±0.43	70.33±44.42	135.92±75.98	89.83±51.69	51.83±30.88
Vellakoil	29.25±1.86	7.16±0.36	64.92±34.22	151.33±86.07	107.25±63.89	50.00±34.65
Pottal	28.83±1.75	7.15±0.35	69.33±29.72	144.75±76.45	108.25±58.12	43.17±33.64
Manapadai Veedu	29.50±1.31	7.22±0.34	83.75±36.63	144.92±75.64	92.83±69.09	40.75±34.95
Seevalaperi	28.83±1.19	7.39±0.39	71.25±28.58	144.83±74.07	108.75±55.21	42.83±34.61
Maruthur Anicut	30.17±1.85	7.43±0.55	94.58±74.62	194.33±130.41	154.92±132.44	47.17±38.60
Karungulam	30.42±1.98	7.41±0.34	49.58±38.32	150.00±87.66	100.00±61.83	59.50±38.79
Srivaikundam	30.17±1.27	7.25±0.31	46.75±32.77	159.75±73.92	113.50±47.69	54.00±40.76
Eral	29.75±0.97	7.70±0.20	106.33±14.74	179.58±89.72	131.17±167.64	62.42±58.51
Aathur	29.75±0.75	7.43±0.42	188.83±37.29	333.83±147.31	280.67±130.84	86.42±124.96

All the values except pH are in mg/l. Each value represents the mean of 12 samples ± STD.

Table 4.5: Physico-chemical Parameters of River Tamirabarani at Different Sampling Stations During 2001

Sampling Site	*Total Hardness*	*Calcium*	*Magnesium*	*Chloride*
Agasthiar Falls	20.92±08.24	4.33±01.92	3.17±03.76	9.00±02.92
V.K. Puram	31.25±11.24	5.25±01.54	5.33±04.14	11.5±04.64
V.K. Puram	50.67±41.34	8.90±06.40	6.17±08.19	19.75±22.71
Ambasamudram	33.33±15.10	9.33±07.84	4.33±02.64	11.67±04.38
Ambasamudram	24.67±11.73	8.08±07.82	2.67±01.92	9.67±02.96
Thirupadi Maruthur	48.17±27.81	11.83±05.86	6.42±05.71	16.25±12.00
Mukkudal	44.50±22.54	9.50±03.37	4.75±02.56	13.83±04.76
Cheranmahadevi	43.33±17.98	11.67±05.05	4.08±02.87	13.33±04.98
Cheranmahadevi	44.50±25.57	11.25±04.85	4.33±03.14	13.75±06.93
Tharuvai	47.33±18.81	12.50±06.50	4.75±02.90	14.33±04.54
Kurukku Thurai	50.71±24.43	15.00±09.79	5.58±03.26	14.00±03.79
Kokkirakulam	68.17±38.59	18.33±10.96	7.17±04.09	19.33±10.92
Vellakoil	68.33±33.99	16.75±08.58	11.92±13.42	16.42±03.50
Pottal	73.55±31.51	15.58±06.79	11.67±08.78	19.75±08.59
Manapadai Veedu	81.75±29.41	17.33±07.69	11.83±07.11	19.42±05.95
Seevalaperi	78.92±33.39	16.75±07.25	12.50±15.00	17.25±04.29
Maruthur Anicut	104.00±90.84	19.17±10.28	16.67±20.15	21.17±30.06
Karungulam	81.83±31.91	15.42±08.05	8.83±02.48	17.42±05.40
Srivaikundam	79.25±29.27	17.08±05.93	14.25±16.36	20.58±03.63
Eral	90.42±29.79	16.42±07.18	13.83±11.17	23.25±07.31
Aathur	122.75±33.26	18.83±12.29	18.33±05.60	62.33±15.60

All the values are in mg/l. Each value represents the mean of 12 samples ± STD.

Studying the water quality of river Tamirabarani during the year 2002, it is evident that the DO is quite high at station 1, Agasthiar falls. However at the other stations the DO is uniformly well above the standard levels. BOD of the river water is slightly high in some of the places. At station 11 and 12, the BOD level exceed the permissible level, indicating that the water is quite unfit for drinking purpose. COD level is also higher at Kokkirakulam and Vellakoil. High organic content in the water is the primary cause for the elevated levels of BOD and COD at these sites. The level of nitrate is comparatively higher at sites Kokkirakulam and Kurukkuthurai

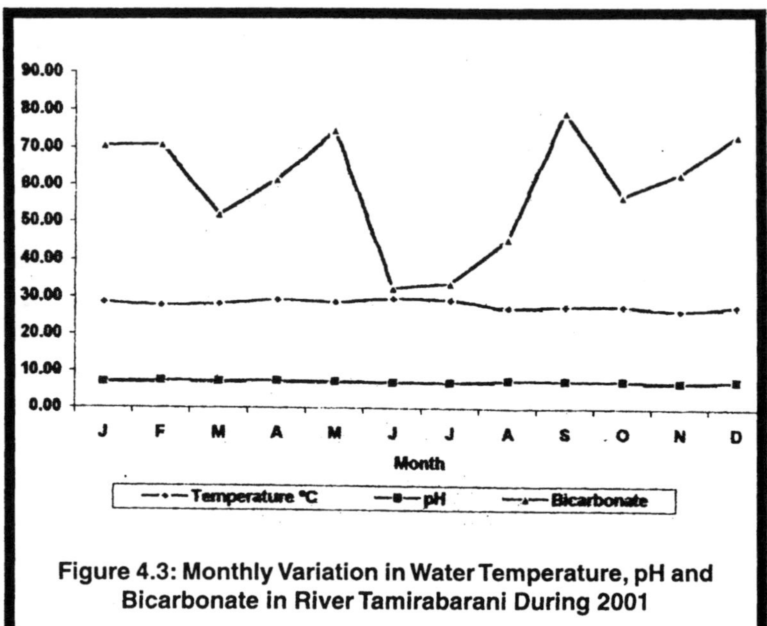

Figure 4.3: Monthly Variation in Water Temperature, pH and Bicarbonate in River Tamirabarani During 2001

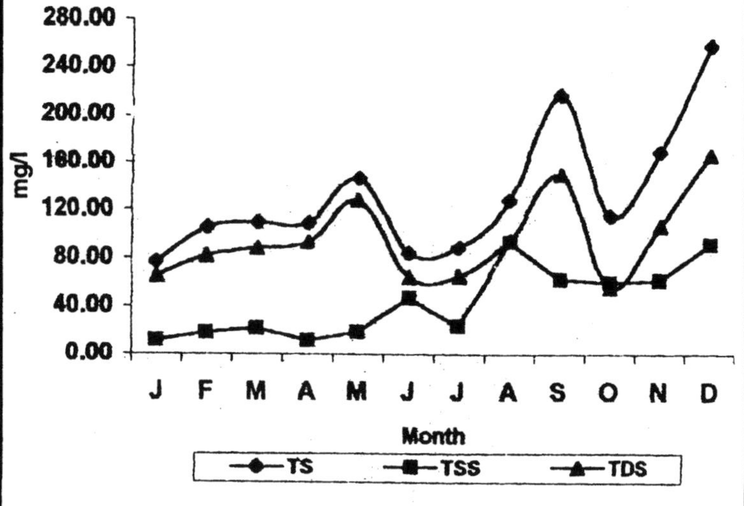

Figure 4.4: Monthly Variation in TS, TDS and TSS in River Tamirabarani During 2001

(Table 4.6). Agricultural field nearby release agrochemicals into the surrounding environment, which finally reaches the river at these sites. Presence of nuisance aquatic weed, water hyacinth in Kurukkuthurai, Kokkirakulam and Vellakoil is an indicative of enrichment of water by nitrates. However, the nitrate level in river Tamirabarani does not exceed the permissible level.

Table 4.6: Physico-chemical Parameters of River Tamirabarani at Different Sampling Stations During 2002

Sampling Sites	*DO*	*BOD*	*COD*	*Nitrate*	*Phosphate*
Agasthiar Falls	6.43±0.50	2.79±1.21	11.33±02.31	1.05±0.48	0.17±0.24
V.K. Puram	5.75±0.50	2.97±1.37	12.75±02.99	1.11±0.46	2.26±0.31
V.K. Puram	5.62±0.29	3.15±1.50	18.17±03.59	1.44±0.52	0.23±0.25
Ambasamudram	5.40±0.46	3.89±2.44	16.50±01.93	1.52±0.78	0.36±0.39
Ambasamudram	5.06±0.56	5.30±1.66	19.67±06.07	1.75±0.68	0.41±0.36
Thirupadi Maruthur	5.65±0.49	3.76±2.19	20.50±06.23	1.66±0.64	0.22±0.19
Mukkudal	5.74±0.53	3.24±1.48	19.42±04.19	1.41±0.51	0.18±0.16
Cheranmaha-devi	5.41±0.59	3.30±0.99	17.67±07.36	1.51±0.73	0.43±0.26
Cheranmaha-devi	5.59±0.32	2.93±1.43	21.42±04.85	1.72±0.71	0.37±0.25
Tharuvai	5.75±0.59	2.34±0.96	12.25±04.52	1.74±0.89	0.22±0.22
Kurukkuthurai	5.07±0.42	5.63±3.79	25.08±14.39	2.20±1.03	0.33±0.29
Kokkirakulam	5.46±0.56	6.20±3.36	32.92±14.88	2.29±1.20	0.43±0.27
Vallakoil	5.37±0.37	3.87±1.94	26.00±13.88	1.63±0.72	0.32±0.17
Pottal	5.23±0.51	2.82±1.39	25.67±21.25	1.36±0.51	0.23±0.15
Manapadai Veedu	5.32±0.25	2.80±1.30	21.00±12.20	1.43±0.52	0.31±0.30
Seevalaperi	5.08±0.65	3.18±0.78	14.83±04.71	1.27±0.40	0.24±0.25
Maruthur Anicut	5.19±0.48	2.94±0.96	15.50±05.20	1.28±0.53	0.16±0.13
Karungulam	5.00±1.17	3.02±0.78	15.83±06.12	1.41±0.52	0.20±0.18
Srivaikundam	5.01±1.06	3.07±1.07	17.00±05.49	1.67±0.65	0.24±0.18
Eral	4.91±0.99	2.75±0.77	15.67±03.39	1.60±0.64	0.30±0.39
Aathur	5.50±0.64	3.06±1.67	19.67±16.84	1.32±0.46	0.33±0.40

All the values are in mg/l. Each value represents the mean of 12 samples ± STD.

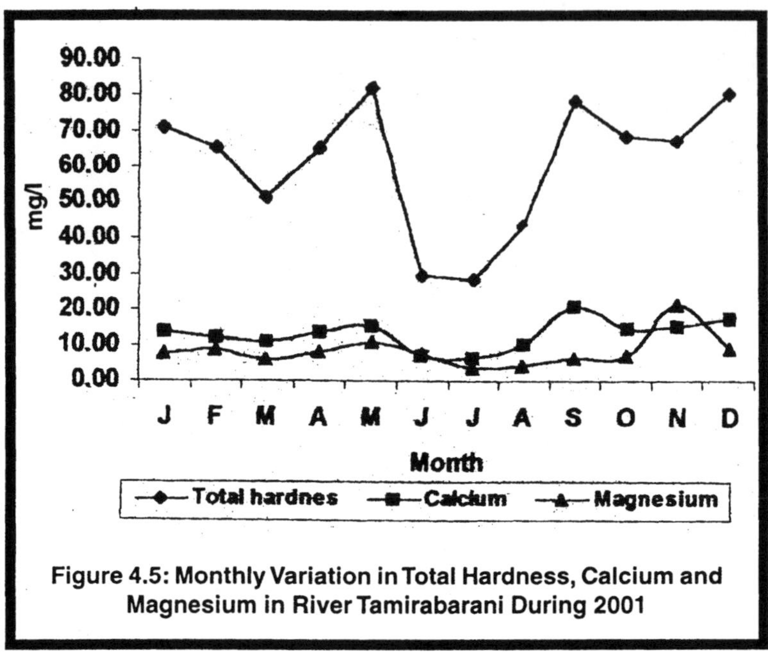

Figure 4.5: Monthly Variation in Total Hardness, Calcium and Magnesium in River Tamirabarani During 2001

Studying the monthly variation of the DO, BOD and COD, it is evident that the DO content has slightly decreased during summer and the BOD level has been raised. But the COD level has shot up during August and September and then gradually come down (Figure 4.6). River Tamirabarani has been found to contain nitrate and Phosphates in minimum amounts. The monthly variation of the nitrate level shows an elevation during June, when the water flow has been lower and during November and December. The level of phosphate has been elevated during November 2002 (Figure 4.7). Influence of agriculture cannot be marked in any specific area but its existence cannot be denied. With the availability of better irrigation facilities more and more land is taken under cultivation, the use of fertilizer and pesticides has also increased.

Water temperature of river Tamirabarani range between 25.25 and 31.25°C, the water pH has been slightly above the neutral pH 7, which is an indicative of alkaline nature of the water, and this portrays that the composition of organic matter received by the river has not suppressed the pH. Total solids in the river have been higher

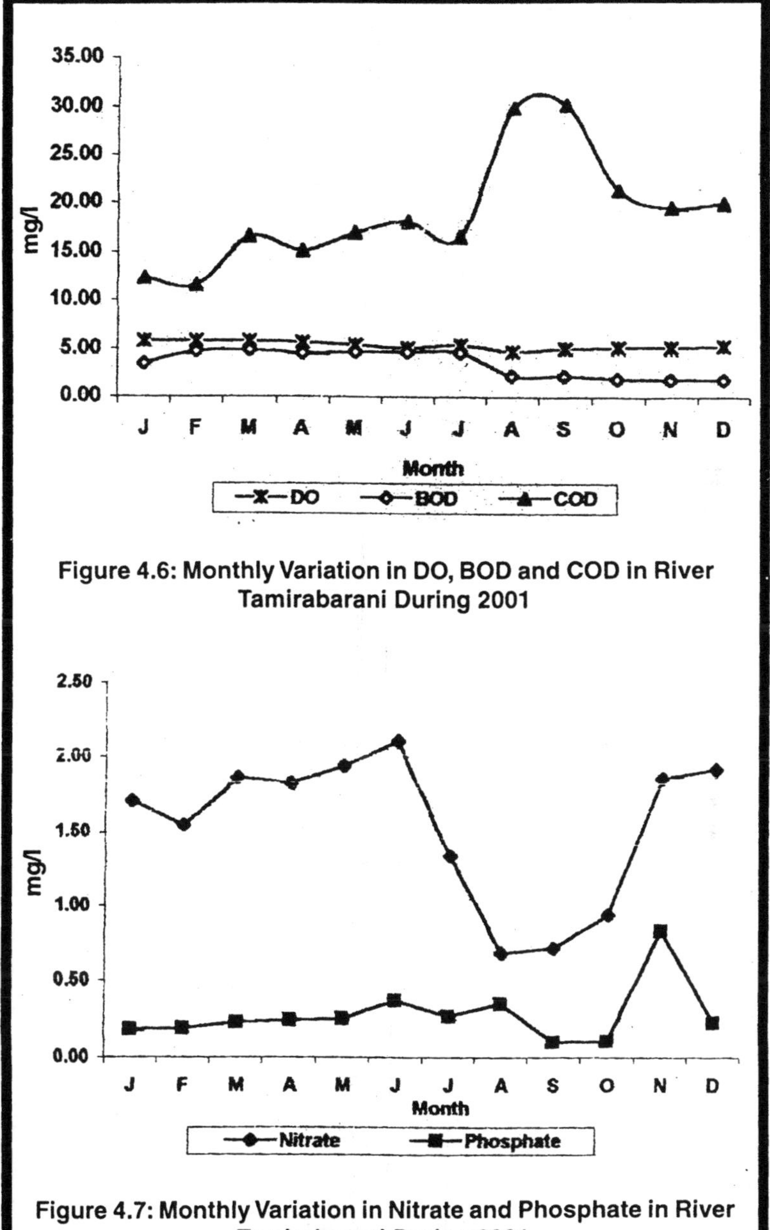

Figure 4.6: Monthly Variation in DO, BOD and COD in River Tamirabarani During 2001

Figure 4.7: Monthly Variation in Nitrate and Phosphate in River Tamirabarani During 2001

when compared to the previous year (Table 4.7). High amount of solids have been recorded in areas such as Vickramasingapuram, Ambasamudram, Kurkkuthurai, Kokkirakulam, Seevalaperi, and Aathur. Organic pollution is the major cause for this elevation. TDS has been higher in Ambasamudram, Tharuvai, Kokkirakulam and Aathur, but the TSS has been high at Cheranmahadevi, Kurukkuthurai, Kokkirakulam and Vellakoil. The level of solids has been comparatively higher in the downstream stations. Figure 4.8 shows the monthly variation of TS, TDS and TSS. It is clear that the solids have been high during the month of June, this may pertain to the decrease in the water flow in the river.

Free carbondioxide generally develops in water bodies when oxygen content is low. High free carbondioxide concentration therefore generally indicates greater pollution. Its value should not exceed 10 mg/l in fresh water bodies. In the upstream water of river Tamirabarani, free carbondioxide was from 2–5 ppm. The value in the downstream water was 4–7 ppm. The free carbondioxide concentration was always high in the downstream water. Alkalinity

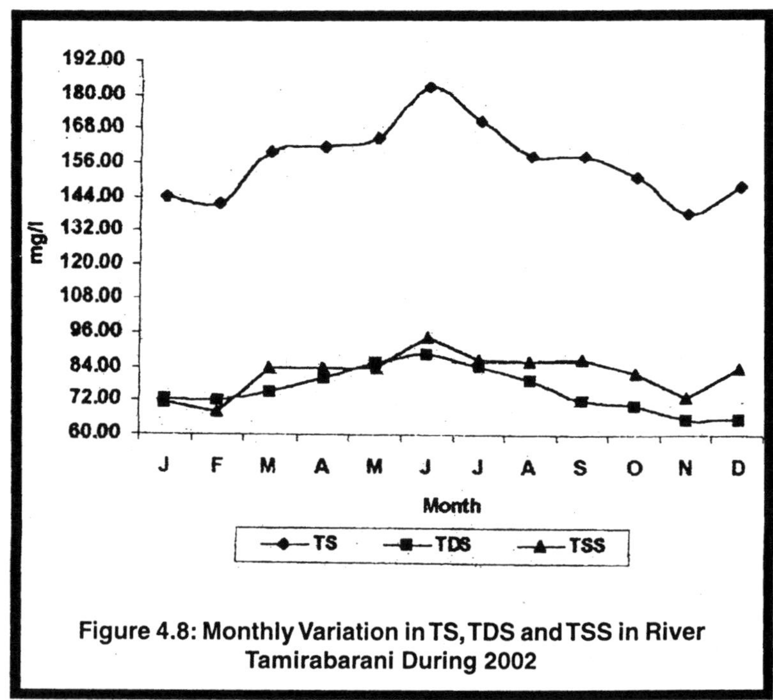

Figure 4.8: Monthly Variation in TS, TDS and TSS in River Tamirabarani During 2002

Table 4.7: Physico-chemical Parameters of River Tamirabarani at Different Sampling Stations During 2002

Sampling Site	*Water Temperature °C*	*pH*	*TS*	*TDS*	*TSS*
Agasthiar Falls	25.25±4.56	7.80±0.21	138.67±19.68	102.58±23.04	36.08±08.40
V.K. Puram	25.63±4.65	7.38±0.26	199.75±70.55	103.92±15.19	95.83±66.09
V.K. Puram	25.96±4.85	7.23±0.25	155.75±54.11	69.75±10.80	87.08±49.26
Ambasamudram	26.79±4.42	7.35±0.14	184.33±23.74	118.92±18.52	67.92±12.92
Ambasamudram	27.13±3.92	7.13±0.36	127.00±32.71	54.67±11.48	72.33±22.80
Thirupadi Maruthur	27.08±3.90	7.14±0.21	127.83±20.01	52.67±08.51	75.17±12.47
Mukkudal	26.71±4.14	7.22±0.20	105.75±22.60	46.92±14.02	62.17±11.65
Cheranmahadevi	27.25±4.32	7.13±0.19	118.00±18.25	49.42±05.76	68.58±16.29
Cheranmahadevi	27.83±4.11	7.18±0.19	145.00±49.51	39.50±10.55	104.25±54.30
Tharuvai	28.08±3.21	7.13±0.30	169.17±17.94	117.33±12.56	51.83±17.68
Kurukku Thurai	28.00±3.83	7.02±0.28	199.42±35.08	83.92±26.78	115.50±21.85
Kokkirakulam	29.13±3.74	7.05±0.34	233.58±49.38	125.08±15.92	125.17±40.72
Vellakoil	29.54±3.45	7.01±0.26	162.67±36.13	46.58±10.35	116.08±33.46
Pottal	29.54±3.31	7.02±0.27	116.75±15.79	40.25±12.56	75.92±8.11
Manapadai Veedu	29.96±3.08	7.07±0.28	137.75±12.31	75.33±07.60	61.83±11.76
Seevalaperi	30.17±3.10	7.14±0.25	185.33±14.87	89.75±06.70	95.67±11.05
Maruthur Anicut	30.42±3.25	7.28±0.28	151.00±23.29	71.58±07.76	79.42±09.10
Karungulam	30.54±3.42	7.22±0.23	163.58±11.96	76.50±05.37	86.92±08.04
Srivaikundam	3.50±3.47	7.14±0.27	155.50±13.82	57.58±05.68	97.92±11.14
Eral	31.08±3.96	7.27±0.28	131.00±21.77	71.08±10.36	57.33±16.77
Aathur	31.25±4.00	7.35±0.31	192.00±29.35	102.50±15.20	90.33±24.73

All the values except pH are in mg/l. Each value represents the mean of 12 samples ± STD.

of water is caused mainly due to OH^-, CO_3 and HCO_3 Ions etc. The alkalinity in the river Tamirabarani was due to the presence of bicarbonate ions in surplus amounts. Surplus amounts of bicarbonates were found in all the stations after Tharuvai (Table 4.8). Human activities release detergents, which contribute to the elevation in the levels of alkalinity. Kurukkuthurai and Kokkirakulam were the sites with higher amounts of acidity. Figure 4.9 shows the monthly variation in the levels of temperature, pH, free carbondioxide and bicarbonate. The level of bicarbonates has been high throughout the year. The other parameters have remained within their permissible limits.

Table 4.8: Physico-chemical Parameters of River Tamirabarani at Different Sampling Stations During 2002

Sampling Site	*Free CO_2*	*Acidity*	*Bicarbonates*
Agasthiar Falls	2.57±0.86	5.00±1.07	26.67±04.44
V.K. Puram	2.93±1.08	4.79±0.72	31.67±03.26
V.K. Puram	2.75±0.99	3.75±1.31	25.42±05.82
Ambasamudram	2.57±0.86	3.54±1.29	35.42±06.89
Ambasamudram	3.82±1.00	5.42±0.97	33.33±04.44
Thirupadi Maruthur	4.42±0.90	6.67±1.23	37.50±05.00
Mukkudal	4.22±0.64	5.21±0.72	34.17±05.57
Cheranmahadevi	5.32±1.13	5.00±1.85	32.92±04.98
Cheranmahadevi	4.60±0.63	5.63±1.13	42.92±05.42
Tharuvai	7.33±1.43	6.04±2.25	61.25±09.80
Kurukku Thurai	6.78±1.13	7.71±1.67	59.17±09.96
Kokkirakulam	6.60±1.62	7.92±2.09	67.92±15.44
Vellakoil	5.87±1.08	6.88±1.13	35.83±06.34
Pottal	4.58±0.64	6.67±1.23	71.25±08.01
Manapadai Veedu	4.22±0.64	6.08±1.78	85.83±11.65
Seevalaperi	4.40±0.94	5.00±1.51	80.00±07.39
Maruthur Anicut	4.40±0.00	5.21±0.72	85.42±05.42
Karungulam	4.58±1.13	4.79±0.72	45.00±03.69
Srivaikundam	4.58±1.13	3.54±1.29	42.08±04.50
Eral	5.13±1.08	4.58±1.44	102.92±13.05
Aathur	4.95±0.99	4.38±1.88	132.92±12.52

All the values except pH are in mg/l. Each value represents the mean of 12 samples ± STD.

The total hardness of the water of river Tamirabarani is within the permissible limits, however some of the stations record high hardness levels when compared to the others. The total hardness of the downstream water has been comparatively higher. Stations Eral and Aathur have recorded the highest hardness levels of 102.92 and 132.92 mg/l respectively (Table 4.9). All the stations after Kurukkuthurai have been found to contain higher amounts of calcium and magnesium when compared to the upstream sites. The chloride content of river Tamirabarani range between 11–33 mg/l. The highest value is recorded at Cheranmahadevi, Kokkirakulam, Vellakoil and Eral come after this. Sulphate level is also quite adequate in some of the places. Station such as Ambasamudram, Thirupadai Maruthur, Cheranmahadevi, Kurukkuthurai, and Aathur, are found to contain comparatively higher amounts of sulphates. Monthly variation in the total hardness, calcium, magnesium, chloride and sulphate shows that there is not much frustrations in this level, except for sulphate which has slightly elevated during summer when there was lean water flow (Figure 4.10).

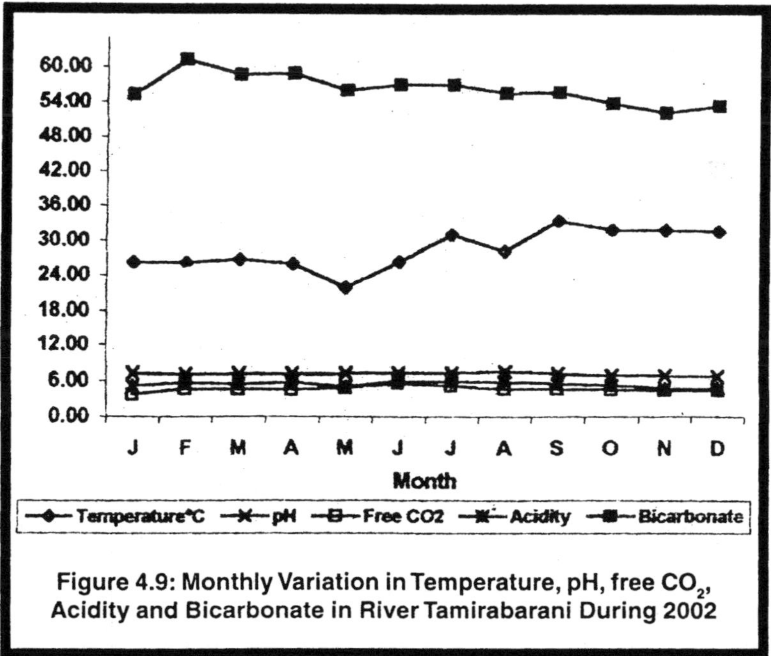

Figure 4.9: Monthly Variation in Temperature, pH, free CO_2, Acidity and Bicarbonate in River Tamirabarani During 2002

Table 4.9: Physico-chemical Parameters of River Tamirabarani at Different Sampling Stations During 2002

Sampling Site	*Total Hardness*	*Calcium*	*Magnesium*	*Chloride*	*Sulphate*
Agasthiar Falls	26.67±04.44	1.59±0.20	12.98±1.36	13.02±1.30	78.83±09.49
V.K. Puram	31.67±03.26	2.92±0.44	10.89±1.18	13.47±1.08	78.25±09.10
V.K. Puram	35.42±05.82	2.96±0.35	5.00±1.20	15.02±1.06	48.08±06.86
Ambasamudram	35.42±06.89	3.57±0.28	4.46±0.95	15.78±1.07	103.83±23.68
Ambasamudram	33.33±04.44	3.53±0.52	4.51±0.54	14.02±2.78	113.08±19.09
Thirupadi Maruthur	37.50±05.00	5.16±0.98	3.92±0.63	15.73±0.89	110.58±14.49
Mukkudal	34.17±05.57	4.95±0.76	4.05±0.37	17.67±0.79	83.08±20.88
Cheranmahadevi	32.92±04.98	5.19±0.79	4.30±0.67	18.02±2.32	100.58±21.99
Cheranmahadevi	42.92±05.42	5.18±0.72	2.86±0.86	33.14±1.74	87.67±14.18
Tharuvai	61.25±09.80	8.86±1.35	2.75±0.72	10.66±2.77	87.17±11.28
Kurukku Thurai	59.17±09.96	9.32±1.09	11.80±2.28	11.05±3.32	102.75±16.78
Kokkirakulam	67.92±15.44	10.56±1.15	11.69±1.76	30.98±1.71	105.50±15.31

Contd...

Table 4.9—Contd...

Sampling Site	*Total Hardness*	*Calcium*	*Magnesium*	*Chloride*	*Sulphate*
Vellakoil	35.83±06.34	12.51±3.31	13.71±2.05	30.21±3.63	86.92±08.70
Pottal	71.25±08.01	6.81±1.50	10.86±2.26	11.77±0.84	86.25±07.34
Manapadai Veedu	85.83±11.65	11.48±3.33	24.33±2.61	12.41±0.70	78.75±06.58
Seevalaperi	80.00±07.39	13.57±3.39	25.26±0.90	15.74±1.25	88.75±11.82
Maruthur Anicut	85.42±05.42	37.67±6.40	42.14±2.09	16.47±0.58	94.83±13.09
Karungulam	45.00±03.69	21.61±1.50	52.64±1.99	15.69±2.38	95.58±10.03
Srivaikundam	42.08±04.50	44.95±8.28	51.60±1.03	17.38±2.39	97.50±11.77
Eral	102.92±13.05	19.41±2.53	22.00±2.86	30.10±3.12	94.33±08.97
Aathur	132.92±12.52	19.21±1.87	21.36±2.32	18.24±1.49	101.83±13.37

All the values except pH are in mg/l. Each value represents the mean of 12 samples ± STD.

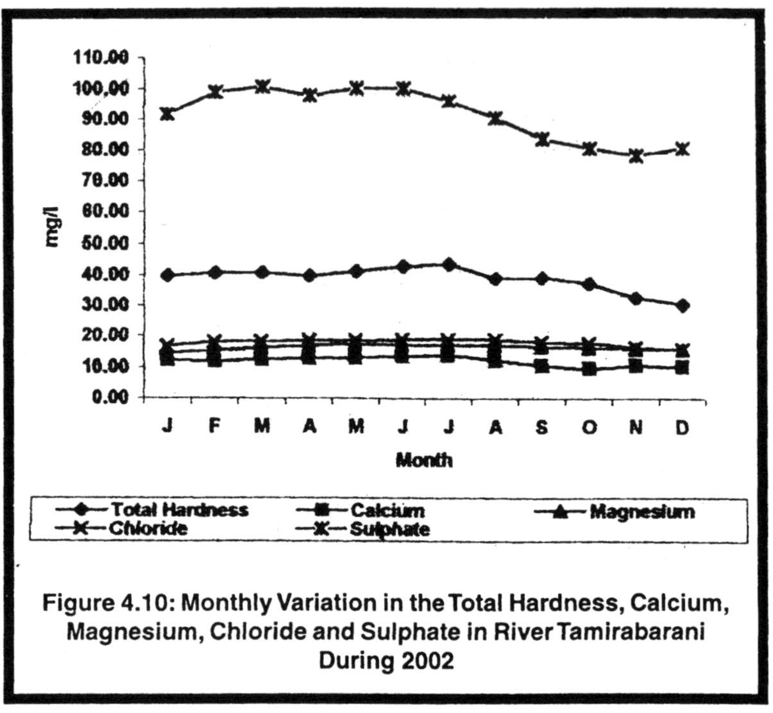

Figure 4.10: Monthly Variation in the Total Hardness, Calcium, Magnesium, Chloride and Sulphate in River Tamirabarani During 2002

The coliform group of organisms meets the criteria for a biological indicator of water contamination (Fair and Geyer, 1961). Present observation showed that highest number of total coliform (238/100 ml) at Tharuvai, followed by Srivaikundam (223/100 ml), Kurukkuthurai (221/100 ml), Mukkudal (191/100 ml), Vellakoil (187/100 ml), Kokkirakulam (176/100 ml) is indicative of higher contamination compared to that of other stations (Table 4.10). The vital cause for this may be due to the higher concentration of organic matter by the entry of domestic sewage. Monthly variation reveals that the total coliforms have been high during July, October but the level has shot up tremendously during December (Figure 4.11). This is because the bacterial contamination caused by human excreta has been excessive during these months. The best example is the contamination of water by the pilgrims who visit the Pabanasam Temple during the month of August for the Aadi Amavasi. The left over food, plastic bags and other polythene are thrown into the river. Defaecation along the river increases the total and faecal coliform levels during this season. All the sampling stations have been found to contain faecal coliform in contrast to the permissible limit of 0/100 ml. Faecal coliform level has not shown much fluctuation in the different seasons, however some of the sampling stations such as Kurukkuthurai, Kokkirakulam, Vellakoil and Srivaikundam have been found to contain faecal coliforms above 10/100 ml.

Table 4.10: Biological Characteristics of River Tamirabarani at Different Sampling Stations During 2002

Sampling Site	*Total Coliforms*	*Faecal Coliforms*
Agasthiar Falls	123.67±046.69	6.00±1.48
V.K. Puram	62.00±034.55	6.50±1.62
V.K. Puram	66.58±020.36	6.50±2.54
Ambasamudram	117.33±247.71	5.42±3.09
Ambasamudram	49.75±041.65	8.08±3.90
Thirupadi Maruthur	164.50±235.88	8.67±4.19
Mukkudal	193.33±445.38	9.08±7.62
Cheranmahadevi	34.42±011.49	8.08±2.35
Cheranmahadevi	59.33±041.17	8.67±3.98
Tharuvai	238.33±439.65	7.82±4.47
Kurukku Thurai	221.42±116.75	15.00±9.11
Kokkirakulam	176.67±073.65	12.83±4.53
Vellakoil	187.33±244.97	10.00±3.25
Pottal	140.00±243.15	9.58±3.29
Manapadai Veedu	63.17±048.96	8.50±2.24
Seevalaperi	85.42±131.75	9.33±7.68
Maruthur Anicut	39.42±015.73	5.33±2.27
Karungulam	70.92±085.44	5.25±3.39
Srivaikundam	223.33±225.68	11.58±7.75
Eral	76.25±134.59	8.17±6.22
Aathur	105.42±250.42	5.75±2.42

All the values except pH are in mg/l. Each value represents the mean of 12 samples ± STD.

Though the magnitude of pollution is not higher the impact of human activity is quite apparent on its water quality. The cultural pollution in the form of mass bathing, cloth and automobile washing,

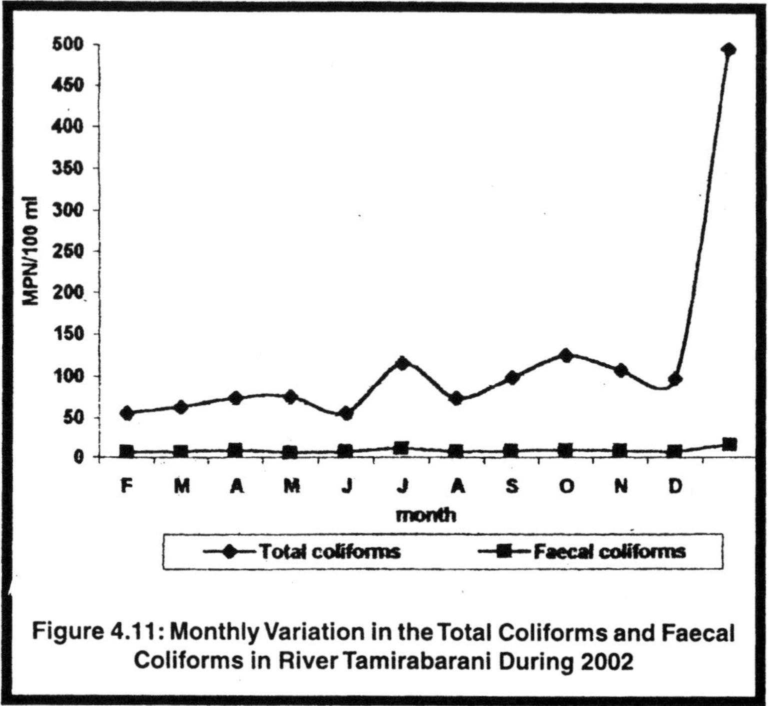

Figure 4.11: Monthly Variation in the Total Coliforms and Faecal Coliforms in River Tamirabarani During 2002

cattle washing and religious activities is evident throughout the river stretch but it is particularly severe at few places. The range of fluctuations in the levels of the various parameters seems to be proportional to the extent of pollution and the number of sources of pollution. The dissolved oxygen measurement generally shows the effect of oxygen consuming substances of the river. Dissolved oxygen content of river Tamirabarani was uniformly well above the prescribed ISI value of 3 mg/l, thus giving support to the contention that, under natural conditions, the flowing waters typically contain a relatively high concentration of DO. Moreover it is evident that the self-purification capacity of riverTamirabarani can assimilate the existing load of pollutants (Lester, 1967).

River Tamirabarani serves as the principal source of water for Ambasamudram, Vickramasingapuram, Cheranmahadevi, Tenkasi, Shenkottai, Kadayanallur, Nanguneri, Sankarankovil, covering Tirunelveli, Thoothukudi and Virudhunagar districts of Southern Tamil Nadu. The water quality profile of river Tamirabarani during 1986 has revealed that the physico-chemical parameters at all the stations confirm to standards (Murugesan *et al.*, 1994). However it was stated that a study of the biological quality was essential to confirm the 100 per cent suitability of the water for drinking purposes. The present study has made an analysis of the physico-chemical and biological characteristics of the water of river Tamirabarani at 21 stations and has come forward to state that the river water is not grossly polluted, however the BOD levels exceed the standards at some of the stations. The other parameters are within the permissible limits. WHO water quality standards, ICMR Standards for drinking water and BIS Standards for Drinking Water (IS: 10500–1983) recommend that no coliforms should be present in the drinking water. The biological quality of the river water is very poor. The water of river Tamirabarani is to be therefore purified well before it is supplied to the community. Moreover necessary steps can be taken to provide proper sanitation to the people in the Tirunelveli and Thoothukudi districts so as to prevent human defecation along the riverbanks. Sewerage systems should be not allowed to confluence with the river, which would furthermore prevent the entry to organic load into the river.

Acknowledgement

We thankfully acknowledge the financial assistance from the World Bank through Water Resources Organization.

References

Abraham Muthukumar, S. and Selvin Samuel, A. (1999). Observation on the effect of pollution and distribution of weeds in the river Tamirabarani, South India. *Proc. Sem. Env. Prob.*, 7-11.

Aditya Tyagi, Sharma, M.K. and Bhatia, K.S. (2003). The study of temporal and spatial trends of water quality of river Kshipra using water quality index. *Indian J. Environ. Hlth.*, **45(1)**: 15-20.

Agarwal, A. and Sharma, R. (1982). *State of India's Environment: A Citizen's Report*. Centre for Science and Environment, New Delhi.

Agarwal, S.K. (1982). Eutrophication of Chambal river at Kota. *Acta. Ecol.*, **4**: 5-9.

Ajmal, M. and Ragi-ud-Din (1988). Studies on the pollution of Hindon river and Kali river (India). In: *Ecology and Pollution of Indian Rivers*, R.K. Trivedy ed.. Ashish Publishing House, New Delhi, India, 87–111.

APHA (1995). *Standard Methods for the Examination of Water and Wastewater, 19th edition*. American Public Health Association, New York.

Badola, S.P. and Singh, H.R. (1981). Hydrobiology of the river. Alaknanda of the Garhwal Himalaya. *Indian. J. Ecol.*, **8**: 269-276.

Balani, M.C. and Sarkar, H.L. (1965). Some observations on pollution of Yamuna river at Okhla water works intake, Delhi. *Environ. Hlth.*, **7**: 84–86.

Chacko, P.I. and Sreenivasan, R. (1955). Observations on hydrobiology of three major rivers of Madras state, South India. *Contr. F. W. Biol. Stn. Madras*, **13**: 1-16.

Chandra, S. and Krishna, G. 1983. Effect of tannery waste disposal on the quality of the river Ganges at Kanpur. *Poll. Res.*, **2**: 63-64.

Chaudhari, N. (1982). *Water and Air Quality Control: The Indian Context for the Prevention and Control of Water Pollution*, New Delhi, India.

Chavadi, V.C. and Nalawadi, A.R. (1989). Hydro-environmental indicators: A review and evaluation of their use in the assessment of environmental impacts on Kali river system. *Manage. Aqu. Ecol.*, pp. 189-205.

Citizen's Report (1982). *The State of India's Environment*. Centre for Science and Environment, New Delhi, India.

Datta, N.C. Bandyopadhyay, B.K., Majumder, A. and Abuja, D. (1988). Hydrobiological profile of Hooghly sector (Bally to Bandel) of river Ganga. In: *Ecology and Pollution of Indian Rivers*, R.K. Trivedy ed. Ashish Publishing House, New Delhi, India, pp. 113-129.

Fair, G.M. and Geyer, J.C. (1961). *Water Supply and Wastewater Disposal*. John Wiley Publication, New York.

Gautam, A. (1992). Conservation of water resources. In: *Aquatic Environment*, ed. Ashutom Gautam, Ashish Publishing House, New Delhi, pp. 100-105.

George, M.G., Qasim, S.Z. and Siddiqui (1966). A limnological survey of the river Kali with reference to fish mortality. *Environ. Hlth.*, 8: 262-265.

Haniffa, M.A., Murugesan, A.G. and M. Porchelvi (1986). Hematological effects of distillery and paper mill effluents of *Heteropneuste fossilis* (Block) and *Sarotherodom Mossambious* (Peters). *Proc. Indian Acad, Sci.*, **95(2)**: 155-161.

Jain, C.K., Bhatia, K.K. and Seth, S.M. (1998). Effect of waste disposals on the water quality of river Kali. *Indian J. Environ. Hlth.*, **40(4)**: 375-376.

Jebanesan, A., Selvanayagam, M. and Thaddeus, A. (1987). Distributing pattern of dissolved oxygen in the selected station of Cooum river and its effect on the aquatic fauna. *Proc. Symp. Environ. Biol.*, pp. 303-311.

Krishnan, N., Jayachandran, A. and Jeyakumar, G. (1994). Faecal pollution in river Vaigai. *Indian. J. Environ. Hlth.*, **36(2)**: 128-129.

Kundra, R., Nagpal, J.L., Verma, S.R. and Srinivastava, S.K. (1977). Rain water quality at Wazirabad and Okhla reservoirs at Delhi. *Indian J. Environ. Hlth.*, **19(3)**: 319-328.

Lester, W.F. (1967). Management of river water quality. In: *River Management*, P.C.G. Isaac, ed. Maolaren and Sons Ltd., London.

Mahadevan, A. and Krishnaswamy, S. (1983). A quality profile of river Vaigai (S. India). *Indian J. Environ. Hlth.*, **25**: 288-299.

Mishra, K.D. (1996). Impact of sewage and industrial pollution on physico-chemical characteristics of water in river Betwa at Vidisha, Madhya Pradesh. In: *Assessment of Water Pollution*, S.R. Mishra ed. APH Publishing Corporation, New Delhi.

Mitra, A.K. (1982). Chemical characteristics of surface water at a selected gauging station in the river Godavari, Krishna and Turngabhadra. *Indian J. Environ. Hlth.*, **24**: 165-179.

Mittal, S. and Sangar, R.M.S. (1991). Studies on the distribution of algal flora in polluted regions of Karwan river at Agra (India). In: *Current Trends in Linmology–I*, Nalin K. Shastree ed. Narendra Publishing House, Delhi, India, pp. 221-230.

Murugesan, A.G. and Sukumaran, N. (1999). Impact of urbanization and industrialization on the river Tamirabarani, the lifeline of Tirunelveli and Thoothukudi districts. *Proc. Sem. Env. Prob.*, pp. 1-6.

Murugesan, A.G. (1988). *Toxicity of Textile Mill Effluent to an Air Breathing Fish*. Ph.D thesis, Madurai Kamaraj University, Madurai, India.

Murugesan, A.G. and Haniffa, M.A. (1992). Histopathological and histochemical changes on the oocytes of the air breathing fish, *Heteropneustes fossilis* (Block) exposed to textile mill effluent. *Bull. Environ. Contam. Toxicol.*, **48**: 929-936.

Murugesan, A.G., Meena, K. and Sukumaran, N. (2000). Biomonitoring of riverine ecosystem: An urgent need to save the pristine habitat from multifarious pollutants. *Mano Research Papers*, pp. 32-38.

Murugesan, A.G., Abdul Hammed, K.M.S.A. and Sukumaran, N. (1994). Water quality profile of the perennial river Tamirabarani. *Indian J. Environ. Protec.*, **14(8)**: 567-572.

Murugesan, A.G. and Haniffa, M.A. (1985). Effects of textile mill effluents an hematological changes of the obligatory air breathing fish, *Anabas Testudineus* (Block). *Proc. Sysm. Assess. Environ. Pollut.*, 121-128.

Nag, B.S. and Kathpalia, G.N. (1975). Water Resources in India. *Proc. Second World Congress on Water Resources, Vol. II*. CBIP, New Delhi.

Nandkumar, V.K., Backyarathy. D.M. and Bhagyalakshmi, K. (1988). Niva river pollution near Chittor town. In: *Ecology and Pollution of Indian Rivers*, R.K. Trivedy ed. Ashish Publishing House, New Delhi, India, pp. 247-254.

Ownbey, C.R. and Kea, D.A. (1967). Chlorides in Lake Eyre. *Proc. Conf. Great Lakes Res.*, **10**: 382-389.

Pate, H.R. (1916). *Tinnevelly District Gazetter*. Govt. Press, Madras.

Patil, M.R. and Patil, A.R. (1983). Water quality of Ulhas river with respect to copper, cadmium and mercury. *Poll. Res.*, **2**: 24-27.

Prakash, C., Rauat, D.C. and Grover, P.F. (1978). Ecological study of the river Jamuna. *IAWPC Tech. Annual*, **5**: 32-45.

Rajagopalan, S.B., Dhaneshwar, A.K. and Raim C.S.G. (1973). *Pollution of river Subarnarekha at Ranchi: A Survey*.

Rajurkar, N.S., Nongbri, S. and Patwardan, A.M. 2003. Physico-chemical and biological investigation of river Umshypri at Shillong, Meghalaya. *Indian J. Environ. Hlth.*, **45(1)**: 83-92.

Rao, S.P. and Rao, K.S. 1986. Pollution due to pilgrim bathers during kumbha mela 1980 in Kshipra river (Ujjain). *J. Hydrobiol.*, **2**: 47-53.

Sangu, R.P.S. and Sharma, K.D. (1985). Studies on water pollution of Yamuna river at Agra. *Ind. J. Environ. Hlth.*, **27**: 257-261.

Sharma, R.C. (1984). Potamological studies on lotic environment of the upland river Bhagirathi of Garhwal Himalaya. *Environ. and Ecol.*, **2**: 239-242.

Sikander, M.H.D and Tripathi, B.D. (1983). The physico-chemical characteristics of Ganga water at Varanasi. *Proc. Nat. Conf. River. Poll. Human Hlth.*, pp. 53-62.

Singh, D.K. and Singh, C.P. (1990). Pollution studies on river Subarnarekha around industrial belt at Ranchi (Bihar). *Indian J Environ. Hlth.*, **32**: 26-33.

Singh, A.K. (1997). Abundance of macrozoobenthic organisms in relation to physico-chemical characteristics of river Ganga at Patna (Bihar). *Indian J. Environ. Biol.*, **18(2)**: 103-110.

Sivasubramani, R. (1999). Water quality of river Periyar (River Suruliyar) in Tamil Nadu. In: *Limnological Research in India*, Mishra, S.K. ed. Daya Publishing House, New Delhi, India, pp. 1-15.

Somasekhar, R.K. (1984). Studies on water pollution of the river Cauvery: Physico-chemical characteristics. *Inter. J. Environ. Studies*, **23**: 209-216.

Sreenivasan, A. and Ray, R.S. (1967). Effect of certain wastes on the water quality and fisheries of river Cauvery and Bhavani. *Environ. Hlth.*, **9**: 13-21.

Thresh, J.C., Suckling, E.V. and Beale, J.E. (1944). *The Examination of Water and Water Supplies*, London.

Unni, K.S., Chauhan, A., Varghese, M. and Naik, L.P. (1992). Preliminary hydro-biological studies of river Narmada from Amarkantak to Jabalpur. In: *Aquatic Ecology*, Mishra, S.R. and Saksena, D.N. eds. Ashish Publishing House, New Delhi, India, pp. 221-229.

Vass, K.K., Raina, H.S., Zutshi, D.P. and Khan, M.A. (1977). Hydrobiological studies on river Jhelum. *Geobios*, **4**: 238-242.

Venkateswarlu, T. and Jayanti, T.V. (1968). Hydrobiological studies on the river Sabarmati to evaluate water quality. *Hydrobiology*, **31**: 443-448.

Verma, S.R., Shukla, G.R. and Dalela, R.C. (1980). Studies on the pollution of Hindon river in relation to fish and fisheries. *Limnologica (Berlin)*, **12**: 33-75.

World Resources (1986). Basic Books, New York, p. 12.

WRO (1996). State framework water resources plan (Annexure–14) Tamirabarani river basin. Institute of Water Studies, IWS Report, 4/97.

5

Pollutional Impact of Insecticides on Fishes: A Review

☆ *Ashok Verma*

Introduction

Modern Indian agriculture heavily depends on the use of insecticides to control insect pests of different crops. Today about 4500 pesticides are in general use all over the world, out of which 25 have high toxicity potential to a wide range of flora and fauna of economic importance (Muirhead-Thomsen, 1971; Matsumra *et al.*, 1972).

The indiscriminate use of insecticides to boost agricultural production affects the aquatic environment to a great extent. It is reported that the insecticide accumulates in fresh water (Anon, 1962; Mount, 1962; Westlake and Gunther, 1966; Westlake *et al.*, 1966; Edwards, 1970; Ruparelia *et al.*, 1984; Haider and Inbaraj, 1988; Datta, 1996; Rao and Rameshwari, 2000; Luther Das *et al.*, 2000 and Ganeshwande *et al.*, 2002), consequently beside target organisms, non-target organisms are also affected (Holden, 1973).

The toxic effects of large number of chemicals and heavy metals and pesticides in fishes have been studied by a number of workers (Konar,1983; Saxena *et al.*,1997; Shanthankumar *et al.*, 2000; Sonawane and Nikam, 2000; Aditya *et al.*, 2002 and Verma and Sharma, 2002a,b).

Material and Methods

Evaluation of Insecticide Toxicity

To analyse the toxicity of the insecticides, static bioassay test is done as per APHA (1985) and Trivedi and Goel (1986) and LC_{50} values is calculated, which describes the concentration of the compound in water to kill 50 per cent of the fish population in a specific time period. Different experiment are done to analyze the impact of insecticides on fishes. Histological studies are done after routine procedures (Humason, 1972).

Discussion

In the present review, an attempt has been made to analyse the impact of insecticides on the fish physiology.

Pesticide hazard on fish mortality, growth and tissue damage have been amply demonstrated (Wildish *et al.*, 1971 and Jackson, 1976). Pesticides not only produce biological or pathological changes but also cause significant biochemical alteration in the living system, which in turn leads to reduced growth. Thus, insecticide application in agricultural pursuits has manifold effects on the fish biology, these are discussed as under.

Insecticides have been classified into three category *viz.*, inorganic, organic and synthetic organic insecticides. The inorganic insecticides include borates, fluorides and mercurial compounds; the natural organic compounds consists of pyrethrum, rotenone and nicotine and the synthetic organic compounds include chlorinated hydrocarbons, organophosphates and carbamates (Patnaik *et al.*, 2002).

Thus, increasing danger of water pollution by pesticides warrants establishment of water quality criteria and determination of safety limits for the fish (Mount and Putnick, 1966).

Generally, pesticides are toxic which may be lethal or sub-lethal concentrations in aquatic environment. Lethal concentrations cause death of the organisms directly. But the sub-lethal concentration effect and disturb the metabolic activities (Tilak and Satyavardhan, 2002).

Effect on Skin

Skin is the first barrier which protects all the living organisms. Structural deformities in chromatophores and their depigmentation has been observed by Mankar and Kulkarni (2000) in *Channa orientalis* (Sch.).

Fish Behaviour

Sprague (1971) has shown that substantial attention is being given to sub-lethal dose which bring about behavioral changes, that can affect the fertility, growth, susceptibility to disease or predation. Indian cat fish, *Heteropneustes fossilis* under the stress of organophosphorus, carbamate and synthetic pyretheroid pesticides showed un-coordinated movements, sluggishness, alternating with hyper-excitability and difficulty in respiration along with excessive secretion of mucus on general body surface (Kumar and Gupta, 1997). These behavioral changes are similar to those reported by earlier workers in fish subjected to other organophosphorus, carbamate, chlorinated and pyretheroid compounds (Naqui and Hawkins, 1988; Rajmanickan and Karpagaganapathy, 1988; Claudio *et al.*, 1993; Kamler *et al.*, 1974 and Haya, 1989). It is believed that behavioral changes are the most sensitive measure of neurotoxicity.

Sonawane and Nikam (2000) have observed high excitation, increased opercular movements, complete loss of equilibrium, dark colouration in *Lepidocephaicthyes thermalis*, when exposed to sumithion.

Similar results on morphological and behavioural changes were observed by Konar and Ghosh, (1983); Trivedi and Saxena (1999); Jugadessan and Vijayalakshmi (1999); Pashin *et al.* (2000) and Santhakumar *et al.* (2000).

Feeding Behaviour

Decrease in the feeding rate of *Mystus vitattus* and *Macropodus cupanus* exposed to carbaryl (Arunachalam *et al.*, 1980 and Arunachalam and Palanichamy, 1982), *Barbus stigma* exposed to sublethal levels of endosulphan (Manoharan and Subiah, 1982) and *Channa striatus* exposed to sublethal concentration of endosulphan, malathion and hexavin (Ramakrishnan *et al.*, 1997) have been studied.

Decrease in feeding rate of the fish exposed to methyl parathion appears to be due to the impairment of impulse transmission (O'Brien, 1975) leading to depressed swimming activity, predatory ability and feeding capacity. Similar observations were recorded by Arunachalam *et al.* (1980) in *Oreochromis mossambicus* with phosphamidon and methyl parathion.

It has been observed that behavioral changes are the most sensitive measure of neurotoxicity. Ranjana *et al.* (1999) have observed that *Channa punctatus* after exposure to quinolphos and cypermethrin, lost aptitude of feeding, it is suggested that the insecticides possibly damage the olfactory epithelium of the nasal capsule, so that the olfactory response is lost.

It has been found that the exposure of *Oreochromis mossambicus* to methyl parathion not only reduced feeding rate but also resulted in considerable increase in the metabolic rate, possibly due to expenditure of more energy to overcome the stress (Nirmala and Vinoliya, 2003).

Oreochromis mossambicus under the stress of a pyretheroid derivative, ethofenprox, showed unususal behaviour showing initial increased gill opercular movement to cope with enhanced physiological activities, followed by a decrease which may be attributed due to accumulation of mucus on the gill (Muniyan and Veeraraghavan, 1999). Koundinya and Ramamurty (1980) and Prabahakar *et al.* (1993) have also reported similar results.

The accumulation and increased secretion of mucus in fishes exposed to pesticides may be an adaptive feature which perhaps provides an additional protection against corrosive nature of the pesticides and to prevent the absorption of the pesticides through the general body surface (Santhakumar and Balaji, 2000).

Santhakumar *et al.* (2001) have shown the optomotor behaviour of *Anabas testudineus* under the stress of monocrotophos.

Anabas testudineus, an air breathing fish, shows erratic swimming, becomes hyper and hypoactive, imbalance in posture, increased surface activity accompanied by gradual decrease in opercular movement, gradual loss in equilibrium, spreading of excess of mucus all over the body surface followed by slugishness and death (Sandhu, 1993; Sabita Borah and Yadav, 1995 and Santhakumar and Balaji, 2000).

Murali Mohan *et al.* (2002) have studied the effect of chloropyrifos on fingerlings of *Tilapia mossambica* and it was observed that fingerlings initially showed rapid movements, which gradually decreased with increasing concentration.

Insecticide toxicity in the fingerlings of *Sartherodon mossambicus* resulted in increased irritability, hyper-excitability and tremors of the whole body (Patnaik *et al.*, 2002).

Santhakumar *et al.* (2000) have shown the optomotor behaviour of *Anabas testudineus* under the stress of monocrotophos and Sonawane and Nikam (2000) observed high excitation, increased opercular movements, complete loss of equilibrium, dark colouration in *Lepidocephaichthyes thermalis* when exposed to sumithion.

Effect on Biochemical Aspects of Fish

Fishes are sensitive to contamination of water and pollutants may significantly damage certain physiological and biochemical processes, when they enter the organs of these animals (Nemcsok *et al.*, 1987). Biochemical studies are therefore more useful since the fish constitute a rich source of nutrition and are of high calorific value.

Insecticides like other environmental pollutants bring about damage to different organs or disturb the physiological and biochemical processes of the organisms following exposure (Thakur and Sahai (1987).

The nutritional quality of fish depends upon the proportional composition of fats, carbohydrates and proteins and these are likely to be affected by the pollutants (Chaudhary *et al.*, 1998; Sakthivel, 2002 and David *et al.*, 2003).

Organophosphate pesticides are known to methylate and phosphorylate cellular protein (Wild, 1975). Ramlingam and Ramlingam (1982) have observed that total protein in liver and muscle of *Sartherodon mossambicus* showed a declining trend following treatment with DDT, malathion and mercury, on the other hand, there was an increase in the lipid content after treatment thereby indicating lipogenesis under phosphamidon intoxication.

Similar trend was observed by Rao *et al.* (1987) in renal lipid contents in *Tialpia mossambica* with hepatchlor, phosphamidon and dichlorovas. Rege and Gaikwad (1989) have also observed an increase in fat content in the liver of *Tilapia mossmbica* treated with Thiodon EC.

Among pyretheroids, fenvalerate is a widely used synthetic toxicant that causes severe metabolic disorders in fish (Coats *et al.*,1989; Reddy *et al.*, 1991; Reddy and Philip, 1994 and Thakur and Bais, 2000).

The acute toxicity of fish may be due to efficient gill uptake, inefficient detoxification and elimination and sensitivity at the site(s) of action (Bradbury *et al.*, 1986; Coats *et al.*, 1989 and Tripathi *et al.*, 2002). The toxicity of any chemical necessarily impairs the metabolic strategy of the fish as reported by Santhamma *et al.* (1999).

Gautam and Gautam (2000) have studied the basic protein contents in the gastrointestinal tract of *Channa punctatus* following treatment with diazinon and endosulphan.

Tilak *et al.* (2001) have studied biochemical changes in the total protein and glycogen content of *Labeo rohita* under stress of chloropyriphos. Sakthivel (2002) have shown that phosphamidon induced low protein content in *Gambusia affinis*. Tripathi *et al.* (2002) have shown inhibitory effects of fenevalerate on DNA, RNA and protein contents in *Clarias batrachus* under fenvalerate toxicity.

Tilak *et al.* (2002) have shown that depletion in the food reserve and enzyme activity was observed in *Catla catla* (Hamilton), *Labeo rohita* (Hamilton) and *Cirrhnius mrigala* (Hamilton).

Zutschi (2003) has observed that the fenthion affects the steroid metabolism by impairing the synthesis of cholesterol and therefore affects the reproductive potential of *G. guiris.*

Effect of synthetic pyretheroids on the biochemical changes in *Cyprinus carpio* has been worked out by Kumaraguru and Beamish (1983) and Saradamni and Kamalaveni (2002).

There is significant decrease in SDH activity and increase in LDH activity possibly due to shift from aerobic metabolism to anaerobic side and it may be because of necrosis and atrophy of secondary gill lamellae as reported by Santhamma *et al.* (1999) in *Tilapia mossambica.*

Loteste *et al.* (2002) has shown inhibition of plasmatic cholinesterase activity in *Prochilodus lineatus* following exposure to organophosphorus pesticides.

Rawat *et al.* (2002) have shown decrease in glycogen content of liver under the stress of endosulphan toxicity in *Heteropneustes fossilis.*

It can be concluded that the biochemical changes seem to be the result of the greater stress, the organ experienced during the process of detoxification of the pesticide and their metabolites (Arasta *et al.*, 1995).

Effect on Fish Physiology

One of the most important manifestations of the toxication of chemical is over stimulation or depression of respiratory activity (Muirhead-Thomsen, 1971).

Depletion in oxygen (O_2) contents occurs in the medium when pesticides, chemicals, sewage and other effluents containing organic matter are discharged into water (Jones, 1973). Pesticides enter into the fish mainly through gills (Holden, 1973).

The damage observed in the gills of the fish exposed to toxicant is due to impaired respiratory metabolism (Hughes and Perry, 1976 and Ramamurthy, 1988).

The oxygen consumption under pesticide stress showed significant decrease resulting in hypoxic conditions (Fergussion *et al.*, 1966). Effect of pesticides on fish metabolism, particularly oxygen consumption has been reviewed by Bradbury *et al.* (1986).

In *Channa punctatus*, fenvalerate altered the opercular activity which in turn influence oxygen uptake (Tilak and Satyavardhan, 2002). Reduction in respiratory metabolic rate on exposure of *Oreochromis mossambicus* to methyl parathion spare more energy for the growth of the tissue (Nirmala and Vinoliya, 2003).

The inhibition in oxygen consumption may be due to disintegration or rupture of respiratory epithelium and coagulation of mucus film over the surface of the gills.

Effect on Haematological Parameters

Observation pertaining to the effects of pesticides on the blood of fish are many (Lone and Javaid, 1976b; Thakur and Sahai, 1987; Joshi *et al.*, 1988; Varadraj *et al.*, 1993).

Significant increase in the total erythrocyte count (TEC) occurred in *Labeo rohita* following exposure to chlordane (Bansal *et al.*, 1979).

Pandey *et al.* (1976) recorded polycythemia under the effect of chlorinated insecticides. On the other hand, Mount and Putnick (1966), Lone and Javaid (1976a) and Kumar *et al.* (1999) have reported significant increase in the total erythrocyrte count following application of endosulphan.

Bansal *et al.* (1979) have also reported a gradual increase in the TLC following treatment with insecticides in *Labeo rohita.* Similar observations were recorded by Abidi and Srivastava (1988) in *Anabas testudineus* due to endosulfan treatment.

Changes in the haemoglobin content of fish following treatment with insecticides have been well documented in fishes by different workers. Thus, Abidi and Srivastava (1988) reported significant increase in haemoglobin content due to endosulphan exposure. Kumar *et al.* (1999) have also reported marked increased in the haemoglobin content in *Anabas testudineus* to different concentrations of endosulfan.

Significant increase in total WBC count occurred, at 30 days exposure to BHC, in *Channa punctatus*. The increase in lymphocyte count can be attributed to an increase in antibody production (Joshi *et al.*, 2002). On the other hand, Mount and Putnick (1966) reported that endrin poisoning decreased the hematocrit values to about half of the affected fish. Similar decrease in the hematocrit values of *Channa punctatus, Clarias batrachus* and *Heteropneustes fossilis* following malathion treatment has been observed by Mukhopadhya and Dehadrai (1980). On the other hand Gopal *et al.* (1982) reported an increase in haematocrit in lower concentration of endosulfan, which decreased at higher concentration.

A significant decrease in TEC, Hb percentage, PCV, MCV, MCH and increase in MCHC, TLC and ESR in *Heteropneustes fossilis* have been noticed following treatment with divithion (Nath and Banerjee, 1995).

Murugappan *et al.* (1999) have shown that in *Cyprinus carpio* the frequency of micronucleated erythrocytes increased with increasing concentration of pollutants. Induction of micronuclei was observed in *Channa striatus* exposed to endosulfan (Murugappan and Gunasundari, 1996).

Thakur and Bais (2000) have suggested that the changes in the haematological parameters of *Heteropneustes fossilis* following treatment with aldrin and fenvalerate might be due to the disruptive action of the pesticides on the erythropoetic tissue.

Gautam and Gautam (2002) have studied the toxic effects of biological and haematological alterations *Channa punctatus* following exposure to endosulfan and diazinon. A reduction in number of leucocytes show immunological response due to these pesticides.

Joshi *et al.* (2002) have observed that haematological parameters like Hb, PCV, RBC, WBC and ESR are affected following exposure of *Channa batrachus* to lindane and malathion.

Histological Changes

Under pesticide stress different tissue of the fish show pathological changes, which are discussed as under:

Nervous Tissue

Vascular dilation of brain tissue has also been reported by earlier workers (Kennedy *et al.*, 1970). Santhakumar *et al.* (2000) have observed the effect of monocrotophos on the brain tissue of *Anabas testiduneus.*

Gills

Gills are liable to be damaged by any irritant material, whether dissolved or suspended in water (Lemke and Mount, 1963). Proliferative thickening of gill epithelium was produced by most kinds of environmental toxicants (Skidmore and Tovel, 1972 and Narain *et al.*, 1990).

Pathological changes like hyperplasia and fusion of gill epithelium due to the separation of epithelium, necrosis of gill epithelium, degeneration of pilaster cells and development of vacuoles in the epithelium are observed in fishes exposed to pesticides (Sunitha and Sahai, 1993).

Mercury toxicity causes damage to the respiratory epithelium of the gills (Paulose, 1987, 1989). Gupta and Dua (2002) have observed mercury induced architectural alterations in the gills surface of *Channa punctatus.*

Santhamma *et al.* (1999) have observed necrosis in respiratory lamellae, degenerative changes in the inter-lamellar space and atrophy of respiratory lamellae in *Tialpia mossambica* exposed to monocrotophos.

Kumar *et al.* (2002) have studied the effect of deltmethrin induced changes in the gills of *Heteropneustes fossilis* and have observed bulging at basal and distal part of the lamellae, hypertrophy and hyperplasia of lamellar and inter-lamellar cells, erosion of epithelial layer, necrosis, pycnosis and fusion of secondary lamellae.

Santhakumar *et al.* (2001) have shown lesion in gill of *Anabas testudineus* exposed to monocrotophos.

Gastrointestinal Tract

Many workers have studied the accumulation of insecticides in fish tissue (Verma *et al.*, 1981; Sandhu and Mukhopdhya, 1985; Sahai, 1992; Jebakumar, *et al.*, 1993; Begum *et al.*, 1994, Aditya and Chattopadhya, 2000 and Susan and Tilak, 2003).

Toxic effect of insecticides on the digestive tracts of fishes have been studied by many workers (Bhatnagar and Srivastava, 1975; Konar, 1983).

Mandal and Kulshrestha (1980) observed enlargement of mucosal cells in *Lepidocephalichthys thermalis* following treatment with HBC and Sumithion. Hyperplasia of epithelial cells with pycnotic nuclei and finally damage to the epithelial cells were observed by Patel *et al.* (2000) in *Nemacheilus botia.*

Liver plays an important role in the detoxification of toxins. Effects of toxicants on the liver has been studied by many workers (Mathur, 1962; Chakrabarty and Kumar, 1974; Bhattacharya *et al.*, 1975; Mathur, 1976).

Mandal and Kulshrestha (1980) observed splitting, necrosis, vacuolation, formation of multi-nucleate cells in liver due to the effect of sub-lethal doses of insecticides in *Clarias batrachus.*

Kulshrestha and Jauhar (1984) obtained similar results due to the effects of thiodon and sevin on the liver of *Channa striatus,* and phenolic compounds on liver of *Notopterus notopterus.*

Sharma *et al.* (2002) have observed the additive effect of DDT and ammonium chloride on the kidney of *Labeo rohita* (Ham.). Further,

Sharma and Sharma (2002) have reported degenerative changes in the liver of *Cirrihinus mrigala* in response to ammonium chloride and DDT.

Susan and Tilak (2003) have reported degeneration of cytoplasm in hepatocytes, atrophy and rupture in the blood vessels causing appearance of blood streaks in *Cirrhinus mrigala* exposed to fenvalerate.

Pesticides are known to cause several histopathological changes in the gonads of the fish (Burdick *et al.*, 1972; Donaldson, 1975; Saxena and Garg, 1978; Mani and Saxena, 1985; Singh and Singh, 1987 and Saxena *et al.*, 1997).

It has been found that pesticides result in the arrest of ovarian recrudence due to the possible inhibition of gonadotropin levels (Ansari and Kumar, 1987; Umarani and Rajendranath, 1988; Rastogi and Kulshrestha, 1990).

Total degeneration and necrosis of ovarian follicle following treatment with carbaryl in *Clarias batrachus* has been observed by Jyothi and Narayan (1999).

Decrease in gonadosomatic index is the most common and important effect in female fish following exposure to pesticides (Mani and Saxena, 1985; Sahai, 1987; Choudhary *et al.*, 1993).

Inhibited growth of oocytes and increased incidence of follicular atresia was evident in case of certain fishes (Mani and Saxena, 1985; Ghose, 1986; Khilare and Wagh, 1987 and Saxena and Garg, 1978).

Ramachandra Mohan (2000) has observed malathion induced changes in the ovary of freshwater fish, *Glossogobius giuris*. Aditya *et al.* (2002) have observed the deleterious effects on the ovaries of *Labeo rohita* following treatment with methyl parathion.

Ramchandra Mohan (2000) has shown that lower dose of malathion brings about a reduction in the ovarian weight and growth of pre-vitellogenic oocyte in *Glossogobius giuris* (Ham.).

Degeneration and necrosis of testicular epithelium occurred in *Clarias batrachus* after treatment with carbaryl (Jyothi and Narayan, 1999).

Delayed sexual maturation by endrin and other toxicants in guppies were reported by Mount (1962). Further, Cairns *et al.* (1967) have reported reduced number of broods after dieldrin treatment, and abortion in *Gambusia affinis* due to endrin and eldrin treatment (Boyd, 1964).

Pesticides have played significant role in bringing out green revolution in India, however, their injudicious and indiscriminate use has resulted in an adverse impact on flora and fauna including human health and environment. The adverse effects of insecticides can be mitigated by its judicious use and developing suitable biodegradation methodologies (Sable and Patel, 2001).

It is suggested that foolproof toxicological tests are performed for each and every pesticide to determine its impact on flora and fauna. The foregoing discussion reveals very interesting results of pollution impact of insecticides on fishes.

References

Abidi, R. and Srivastava, U.S. (1988). Effect of endosulfan on certain aspect of haematology of the fish, *Channa punctatus* (Bloch). *Proc. Nat. Aca. Sci. India.* **58(8)**: 55-65.

Aditya, A.K., Chattopadhyay, S. and Mitra, S. (2002). Effect of mercury and methyl parathion on the ovaries of *Labeo rohita* (Ham). *J. Environ. Biol.*, **23(1)**: 61-64.

Aditya, A.K. and Chattopadhyay, S. (2000). Accumulation of methyl parathion in the muscle and gonad of *Labeo rohita. J. Environ. Biol.*, **21(1)**: 55-52.

Annon. (1962). *The Wealth of India: Raw Material,* Volume IV, Supplement Fish and Fisheries, CSIR, New Delhi, p. 132.

Ansari, B.A. and Kumar, K. (1987). Malathion toxicity: Effect on ovary of zebra fish, *Brachydanio rario* (Cyprinidae). *Int. Rev. Gasamten. Hydrobiol.* **72(4)**: 517-528.

APHA (1985). *Standard Methods for the Examination of Water and Wastewater.* APHA, AWWA, WPCF. Amer. Publ. Health Assoc., Washington, DC.

Arsta, T., V.S. Bais and N.C. Agarwal (1994). Relative toxicity of organochlorine compound (Aldrex) on *Mystus vittatus* under pH conditions. *Mendel,* **2(3-4)**: 185-186.

Arunachalam, S. and Palanichamy, S. (1982). Sublethal effects of carbaryl on surfacing behaviour and food utilization in the air breathing, *Macropodus cupanus. Physiol. Behav.*, **25**: 23-27.

Arunachalam, S., Jayalakshmi, K. and S. Aboobucker (1980). Toxic and sub-lethal effects of carbaryl in a freshwater catfish, *Mystus vittatus* (Bloch.) *Arch. Environ. Contam. Toxicol.*, **9**: 307-316.

Bansal, P.K., S.R. Varma, A.K. Gupta, S. Rani and R.C. Dalella (1979). Pesticide induced alterations in oxygen uptake rate of a freshwater major carp *Labeo rohita. Ecotoxicol. Environ. Safety*, **3**: 374-382.

Begum, G.S., V. Raghaban, P.M. Sharma and S. Hussain (1994). Study of dimethoate bioaccumulation in liver and muscle tissue of *Clarias batrachus* and in the elimination following cessation of exposure. *Pesti. Sci.*, **49(5)**: 201-205.

Bhatnagar, S.L. and R.S. Srivastava (1975). Histopathological changes due to copper in *Heteropneustes fossilis. Proc. 62nd Ind Sci. Cang.*, **3:** 118.

Bhattacharya, S., S. Mukherjee and S. Bhattacharya (1975). Toxic effects of endrin on hepatopancreas of the fish *Clarias batrachus* (Linn.) *Ind. J. Exp. Biol.*, **13**: 185-186.

Boyd, C.E. (1964). Insecticide causes mosquito fish to abort. *Fish. Cult.*, **26**:138.

Bradbury, S.P., J.R. Coats and J.M. Mckim (1986). Toxico kinetics of fenvalerate in rainbow trout (*Salmo gairdneri*). *Environ. Toxicol Chem.*, **6**: 555-567.

Burdick, C.K., A.J. Dean, E.J. Marris, J. Seka, R. Darchar and C. Frisca. (1972). Effect of rate and duration of feeding DDT on the reproduction of salmonoid fish reared and held under control condition. *N.Y. Fish and Game Jour.*, **19**: 97-115.

Cairns, J. Jr., N.R. Foster and J.J. Loose (1967). Effects of sublethal concentrations of Dieldrin on laboratory populations of guppies, *Poecilla reticulata* Pelias. *Proc. Nat. Acad., Sci. Phil.*, **119**: 75-91.

Chakrabarty, G. and S.K. Kumar (1974). Chronic effect of sublethal levels of pesticides on fish. *Proc. Natl. Acad Sci. India.* **4**: 241-246.

Chaudhary, B., R. Sharma and G.R. Shukla (1998). Alteration in tissues composition of *Mystus vittatus* exposed to toxaphone and maled. *J. Natcon*, **10(2)**: 225-229.

Choudhary, C., A. Ray, S. Bhattacharya and Shelley Bhattacharya

(1993). Non-lethal concentration of pesticide impairs ovarian function in the fresh water perch, *Anabas testudineus. Environ. Biol. Fishes*, **36**: 319-324.

Claudio, M., Johnson and C. Maric, F. Teledo (1993). Acute toxicity of endosulfan to the fish, *Hyphessobrycon bifaciatus* and *Brachydanio rario. Arch. Environ. Contam. Toxicol.*, **24**:151-155.

Coats, J.R., D.M. Symonik, S.P. Bradbury, S.D. Dyer, L.K. Timson and G.J. Atchison (1989). Toxicology of synthetic pyrethroides in aquatic organisms: An overview. *Environ. Toxicol. Chem.*, **6**: 671-679.

Datta, H.M. (1996). A composite approach for evaluation of the effects of pesticides on fish. *Fish Morphology*, pp. 149-277.

David, M., S.B. Mushigeri, M.S. Prashant and S.G. Mathad (2003). Role of phosphates during transport and energy metabolism in freshwater fish, *Cyprinus carpio* exposed to cypermethrin. *Poll. Res.*, **22(2)**: 277-281.

Donaldson, E.M. (1975). Physiological and physico-chemical factors associated with maturation and spawning. *EIFAC, Technical Paper*, **25**: 53-71.

Edwards, C.A. (1970). *Environmental Pollution by Pesticides*. Plenum Press, London and New York.

Ferguson, D.E., J.L. Ludke and G.C. Murphy (1966). *Transactions of American Fisheries Society*, **95**: 315.

Ganeshwade, R.M., D.R. Deshmukh and S.R. Sonawane (2002). Ethology of the fish, *Barbus ticto* under dimethoate toxicity. *J. Aqua. Biol.*, **17(2)**: 77-80.

Gautam, R.K. and K. Gautam (2002). Biological and haematological alterations in *Channa punctatus. Aquacult.*, **3(1)**: 35-36.

Ghosh, T.K. (1986). Compensative toxicological evaluation of two commonly used pesticides, Ekalux (EC-25) and Roger (Dimethoate) on the ovarian recrudescence in a, teleost, *Sarotheroden mossambicus. Uttar Pradesh J. Zool.*, **6(2)**: 224-227.

Gopal, K., R.M. Khanna, M. Anand and G.S.D. Gupta (1982). Haematological changes in fish exposed to endosulfan. *Indian J. Environ Hlth.*, **20**: 157-159.

Gupta, N. and A. Dua (2002). Mercury induced architectural alterations in the gills surface of a fresh water fish, *Channa punctatus*. *J. Env. Biol.* **23**: 383-386.

Haider, S. and R.M. Inbaraj (1988). *In vitro* effect of malathion and endosulfan on the LH induced oocyte maturation in the common carp, *Cyprinus carpio* (L). *Water and Soil Pollution*, **39**: 27-31.

Haya, K. (1989). Toxicity of pyretheroid insecticides to fish. *Environ. Toxicol. Chem.* **8**: 381-391.

Holdon, A.V. (1973). Effects of pesticides on fish. In: *Environmental Pollution by Pesticides*, Edwards, C.A. ed. Plenum Press, New York, pp. 213-253.

Hughes, G.M. and S.F. Perry (1976). Polluted fish respiratory physiology on lock wood. In: *Effect on Pollutants on Aquatic Organisms*, APM, ed. Cambridge University Press, Cambridge, pp. 163-183.

Humason, G.L. (1972). *Animal Tissues Technique*. W.H. Permand and Co., San Francisco.

Jackson, G.A. (1976). Biochemical half life of endrin in channel catfish tissue. *Bull. Environ. Contain. Toxicol.*, **16**: 505-507.

Jeba Kumar, S.R., D.A.K. Kumarguru and J. Jayaraman (1993). Accumulation and dissipation of phosphamidon in the tissues of fresh water fish. *Oreochromis mossambicus*. *Comp. Physiol. Ecol.*, **18**: 12-17.

Jones, J.R. (1973). *Fish and Oxygen Pollution by Oxygen Reducing Effluents: Fish and River Pollution*. Butterworth and Co., pp. 5-26.

Joshi, P., D. Harish and M. Bose (2002). Effect of lindane and malathion exposure to certain blood parameters in a fresh water taleost fish *Clarias batrachus*. *Poll. Res.*, **21(1)**: 55-57.

Joshi, S.K., A.K. Pandey and R.K. Nagar (1982). Electro-phoretic variation in the blood protein fraction in malathion treated fish *Channa punctatus* (Blch). *Drre. Mat. Symp. Anim. Meta. and Poll. J.*, pp. 147-149.

Jugadeesan, G. and S. Vijayalakshmi (1999). Alterations in the behaviour pattern in *Labeo rohita* (Ham.) fingerlings induced by mercury. *Ind. J. Env. Toxicol.*, **9(1)**: 45-52.

Jugadeesan, G. and S. Vijayalakshmi (1999). Alteration in the behavioural pattern in *Labeo rohita* (Ham.). *Ind. J. Env. Toxicol.*, **9(1)**: 45-52.

Jyothi, B. and G. Naryan (1999). Toxic effects of carbaryl on gonads of freshwater fish, *Clarias batrachus* (Linnaeus). *J. Environ Biol.*, **(20(1)**: 75-76.

Kamler, E., O. Matlak and K. Srodusz (1974). Further observation on the effect of sodium salt of 2,4-D on early developmental stages of carp (*Cyprinus carpio*). *Pol. Arch. Hydrobiol.*, **21**: 481-502.

Kennedy, H.D., Eller, L.L. and D.F. Walsh (1970). Chronic effect of methoxychlor on blue gills and aquatic invertebrates. *U.S. Bull. Report Fish Wild. Tech. Pap.*, **53**.

Khillare, Y.K. and S.B. Wagh (1987). Autotoxicity of the pesticide endosulfan to fishes. *Environ. and Ecology*, **5(4)**: 805-806.

Konar, S.K. (1983). Lethal effects of the insecticides thiometon on the carp, *Labeo rohita and the fish, Heteropneustes fossilis. Proc. Nat. Acad. Sci.*, **53(130)**: 178-181.

Konar, S.K. and T.K. Ghosh (1983). Lethal effects of carbamate pesticides, dithane M-45 on fish, plankton and worm. *Geobios.*, **10**: 222-226.

Koundinyer, P.R. and R. Ramamurthy (1982). Effect of organophosphorus pesticide sumithion (fenitrothian) on alkaline phosphatase activity of freshwater teleost *Sarotherodon mossambicus* (Peters). *Current Science*, **15**: 503-505.

Kulshreshtha, S.K. and L. Jauhar (1984). Effects of sublethal dose of thiodin and serin on liver of *Channa striatus*. In: *Effects of Pesticides on Aquatic Farms*, S.K. Kulshreshtha, V. Kumar and M.C. Bhatnagar, eds. Academy of Environmental Biology, Muzaffarnagar, India, pp. 71-78.

Kumar, H. and A.B. Gupta (1997). Toxicity of organophosphorus, carbamates and synthetic pyretheroid pesticides to the Indian catfish, *Heteropneutes fossilis. J. Natcon.*, **9(1)**: 111-114.

Kumar, K., Y.K.P. Sinha, and A.K. Pandey (1999). Endosulfan induced haematological alterations in the freshwater air-breathing fish, *Anabas testudineus* (Bloch). *J. Natcon.*, **11(1)**: 125-135.

Kumar, S. Swarnlata and Krishna Gopal (2002). Deltamethrin induced changes in gills of *Heteropneustes fossilis. Proc. Acad. Environ. Biol.*, **9(2)**: 159-163.

Kumaraguru, A.K. and F.W.H. Beamish (1983). Bioenergetics of acclimation to permethrin. (MLCl-143) by rainbow Trout. *Comp. Biochem. Physiol.*, **75 (2)**: 247-252.

Lemke, E. and J. Mount (1963). Some effects of alkyl benzene sulfonate on the blue gill, *Lepomis macrochirus. Trans. Ame. Fish. Soc.*, **92**: 372-378.

Lone, K.P. and M.Y. Javaid (1976a). Effect of sublethal doses of DDT and dieldrin on blood of *Channa punctatus* (Bloch). *Pak. J. Zool.*, **8**: 143-150.

Lone, K.P. and M.Y. Javaid (1976b). Effect of sublethal dose of three organophosphorus insecticides on the haematology of *Channa punctatus* (Bloch). *Pak. J. Zool.*, **8**: 77-84.

Loteste, Alicici, Cazenave, Jimena, Parma, de Crouy and M. Julieta (2002). Recovery of plasmatic cholinesterase activity in a neotropical fish. *Prochilodus lineatus* (Pisces, Curimatidae) exposed to organophosphorus pesticides. *J. Environ Biol.*, **23(3)**: 225-229.

Luther Das, V., K. Samedhama Reyn and K. Kondaish (2002). Toxicity and effect of cypermethrin to freshwater fish *Labeo rohita* (*Hamilton*). *Proc. Acad. Environ Biol.*, **9(1)**: 41-48.

Mandal, P.K. and A.K. Kulshreshtha (1970). Histopathological changes induced by the sub-lethal submission in *Clarias batrachus* (Linn.). *Ind. J. Expt. Biol.* **18**: 547-548.

Mani, K. and P.K. Saxena (1985). Effect of safe concentration of some pesticides on ovarian recrudescence in the fresh water murrel, *Channa punctatus* (B). A quantitative study. *Ecotoxicol. Environ. Saf.*, **9**: 241-249.

Mankar, C.R. and K.M. Kulkarni (2000). Endocil impact on chromotophores of freshwater fish, *Channa orientalis* (Sch). *J. Aqua. Biol.*, **15(192)**: 102-109.

Manoharan, T. and G.N. Subiah (1982). Toxic and sublethal effects on *Barbus stigma. Proc. Indian Acad Sci.*, **91**: 523-532.

Mathur, D.S. (1962). Studies on the histopathological changes induced by the liver, kidney and intestine of certain fishes. *Ind. J. Exp. Biol.*, **18**: 506-509.

Mathur, D.S. (1976). Histopathological changes in the liver of fishes resulting from exposure to Dieldrin and Lindane. *Annual Plant and Microbial Toxins*, pp. 546-552.

Matsumura, F.G., M. Gouch and T. Mirate (1972). *Environmental Toxicity of Pesticides*. Academic Press, New York.

Mount, D.I. and G.J. Putuick (1966). Summary report of the 1963-Mississippi river fish kill. *Trans 31st M. Amer. Wild Nature Resources Conf. Trans.*, pp. 177-231.

Mount, D.J. (1962). Chronic effects of endrin on blunt nose minnows and *guppies. USA and Wildlife Services Report*, **58**: 1-38.

Mount, D.J. and C.E. Stephen (1967). A method of establishing acceptable toxicant limits for fish malathion and butyoxythanol ester at 2,4-D. *Trans. Am. Fish. Sci.*, **95**: 185-193.

Muirhead-Thomson, R.C. (1971). *Pesticides and Freshwater Farms*. Academic Press, New York.

Mukhopadhya, P.K. and Dehadrai, P.V. (1980). Biochemical changes in air breathing cat fish *Clarias batrachus* (Linn.) exposed to malathion. *Env. Poll.*, 22: 149-158.

Muniyan, M. and K. Veeraraghavan (1999). Acute toxicity of ethofenprox to the freshwater fish *Oreochromis mossambicus* (Peters). *Journal of Environ. Biol.* **20(2)**: 153-155.

Murali-Mohan, E., Banu Priya, C.A.Y., Amritha Ganesh, D., Arun Sana Lal, K.S. Pillai and Balakrishna Murthy (2002). Alternative to LC_{50} of organophosphates an attempt. *J. Aqua Biol.*, **15(192)**: 74-76.

Murugappan, R.M. and M. Gunasundari (1996). Evaluation of genotoxic effects of endosulfan in the *Channa striatus. J. Ecotoxicol, Environ. Monit.*, **6(2)**: 131-134.

Murugappan, R.M., M. Muthusamy and M. Gunasundari (1999). Genotoxicity and bio accumulation of BHC and cadmium on the freshwater fish *Cyprinus carpio. J. Environ Biol.*, **20(4)**: 313-316.

Naqvi, S.M. and R. Hawkins (1988). Toxicity of selected insecticides to mosquito fish, *Gambusia affnis*. *Bull. Environ. Contam. Toxicol.*, **40**: 779-781.

Narain, A.S., A.K. Srivastava and B.B. Singh (1990). Gill lesions in the perch, *Anabas testudineus* subjected to sewage toxicity. *Bull. Environ. Contain. Toxicol.*, **45**: 235-242.

Nath and V. Banerjee (1995). Sub-lethal effect of elevation on the haematological parameters in *Heteropneustes fossilis*. *J. Freshwater Biol.*, **7(4)**: 261-264.

Nemcsok, J., L. Orbran., B. Asztalos and E. Vig (1987). Accumulation of pesticides in the organs of carp *Cyprinus carpio* L. at 4°C and 20°C. *Bull. Environ. Contam. Toxicol.*, **39**: 370-378.

Nirmala, A.R.C. and J. Vinoliya (2003). Effect of an organophosphorus pesticide methyl parathion on the respiratory metabolism and food utilization in fingerlings of *Cyprinus*. *Poll. Res.*, **22(1)**: 101-105.

O'Brian, R.D. (1967). *Insecticides, Action and Metabolism*. Academic Press, New York.

Pandey, B.N., A.K. Chandel and M.P. Singh (1976). Effect of malathion on oxygen consumption and blood of *Channa punctatus* (*Bloch.*). *Indian J. Zool.*, **16**: 95-100.

Pashine, R.G. and Kurve, S.S. (2000). Estimation of LC_{50} values for *Lebistes reticulatus* with toxicants zinc sulphate and mercuric chloride. *J. Aqua. Biol.*, **15(1&2)**: 84-85.

Patel, M.G., K. Shareel, L.B. Piwar and P.S. Loher (2000). Histopathological changes in stomach and liver of *Nemacheilus botia* (Ham). on acute exposure to dimecron. *J. Aqua Biol.*, **15(1&2)**: 105-107.

Patnaik, B.B., K. Abdul Subhan and M. Selvanayagam (2002). Toxicity to the fingerlings of a carp. *Sarotherodon mossambicus* by a synthetic pyretheroid fenvalerate. *Indian J. Environ and Ecoplan.*, **6(3)**: 551-558.

Paulose, P.V. (1987). Accumulation of organic and inorganic mercury and histological changes in the gills of *Gambusia affinis*. *Proc. Ind. Natl. Sc. Acad.*, **53(B)**: 235-237.

Paulose, P.V. (1989). Histological changes in relation to accumulation and elimination of inorganic and methyl mercury in gills of *Labeo rohita. Indian J. Exp. Biol.*, **27**: 146-150.

Prabhakar, J.D., R.E. Martin, M.R. Jagtap and Kshemkalayan (1993). Effect of endosulfan and hexa-chlorocycloehxane on the behaviour of fish *Labeo rohita. J. Ecobio.*, **5(2)**: 149-150.

Rajamanickan, C., P.R. Karpagaganapathy (1988). Effect of Lindane on the behavioral changes in the freshwater fish *Tilapia mossambica. Proc. 2nd Nat. Symp. Ecotoxicol.*, pp. 143-145.

Ramakrishnan, M., S. Arunachalam and S. Palanichamy (1997). Sublethal effects of pesticides on feeding energetics in the air breathing fish *Channa striatus. J Ecotoxicol. Environ. Monit.*, **7**: 169-175.

Ramamurthy, K. (1988). Impact of hepatchlor on haematalogical, histological and selected biochemical parameters in the freshwater edible fish *Channa punctatus*. Ph.D. Thesis, S.V. University, Tirupathi, India.

Ramalingam, K. and Ramalingam, K. (1982). Effects of sublethal levels of DDT, malathion, and mercury on tissue proteins of *Sartherodon mossambicus* (Peters). *Proc. Indian Acad. Sci.* (*Animal Sci.*), **19(6)**: 501-505.

Ramchandra, Mohan, M. (2000). Malathion induced changes in the every of freshwater teleost fish, *Glossogobius giuris* (Ham). *Poll. Res.*, **19(1)**: 73-75.

Ranjana, A.M., R. Kumar, H.S. Singh and A.K. Gupta. (1999). Behavioral studies of *Channa punctatus* after exposure to pesticides I-feeding behaviour. *J. Natcon.*, **11(2)**: 169-173.

Rao, Jayatha, K. Md. Azar Baig, V. Raddaich and K. Ramamurthy (1987). Pesticidal impact on fresh water teleost, *Tilapia mossambica* retina. *Curr. Sci.*, **5(56)**: 17.

Rao, L.M. and K. Ramaheswari (2000). Variations in acute toxicity of endosulfan and monocrotophos to *Labeo rohita, Mystus vittatus* and *Channa punctatus. Poll. Res.*, **19(3)**: 461-465.

Rastogi, A. and J. Kulshreshta (1990). Effect of sublethal doses of three pesticides on the ovary of carp minnow, *Rasbora doniconius. Bull. Environ, Contam. Toxicol.*, **45**: 742-747.

Rawat, D.K., N.C. Agarwal and Bais, V.S. (2000). Alterations in the endosulfan and triazophos toxicity on *Heteropneustes fossilis* (Bloch) in relation to pH. *J. Natcon.* **12(1)**: 51-56.

Reddy, P.M. and C.G. Philip (1994). *In vivo* inhibition of ACHE and ATPase activities in the tissue of freshwater fish, *Cyprinus carpio* exposed to technical grade, Cypermethrin. *Bull. Environ, Contm. Toxicol.,* **52**: 619-626.

Reddy, P.M., G.G. Philip and M.D. Bahamohidea (1991). Fenvalerate induced biochemical changes in the selected tissues of the freshwater fish, *Cyprinus carpio. Biochem. Int.,* **23**: 1087-1096.

Rege, M.S. and S.A. Gaikwad (1990). Effect of different concentrations of Thiodon 35 EC and phenyl mercuric acetate (PMA) on proximate composition of *T. mossambica* (Peters). In: *Recent Trends in Toxicol Environment Services,* **3**: 199-202.

Ruparelia, S.G., Y. Verma, J.B. Kashyap and B.D. Chatterjee (1984). Status report on acute toxicity of pesticides in fishes in India. *Proc. Sem. Eff Pest. Ag. Fan.,* pp. 107-114.

Sabita Borah and R.N.S. Yadav (1995). Static Bioassay and toxicity of two pesticides, rogor and endosulfan to the air breathing fish, *Heteropneustes fossilis* with special references to behaviour. *Poll. Res.,* **14(4)**: 435-438.

Sable, S.S. and K. Patel (2001). Biodegradation of fenvalerate by *Pseudomonas* sp. *Indian J. Env. Toxicol.,* **11(1)**: 12-21.

Sahai, S. (1987). Toxicological effects of some pesticides on the ovaries of *Puntius ticto* (Teleostei). *Proc. VII A.E.B. SES and Symp.,* pp. 53-54.

Sahai, S. (1992). Accumulation and induced histopathological effects of pesticides in the testis of *Puntius ticto* (Teleostei). *J. Ecobiol.,* **4(4)**: 303-308.

Sakthivel, V. (2002). Effect of phosphamidon on glycogen, proteins and lipids in the fish. *Gambusia affinis* (Baird and Girand). *J. Aqua Biol.,* **17(2)**: 67-90.

Sandhu, A.K. and P.N. Mukhopadhya (1985). Comparative effect of two pesticide malathion and carbafuran, on testis of *Clarias batrachus* Linn. *J. Environ. Biol.,* **6(3)**: 217-222.

Sandhu, D.N. (1993). Toxicity of an organophosphorus insecticide monocil to the air breathing fish, *Channa punctatus. J. Ecotoxicol. Monit.*, **3(2)**: 133-136.

Santhakumar, M., M. Balaji and K. Ramudu (2001). Gill lesions the perch, *Anabas testudineus* exposed to monocrotophos. *J. Environ Biol.*, **22(2)**: 87-90.

Santhakumar, M., M. Balaji and K. Ramudu (2000). Effect of sub-lethal concentrations of monocrotophos on the ethological response of an air breathing fish, *Anabas testudineus* (Bloch). *Ecol. Env. Cons.*, **6(2)**: 175-177.

Santhakumar, M., M. Balaji (2000). Acute toxicity of an organophosphorus insecticide monocrotophos and its effects on behaviour of an air breathing fish, *Anabas testudineus* (Bloch). *J. Environ Biol.*, **2(2)**: 121-123.

Santhakumar, M., M. Balaji and K. Ramudu (2001). Effect of monocrotophos on plasma phosphatase activity of a freshwater fish, *Anabas testudineus* (Bloch) *Poll. Res.*, **19(2)**: 257-259.

Santhakumar, M., K.R. Aarvanan, D. Sumadry, M. Balaji and Ramudu, K. (1999). Effect of monocrotophos on the activity of acid and alkaline phosphatase in brain and gill of an air breathing fish, *Anabas testudineus. Ecol. Env. and Cons.*, **5(3)**: 363-371.

Santhakumar, M., M. Balaji and K. Ramudu (2000). Histopathological lesions in the brain of a teleost *Anabas testudineus* exposed to monocrotophos, *J. Environ Biol.*, **21(4)**: 297-299.

Santhakumar, M., M. Balaji, K.R. Sarvananan, K. Sowmadry and K. Ramudu (2000). Effect of monocrotophos on the optomotor behaviour of an air breathing fish, *Anabas testudineus* (Bloch). *J. Environ Biol.*, **71(1)**: 65-68.

Santhamma, D., N.J. Sushma and Jayaitha Rao, K. (1999). Effect of monocrotophos on structure and function of certain vital tissues of fish, *Tilapia mossambica. J. Natcon*, **11(1)**: 91-100.

Sardamni,R. and Kamlaveni,K. (2002). Sublethal effect of fenvalerate on bioenergetics and biochemical changes in *Cyprinus carpio. Indian J. Environ. and Ecoplan.*, **6(2)**: 315-318.

Saxena, P.K. and M. Garg (1978). Effect of insecticide pollution on ovarian recrudescence in the freshwater teleost, *Channa punctatus, Indian. J. Exp., Biol.*, **16**: 689-691.

Saxena, V., S. Trivedi and D.N. Saxena (1997). Acute toxicity of two organophosphorus pesticides, Nuvan and Dimecron to a freshwater murrel, *Channa orientalis* (Schneider). *J. Environ Poll.*, **4**: 79-83.

Sharma, Y., A. Verma and R. Sharma (2002). Additive effect of DDT and Ammonium chloride on the kidney of *Labeo rohita* (Ham.). *Proc. 4th Ind. Agric. Scientist and Farmers Congress*, 16-17th Feb., Meerut.

Sharma, Y., and R. Sharma (2002). Synergistic effect of pesticides DDT and fertilizer ammonium chloride on the liver of *Cirrhinus mrigala* (Ham.). *Proc. Poorvanchal Academy of Sci.*, Jaunpur, p. 3.

Singh, M.N. and A.K. Srivastava (1982). Toxicity of a mixture of aldrin and farmothion and other organophosphorus, organochlorine and carbamate pesticides to the Indian catfish, *Heteropneustes fossilis. Comp. Physiol. Ecol.*, **4**: 115-118.

Singh, S.P. and T.P. Singh (1987). Impact of malathion and hexachlorocyclolexane on phases of reproductive cycle in *Clarias batrachus* (Insecticide testosterone Estradiol 17 - Beta-Estrone *vitellogenesis. Pesticide Biochem. Physiol.*, **27(3)**: EM 301-308.

Skidmore, J.F. and P.W.A. Tovel (1972). The toxic effect of zinc sulfate on the gills of rainbow trout. *Water Res.*, **6**: 217-230.

Sonawane, S.R. and Nikam, S.R. (2000). Effect of pesticides, benzene hexachloride (BHC) and Sumithion on Lepidocephalichthys thermalis (C&V). In: *Environmental Issues and Sustainable Development*. Vineet Publications, pp. 30-32.

Sprague, J.B. (1971). Measurement of pollutant toxicity to fish sub-lethal effect and safe concentration. *Klat. Res.*, **5**: 245-266.

Sunitha, S. and S. Sahai (1993). Histopathological change in the gills of *Rasbora daniconius* induced by γ-BHC. *J. Environ Biol.*, **5**: 65-59.

Susan, A. and K.S. Tilak (2003). Histopathological changes in the vital tissues of the fish *Cirrhinus mrigala* exposed to fenvalerate technical grade. *Poll. Res.*, **22(2)**: 179-184.

Thakur, D.B. and V.S. Bais (2000). Toxic effects of aldrin and fenvalerate on certain haematological parameters of a freshwater teleost, *Heteropneustes fossilis* (Bloch.). *J. Environ Biol.*, **21(2)**: 161-163.

Thakur, Nutan and S. Sahai (1987). Changes in blood serum proteins and other hematological parameters in the teleost, *Channa punctatus* on exposure to organochlorine pesticides. *Proc. Acad. of Env. Biol. Environ. Pestic. Toxicol.*, pp. 263-271.

Tilak, K.S. and Satyavardhan (2002). Effect of fenvalerate on oxygen consumption and haematological parameters in the fish *Channa punctatus* (Bloch). *J. Aqua. Biol.*, **17(2)**: 81- 88.

Tilak, K.S., K. Veeraiah and S. Jhansi Lakshmi (2002). Studies of some biochemical changes in the tissues of *Catla catla* (Hamilton), *Labeo rohita* (Hamilton) and *Cirrhinus mrigala* (Hamilton) exposed to NH_3-N, No_2-N and No_3-N. *J Environ. Biol.*, **23(4)**: 377- 387.

Tilak, U.S., K. Veeraiah and G.V. Ramana Kumar (2001). Studies on histopathological changes in the gill, liver, kidney of *Ctenopharynodon idella* exposed to technical fenvalerate and EC 20 per cent. *Poll. Res.*, **20(3)**: 387-393.

Tripathi, G., S. Harsh and P. Verma (2002). Fenvalerate induced macromolecular changes in the catfish, *Clarias batrachus*. *J. Environ Biol.*, **23(2)**: 143-146.

Trivedi S. and D.N. Saxena (1999). Acute toxicity and behavioural response of an organophosphorus Nuvan to *Clarias batrachus* (Linnaeus). *J. Env. Poll.*, **6(1)**: 53-57.

Trivedy, R.K. and P.K. Goel (1986). *Chemical and Biological Methods for Water Pollution Studies*. Environmental Publications, Karad (Mh.), p. 251.

Uma Rani, S. and T. Rajendranath (1988). Toxicity of kitazin on the oocytes of the fish *Channa punctatus* (Bloch). *Indian J. Comp. Anim. Physiol.*, **6(2)**: 140-143.

Varadraj, G., M.A. Subramaniam and B. Nagarajun (1993). The effect of sub-lethal concentrations of paper and pulp mill effluents on the haematological parameters of *Oreochromis mossambicus* (Peters). *J. Environ. Biol.*, **14(4)**: 321-323.

Verma, A. and Y. Sharma (2002a). Pollutional status of fresh water biota: A review. In: *Ecology of Polluted Waters, Vol. 2,* Arvind Kumar, ed. APH Publishing House, New Delhi, pp. 835-845.

Verma, A. and Y. Sharma (2002b). Hydrobiological studies on fresh water ponds. In: *Wetland Conservation and Mangement,* B.B. Hosetti, ed. Pointer Publications, Jaipur, pp. 254-265.

Verma, S.R., V. Kumar and R.C. Dalela (1981). Studies on the accumulation and elimination of three pesticides in the gonads of *Notopterus notopterus* and *Colisa fasciatus. Indian J. Environ. Hlth.,* **23(4)**: 275-281.

Westlake, V. and F.M. Gunther (1966). Organic pesticides in Environment. In: *Advances in the Chemistry,* R.P. Gouled, ed. American Chemical Society, Washington D.C.

Wild, D. (1979). Mutagenecity studies on organophosphorus insecticides. *Mutation Res.,* **32**: 133-150.

Wildish, D.J, W.J. Cartson, T. Cunningham and I.J. Lister (1971). Toxicological effects of some organophosphorus insecticide to Atlantic salmon fish. *Res. Board Can. Mans. Reprod. Ser. Fish Andmans. Ser.,* West Vancouver, British Columbia. **1157**: 1-20.

Zutshi, B. (2003). Effect of fenthion on the testis of freshwater fish, *G. giuris (Histochemical and Biochemical). Poll. Res.,* **22(20)**: 231-236.

6

Physico-chemical Characteristics in Relation to Abundance of Plankton of Jagat Sagar Pond, Chhatarpur, India

✰ *Pushpendra Kumar Khare*

Introduction

Plankton population, related to physico-chemical parameters of water generally studied with reference to its periodicity and abundance. Some notable work on this has been recently by Abbasi *et al.*, 1996; Rao *et al.*, 1996; Singh and Sinha, 1995; Arvind Kumar, 1995; Mansori and Tiwari *et al.*, 1993; Mishra and Saksena, 1993; Khare, 1999; Singh and Singh, 1993; Verma and Dattamunshi, 1987; Kant and Anand, 1978; have studied the inter-relationships of Phytoplankton and Physical factors, the present study was undertaken to assess the various planktonic groups and their seasonal abundance in relation to certain physico-chemical factors in the Jagar Sagar pond during 1996.

Study Area

Jagar Sagar Pond lies at 25°01′ N and 79° 29′ E. It is an old pond on Chhatarpur-Jhansi national highway which is 15 km from Chhatarpur District. The pond is 2–2.5 km long from north to south and its width from east to west 1–1.5 km. The water level showed variation at four experimental stations from 2.05 to 4.15 meters during the period of investigation. It is a perennial pond which is annually filled by Monsoon water. It is used for irrigation purpose and fish farming by the fisheries department of M.P. Government.

Material and Methods

Samples of water were collected and analysed for various physico-chemical parameters by standard methods referred by APHA, 1976. The plankton samples were collected at monthly with a plankton net of bolting silk no. 21 having a diameter of 30 cm and length of 60 cm in lower narrow end of net a glass tube of 50 ml capacity was fixed. Fifty liters of pond water was carefully passed through net and samples collected were preserved 5 per cent Formalin and some drops of glycerine. Number of plankton were counted with the help of Sedgwick-Rafter plankton counting cell and were expressed as organism per liter. Plankton identification was done with the help of standard books and published taxonomic articles (Smith, 1950; Tonapi, 1959; APHA, 1976; Adoni, 1985 and Alfered *et al.*, 1973).

Results

The seasonal variation in physico-chemical characteristics of the pond water are represented in Table 6.1. The species composition of phytoplankton and zooplankton are giving in Tables 6.2 and 6.3 respectively. The physico-chemical characteristics reveal that the water temperature ranged from 16.9 to 31.8 °C whereas transparency of the water varied from 13.10 to 32.67 cm. The pH value showed a variation from 7.1 to 8.1. Dissolved oxygen content varied from 3.71 to 17.63 ppm. Free carbondioxide showed a variation from 2.95 to 7.05 ppm. The chloride content also varied from 19.1 to 22.6 ppm. The nutrient contents live phosphate, nitrate showed a variation from 0.10 to 0.26 ppm and from 0.20 to 0.57 ppm respectively.

Figure 6.1 shows the total number of phytoplankton in three groups of different months; Chlorophyceae dominated in the

Table 6.1: Monthly Variation in Physico-chemical Parameters (Averages) of Jagat Sagar Pond, Chhatarpur During 1996

Parameters	*Jan.*	*Feb.*	*Mar.*	*Apr.*	*May*	*June*	*Jul.*	*Aug.*	*Sep.*	*Oct.*	*Nov.*	*Dec.*
Atmos. Temp. (°C)	14.8	20.3	26.3	32.7	37.9	36.2	30.7	27.3	28.7	26.0	19.0	16.7
Water Temp. (°C)	16.9	19.5	23.9	26.9	30.3	31.8	28.6	26.1	27.4	24.6	23.5	22.7
Light Transparency (cm)	32.67	30.30	26.10	20.35	17.37	16.11	13.10	17.77	20.62	22.51	29.25	31.32
pH	7.1	7.4	7.7	7.8	8.1	8.0	7.8	7.5	7.3	7.3	7.2	7.1
Dissolved O_2 (ppm)	17.63	15.41	5.53	4.77	4.55	4.43	3.71	4.54	5.05	5.58	14.53	16.41
Dissolved CO_2 (ppm)	6.20	7.05	5.65	4.32	4.12	4.75	6.87	6.15	5.10	3.87	4.25	2.95
Phosphate (ppm)	0.26	0.23	0.24	0.18	0.21	0.12	0.10	0.13	0.22	0.22	0.25	0.24
Nitrate (ppm)	0.26	0.20	0.41	0.40	0.53	0.57	0.31	0.20	0.26	0.30	0.27	0.29
Chloride (ppm)	19.2	20.6	21.6	22.0	20.3	21.6	22.6	21.8	18.6	19.1	20.7	19.4

Table 6.2: Seasonal Variation of Phytoplankton in Jagat Sagar Pond During 1996

Groups	*Jan.*	*Feb.*	*Mar.*	*Apr.*	*May*	*June*	*Jul.*	*Aug.*	*Sep.*	*Oct.*	*Nov.*	*Dec.*
Microcyst sp.	43	185	223	382	265	345	225	45	62	32	198	42
Anabena sp.	380	566	625	270	530	580	385	150	132	142	212	153
Gleotrichia sp.	Nil	75	50	108	145	110	Nil	Nil	40	85	75	Nil
Rivularia sp.	42	44	100	86	51	75	70	37	20	48	30	22
Calothrix sp.	28	13	70	28	44	65	Nil	20	15	35	Nil	13
Spirulina sp.	25	65	95	55	35	80	48	Nil	51	78	Nil	Nil
Total Myxophyceae	**518**	**948**	**1163**	**929**	**1070**	**1255**	**728**	**252**	**320**	**420**	**515**	**230**
Melosira sp.	28	35	74	84	112	92	67	65	64	31	48	54
Cymbella sp.	38	40	75	90	85	202	85	86	52	47	72	75
Diatoms sp.	Nil	Nil	22	42	58	93	73	35	Nil	Nil	64	82
Navicula sp.	80	84	80	112	168	348	134	145	114	108	110	142
Pinnularia sp.	22	39	42	90	120	145	41	37	39	27	30	Nil
Synedra sp.	Nil	Nil	66	64	105	38	16	Nil	Nil	Nil	45	36
Total Bacilariophyceae	**168**	**198**	**359**	**482**	**648**	**918**	**416**	**368**	**269**	**213**	**369**	**389**
Eudorina sp.	16	70	105	108	99	86	92	29	30	38	40	44
Volvox sp.	20	84	84	92	75	80	76	03	09	15	14	12
Botryococcus sp.	04	12	17	20	30	25	24	05	06	08	12	13
Chlorella sp.	05	15	22	25	28	25	21	Nil	Nil	06	11	12
Dictyosphaerium	01	10	15	23	22	20	23	Nil	08	12	12	14

Contd...

Table 6.2–Contd...

Groups	*Jan.*	*Feb.*	*Mar.*	*Apr.*	*May*	*June*	*Jul.*	*Aug.*	*Sep.*	*Oct.*	*Nov.*	*Dec.*
Oocystis sp.	06	13	18	22	31	32	29	15	18	14	16	15
Pediastrum sp.	12	18	20	32	30	20	10	Nil	Nil	16	18	19
Scendsmus sp.	11	35	37	45	40	32	30	08	14	12	14	15
Microspora sp.	10	15	18	27	32	26	25	10	12	16	18	20
Ulothrix sp.	18	48	52	64	62	45	42	25	28	20	22	25
Uronema sp.	05	14	18	21	27	21	22	05	08	11	12	14
Chaetophora sp.	03	16	21	25	24	20	25	11	12	18	16	17
Chaetonema sp.	06	17	20	31	28	22	20	08	10	17	10	12
Coleochaete sp.	08	12	42	65	62	48	42	09	14	12	18	22
Spirogyra sp.	20	80	202	620	530	498	460	65	82	87	89	100
Zygnima sp.	28	56	134	294	275	266	251	32	35	49	52	62
Mougeotia sp.	20	52	66	70	76	92	102	20	22	45	47	56
Closterium sp.	15	18	32	80	74	69	68	24	22	22	22	24
Cosmarium sp.	Nil	24	25	20	35	40	39	28	25	24	26	28
Eusstrum sp.	06	10	15	14	20	18	04	Nil	Nil	12	18	16
Pleurotaenium sp.	11	15	18	21	28	30	27	16	12	08	12	14
Micrasterias sp.	08	10	21	26	36	38	25	06	08	11	17	18
Total Chlorophyceae	**233**	**644**	**1002**	**1745**	**1664**	**1553**	**1457**	**319**	**375**	**473**	**516**	**572**
Phacus Total Euglenophyceae	6.0	20	52	37	124	151	134	84	44	33	45	12
Total Phytoplankton	**930.01**	**1847**	**2624**	**3243**	**3536**	**3837**	**2765**	**1044**	**1031**	**1100**	**1463**	**1228**

Table 6.3: Seasonal Variation of Zooplankton in Jagat Sagar Pond During 1996

Groups	*Jan.*	*Feb.*	*Mar.*	*Apr.*	*May*	*June*	*Jul.*	*Aug.*	*Sep.*	*Oct.*	*Nov.*	*Dec.*
Amoeba sp.	-	-	130	160	64	80	55	15	133	17	-	37
Paramecium sp.	65	195	225	52	15	-	30	191	42	-	40	40
Verticella sp.	15	10	149	75	50	28	20	85	40	50	356	52
Telotrichidium sp.	680	355	128	455	237	47	60	-	73	165	173	182
Euglena sp.	127	113	128	224	180	-	-	30	11	90	70	138
Total Protozoans	**887**	**673**	**760**	**966**	**546**	**155**	**165**	**321**	**299**	**322**	**639**	**449**
Asplanchna sp.	-	-	32	40	28	-	-	-	15	19	-	-
Brachionus falcatum	168	149	280	650	290	90	40	10	60	82	465	250
B. angularias	160	185	240	295	360	78	62	-	45	60	85	112
B. calafortus	210	246	180	280	204	86	-	-	48	76	70	128
Brachionus sp.	196	105	150	349	250	102	30	18	108	80	98	280
Keratella tropica	85	-	132	300	85	-	-	-	49	18	08	36
K. tecta	76	49	85	150	70	-	-	-	5	12	27	44
Lecane sp.	97	68	30	76	18	24	20	40	75	30	42	50
Filinia longiseta	73	45	30	80	42	33	28	52	50	100	10	37
Pterodina patina	128	84	128	226	30	06	70	-	224	160	180	115
Total Rotifers	**1193**	**931**	**1287**	**2446**	**1377**	**419**	**250**	**118**	**681**	**637**	**985**	**1052**

Contd...

Table 6.3–Contd...

Groups	*Jan.*	*Feb.*	*Mar.*	*Apr.*	*May*	*June*	*Jul.*	*Aug.*	*Sep.*	*Oct.*	*Nov.*	*Dec.*
Bosmania longispina	–	–	30	–	18	–	–	–	–	–	–	–
Bosmania sp.	–	–	–	38	15	–	–	–	–	–	–	–
Daphnia similis	–	–	270	940	467	33	–	55	37	17	15	270
D. longispina	410	230	786	220	25	–	–	15	30	20	9	220
Moina sp.	68	20	170	–	–	–	–	79	–	–	–	111
Monostyla sp.	49	60	13	103	29	22	15	40	112	27	23	21
Nauplius larvae	134	75	122	184	41	25	–	–	07	10	25	87
Cyclops viridis	17	20	30	104	92	75	–	42	50	97	76	212
C. bicuspidatus	52	75	112	134	42	–	–	–	32	14	60	96
C. sp.	30	62	120	42	41	62	28	–	–	25	9	40
Mesocyclops sp.	27	100	168	140	–	–	15	65	10	–	10	42
Thermocyclops	61	175	184	232	61	7	3	10	40	11	25	61
Diaptomus sp.	40	12	117	–	–	–	–	–	–	–	–	–
Total Crustacea	**888**	**829**	**2122**	**2137**	**831**	**224**	**61**	**306**	**318**	**221**	**252**	**1160**
Total Zooplankton	**2968**	**2433**	**4169**	**5549**	**2754**	**798**	**476**	**745**	**1298**	**1180**	**1876**	**2261**

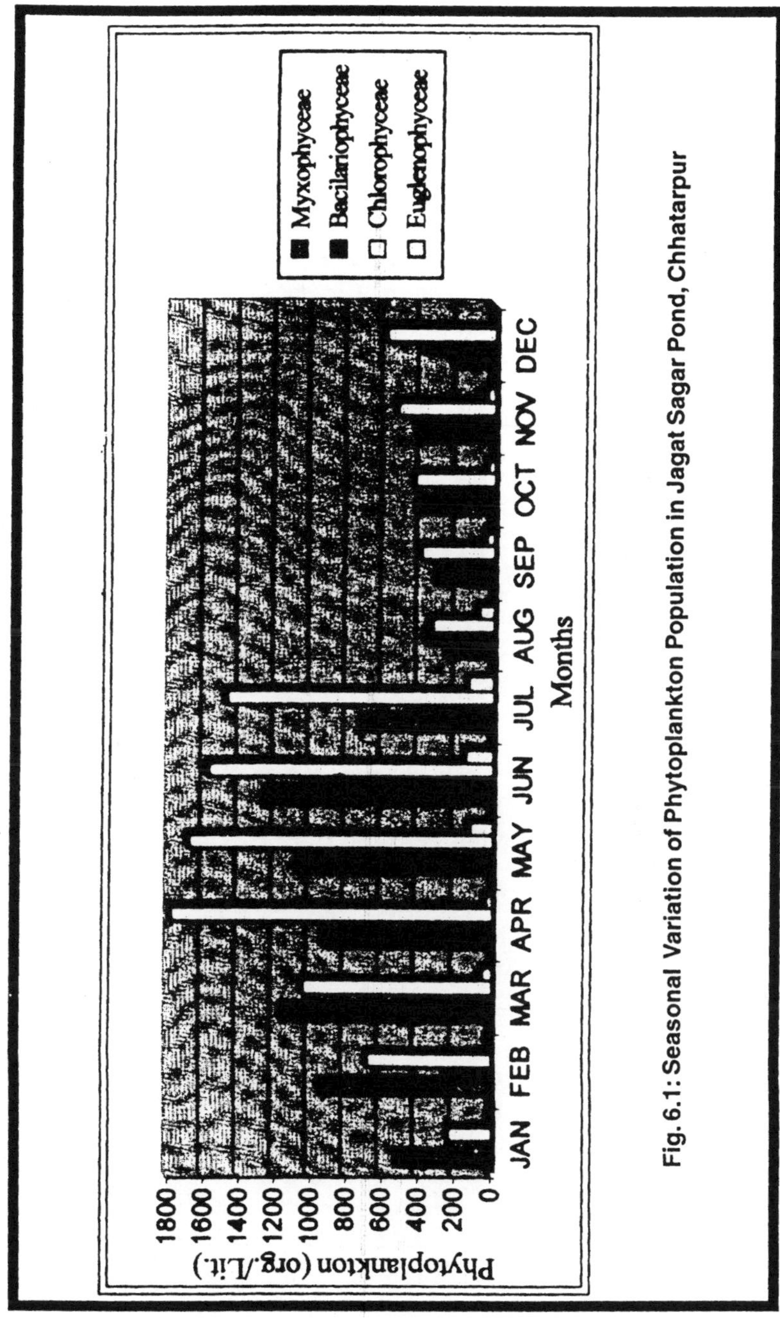

Fig. 6.1: Seasonal Variation of Phytoplankton Population in Jagat Sagar Pond, Chhatarpur

phytoplanktonic flora both in summer and winter months. Myxophyceae showed dominance in the month of February, March and June. The total phytoplankton showed three peaks in the month of April, May and June. The important genera of the phytoplankton recorded was *Eudorina* sp., *Pandorina* sp., *Volvox* sp., *Closterium* sp., *Micrasterias* sp., *Ulothrix* sp., *Zygnema* sp., *Spirogyra* sp., *Anabena* sp., *Microcystis* sp., *Melosira* sp. and *Navicola* sp. etc.

The important zooplankton recorded were–*Brachionus* sp., *Keratella* sp., *Daphnia* sp., *Moina* sp., *Cyclops* sp. and *Asplanchna* sp. etc. The total zooplankton showed a single peak in the month of April. The Rotifers as a group formed 43.52 per cent of the total zooplankton population, 29.66 per cent of crustacean and 26.82 per cent of protozoans. The maximum number of zooplankton (5576 organism/liter) was recorded in the month of April.

Discussion

The result of plankton population are shown in Tables 6.2 and 6.3 and Figures 6.1 and 6.2. The temperature was higher in summer, lower in winter and medium in rainy season, the range being between 14.8 to 37.9 °C. It has an indirect effect on the viscosity of water effecting the toxicity. Intensifying deoxygenation and finally increasing the biomagnification, that it why, the DO depletion and plankton community intensity their span in summer. So it is generally believed that temperature is the most important factor for planktonic growth and variation. But Chako *et al.* (1954) could not establish any correlation between temperature and planktonic growth and variation. The present investigation is similar with the findings of Rao *et al.*, 1996; Kumar, 1995 and Nasar, 1977. They one reported that only temperature is not a limiting factor for the growth of zooplankton through it plays an important role either directly or indirectly in their growth and production.

Secchi transparency was also found high during winter season (32.67 cm in January) at limnetic zones. During rainy, the sacchi transparency was minimum (13.10 cm in July). The range of variation during rainy season was less than that in winter which may be attributed to the constant disturbance caused due to frequent shower which present the solid from settling to the bottom. The present findings are confirmed by the observation of Singh and Sihna, 1995; Datta *et al.*, 1982, but differ with that of Micheal, 1966 and Agrawal *et al.*, 1995.

The value of phosphate recorded in the present investigation is 0.10 to 0.26 ppm. According to Jingran (1971), the phosphate content of more than 0.2 mg/liter may be considered as productive nature of water so the present investigation indicates the good productive nature of the pond.

In the present study, the major groups of phytoplanktonic species dominated as follows:

Chlorophyceae > Cyanophyceae > Bacillariophyceae > Euglenophyceae

The total phytoplanktonic population showed variation according to season as well as sites. The phytoplankton population had been observed to show an increasing trend in the months of winter season with peak during the summer month March to June. The minimum population had been noticed in the rainy season. Different species of phytoplankton and zooplankton showed marked variation in their seasonal population dynamics. Yet, phytoplankton are very poor than zooplankton which is clear cut sign of Eutrophication (Nirmal Kumar, 1997). Goldman and Horne (1983) states that the Myxophycean bloom in a pond is often the first clear-cut sign of cultural Eutrophication. They have very efficient mechanism for the uptake of nutrients at lower concentration.

The population of Rotifers were found higher from November to January and showed a single peak in April. This confirm the findings of summer periodicity of Rotifers as also observed by Malik and Bose, 1988; Sinha and Sinha, 1993 and Arvind Kumar, 1995. The Crustacean did not show any marked changes in the seasonal abundance. However Daphnia longispina was found abundance in March. The other crustacean showed throughout the study period in considerable quantities and its increase in number coincided with higher temperature (32.7 to 37.9°C).

The result concerning the distribution and abundance of plankton in Jagar Sagar pond shows in Tables 6.1 and 6.2 and Figures 6.1 and 6.2. Among the phytoplankton chlorophyceae and cyanophyceae were dominant in the population. The presence of certain planktonic genera like Cladophora, Pithophora, Spirogyra, Mougeotia, Pediastrum, Eudorina, Pandorina, Closterium, Volvox, Anabena, Microcystics as well as some diatoms were noticeable. The data indicates that the Jagat Sagar pond is tending towards eutrophication, and total plankton density showed marked and

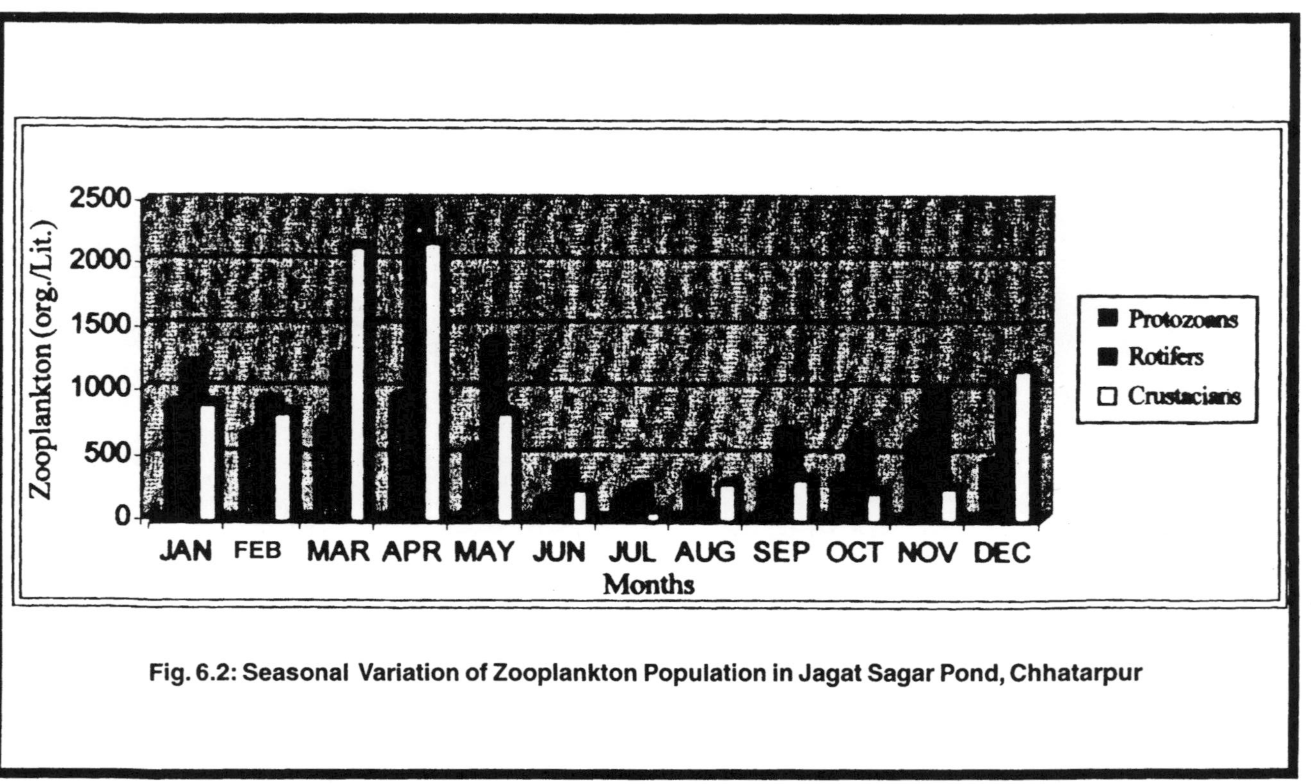

Fig. 6.2: Seasonal Variation of Zooplankton Population in Jagat Sagar Pond, Chhatarpur

significant correlation with water temperature, DO_2, DCO_2, Nitrate and Phosphate.

Acknowledgement

The author is grateful to the Prof. B.L. Vyas, Head, Department of Botany for providing necessary research facilities, encouragement and valuable suggestion.

References

Abbasi, S.A., Bhatia, K.K.S., Kunhi, A.V.M. and Soni, R. (1996). Studies on the Limnology of Kuttiadi lake (North Kerala). *Ecol Env. and Cons.*, **2**: 17-27.

Adoni, A.D. (1985). *Work Book on Limnology*. Pritibha Publication, Sagar (M.P.), India.

Agrawal, N.C., Bais, V.S. and Shukla, S.H. (1995). Contrast effects of temperature and redox potential on the biology of the Sagar lake. *Eco. Env. and Cons.* **1(1-4)**: 31-33.

APHA (1989). *Standard Methods for the Examination of Water and Wastewater, 14th edition*. American Public Health Association, Washington D.C.

Chacko, P.L. and Krishnamurthy J.B. (1954). On the plankton of three freshwater ponds in Madras City. *Ind. Symp. Mar. Freshwater Plankton*. Indo-Pacific Fish Coun., UNESCO.

Datta, N.C., Bandyopadhyay, B.K., Das, M.K. and Bandyopadhyay, S.B. (1982). Diurnal rythem of some physico-chemical properties and zooplankton in a tropical freshwater pond in Calcutta, West Bengal. *Ind., J. Phy. Nat. Sc.*, **2**: 22-27.

Goldman, C.R. and Home, A.J. (1983). *Limnology*. McGraw-Hill International Book Company, New York, p. 464.

Kumar Nirmal, J.I. (1997). Impact of pollution on aquatic biodiversity. *Ecol. Env. and Cons., 3(3-4)*: 209-217.

Kumar, Arvind (1995). Periodicity and abundance of plankton in relation to physico-chemical characteristics of a tropical wetland of South Bihar. *Eco. Env. and Cons.*, **1(1-4)**: 47-51.

Mallik, S. and Bose, S.K. (1988). Effect to some Physico-chemical factors on the rotifers population in a fresh water pond of Ranchi, Bihar, India. *The India. Zoologist*, **12(1&2)**: 21-23.

Michael, R.G. (1966). Diurnal variation in physico-chemical factors and zooplankton in the surface layer of three fresh water ponds. *Indian J. Fish.*, **13**: 48-82.

Nasar, S.A.K. (1977). Investigation on the seasonal periodicity of zooplankton in a freshwater pond in Bhagalpur. *Acta Hydrochem. Hydrobiol.*, **5(6)**: 577-584.

Khare, P.K. (1999). Phytoplankton as indicator of water quality and pollution status of Jagat Sagar Pond, Chhattarpur (M.P.). *Geobios. New Reports*, **18(2)**: 107-110.

Rao, M.M.A., Rao V.N. and Mahmood, S.K. (1996). Assessment of water quality and pollution of Narsingi pond. *Eco. Env. and Cons.*, **2**: 45-49.

Sinha, K.K. and Sinha, D.K. (1993). Seasonal trends in physico-chemical factors and zooplankton in a freshwater pond of Munger, Bihar. *J. Ecobiol.*, **5(4)**: 299-302.

Singh, Meena and Sinha, R.K. (1995). Diel variation of water quality and zooplankton community in a fresh water pond of Patna, Bihar. *Eco. Env. and Cons.*, **1(1-4)**: 57-64.

Verma, P.K. and Dattamunshi, J.S. (1987). Plankton community structure of Banda reservoir, Bhagalpur. *Tropic. Ecol.*, **28**: 200-207.

7

Benthic Food Web Patterns of Tropical Freshwater Ponds

✰ *Mrinal Kanti Ghosh & Samir Banerjee*

Introduction

Food webs summarize predator-prey relationships in ecological communities. Analysis of food web structure and its temporal dynamics is essential for understanding energy flow and population dynamics of species and may contribute to other ecological processes. Conclusions from food web analysis have been much discussed (May, 1983a; Strong, 1988; Lawton, 1989; Martinez, 1991) and structural food web studies represent an active area of theoretical ecology. A fascinating aspect of this work is the capability to compare quantitatively whole communities and ecosystems (Cohen, 1989; Martinez, 1991). Such comparisons may delineate constraints and similarities in all ecosystems (Pimm, 1982; Briand and Cohen, 1984; Cohen and Briand, 1984; Cohen and Newman, 1985; Cohen *et al.*, 1985) and also in certain aspects of ecosystems (Briand, 1983a,b, 1985; Briand and Cohen, 1987). Several food web features were established during 1980's and onwards (Pimm, 1982; Briand and Cohen *et al.*, 1986, 1990), which have been criticized due to the lack of several methodological procedures (Paine, 1988; Schoener, 1989;

Winemiller, 1989; Polis, 1991). Martinez (1992) included two comprehensive and evenly resolved webs in his set of data. In Havens (1992), webs have no detritus and very few links between the top and basal species, which is severely unrealistic (Deb, 1995).

To overcome these shortcomings, the existing trophic links between all trophic levels (including detritus) are required to be estimated to the best possible extent (Deb, 1995). So trophic interactions of benthic communities of two freshwater ponds are described in detail and the relationship between community size has also been studied. In the present study, food-webs were graphically constructed on the basis of established literature (Titmus and Badcock, 1981; Chattopadhyay and Dutta, 1987–88; Alfred, 1974; Jhingran, 1983; Sprules and Bowerman, 1988; Hartig *et al.*, 1962) and statistical procedures were done to test the proposed scaling laws (Briand and Cohen, 1984; Cohen and Briand, 1984; Cohen and Newman, 1985; Martinez, 1992) naming:

1. "Species scaling law" is that the basal, intermediate and top fractions of species do not vary with the total number of species (S) in the web.
2. "Link scaling law" proposes that the fractions of top-intermediate (T–I), top-basal (T–B), intermediate-intermediate (I–I) and intermediate-basal (I–B) links do not vary with S.
3. "Link-species scaling law" proposes that the total number of links (L) is proportional to S and that linkage density (d = L/S) does not vary with S (Briand and Cohen, 1984; Cohen and Briand, 1984; Cohen and Newman, 1985).
4. Constant connectance hypothesis (Martinez, 1992), and predator–prey relationship in a benthic community of both the managed and unmanaged pond.

Materials and Methds

Study Site and Pond Preparation

Two ponds were selected for study at Sonarpur (22.26°N–88.25°E), West Bengal, India. One of the ponds was scientifically managed and maintained for pisciculture, and is referred to herein as the "managed pond." The other pond, was an uncared one, treated as the 'unmanaged pond'. The managed pond was prepared for pisciculture in the year 1985.

Before stocking, the managed pond was treated with mahua oil cake (MOC), prepared from the saponin–containing bark of a plant mahua (*Bassia latifolia* Roxb.) at a rate of 2500 kg/ha to eradicate 'weed' fish species on 12th October, 1985. The known lethal dose of MOC is 250 ppm (Homechaudhury *et al.*, 1986). The pond was then treated with lime at a rate of 1500 kg/ha at two installments on 19th October, 1985 and 20th October, 1985 for alkalization. After 15 days of application of MOC, mustard oil cake and cow dung were administered to the pond at a rate of 3250 kg/ha and 3500 kg/ha at two installments respectively for manuring (Jhingran, 1983) on 28th October, 1985. The fry (< 20 mm) of Indian major carps (IMC), *viz.*, *labeo rohita* Ham., *Catla catla* Ham., and *Cirrhinus mrigala* Ham. in the ratio of 3 : 4 : 3 were stocked at a density of 2,50,000 individuals/ ha on 5th November, 1985 for rearing. The fishes were fed according to the standard feeding schedule (Sinha, 1979) for better growth of fishes. Frequent nettings (once in a month) were done to stir the soil-water interface and to run the cultured fishes.

The unmanaged pond was not treated with any material and the fry of IMC were administered at the same stocking density as in the managed pond.

Sample Collection

As biological interactions in food webs change over seasonal cycles, a series of time slices is needed to make a picture of the seasonal dynamics of the community (Winemiller, 1990; Schoenly and Cohen, 1991).

Benthic organisms were sampled fortnightly at around 08.00 hrs over a period of two years in both the ponds. The phytoplankton and zooplanktonic organisms were sampled over a period of 18 months in the case of unmanaged pond and of 24 months in case of managed pond.

Planktons were collected fortnightly by filtering 100 litres of random samples of surface water using a plankton net with a mesh of 76 μm. The samples were preserved in 4 per cent formalin and counted in triplicates on a Sedgewick-Rafter chamber. Macrozoobenthic organisms were also collected fortnightly by Ekman Grab (measuring 22.9 cm × 22.9 cm) at eight random sites and were mixed thoroughly in a large bucket. The samples were then seived through standard seive (0.595 mm). The macrozoobenthic specimens of each collection were fixed in 4 per cent formalin, indentified and counted manually. The fishes were netted out from the ponds for measuring their size. Each of such collections was treated as a time slice of the system.

Food Web Pattern

Dietary interaction between each pair of species was determined from the literature. Gut content analyses has been avoided here as because the links between species reported in the literature would hold true for the studied species in the pond communities.

Food webs were graphically constructed using the presence absence data for species in the ponds. In constructing the webs the following considerations were undertaken:

1. All the observed algal species, the protozoans (which are primary producers) like Euglenoids and detritus are taken as basal species. Basal species are considered here individually and separately. Detritus is also taken as a separate species.
2. Among the zooplankters, mesocyclops species feed on their own and on each other nauplii (Sprules and Bowerman, 1988) and also feed on the fry of *Cirrhinus mrigala* (Hartig *et al.*, 1962). Among the herbivore rotifers only the species of *Brachionus* are considered here which are fed upon by the *Chironomus* species (Alfred, 1974).
3. Chironomus larvae feed chiefly on detritus (Titmus and Badcock, 1981), algal species and protozoa (Chattopadhyay and Dutta, 1987–88) and species of *Brachionus* (Alfred, 1974). The larvae are fed upon by the *Cirrhinus mrigala* (Jhingran, 1983).
4. *Tanypus* larvae feed on algae, detritus, larval chironomids, oligochaetes, nematods (Chattopadhyay and Dutta, 1987–88) and are fed upon by *Cirrhinus mrigala* (Jhingran, 1983).
5. Oligochaeta species and larvae of *Culicoides* sp. and *Hydrometra* sp. are primarily detritivore and are fed upon by fingerlings and adults of *Cirrhinus mrigala*.
6. Molluscan species are mainly herbivorous and detritivorous because the gut contents of the species contain the basal species referred to here.

7. The fry of *Cirrhinus mrigala* feeds chiefly on zooplankters (here only species of *Brachionus* are considered), but after attaining fingerling size, feeds on detritus and small benthic organisms (Jhingran, 1983). There was no interspecific competition between the cultured Indian major craps (IMC) because of their different food niches. Here only the benthivore and detritivore species, namely *Cirrhinus mrigala*, are taken into consideration because the macrozoobenthic organisms are mainly emphasized here. The phytoplankton and zooplankton population are also considered here as they have a direct or indirect influence on benthic organisms as well as on fish species.

Each food web was analysed and parameters of food web structure were determined following Pimm (1982), Cohen (1978), Schoener (1989), Sprules and Bowerman (1988), Martinez (1991, 1992) and Havens (1991):

Species Richness

Total number of species including phytoplankton, and detritus also considered as a species.

Modal Number of Trophic Levels (Cohen's Maximum Chain Length)

Trophic species refer to clusters containing all taxa with the exact same set of predators and prey (Briand and Cohen, 1984). All possible pathways except cannibalism were traced between a given 'basal' [having no trophic species as prey (Pimm, 1982; Pimm *et al.*, 1991; Briand and Cohen, 1984)], a given "top" [having no trophic species as predator (Briand and Cohen, 1984)] and a given "intermediate" [having both predators and prey (Briand and Cohen, 1984)] trophic species. The length of each path was recorded as the number of links plus one. The modal length of the paths is the modal number of trophic levels.

Total Number of Predator-Prey Interactions

These interactions do not included cannibalism and counted each cycle (species A preys on B and vice versa) as a single interaction.

Food Chains and Chain Length

Food chains are directional paths of trophic energy or equivalently sequences of links that start with basal species such as producers or detritus materials (fine organic matter) and end with consumer organisms. Links are trophic interactions directed from prey to predator. Chain length is measured by the number of links in the sequence from producers or fine organic matter to a consumer. A web's average chain length is the average number of links in every chain connecting all consumers to basal species.

Fractions of Food Links

Food link fractions were analysed between top and intermediate species (T–I link fractions), top and basal species (T–B link fractions), intermediate and intermediate species (I–I link fractions), and intermediate and basal species (I–B link fractions).

Number of Interaction Per Species

Total number of interactions divided by the species richness.

Predator-Prey Ratio

The sum of top and intermediate species divided by the sum of intermediate and basal species.

Linkage Density (d)

It is the number of trophic linkages (L) per trophic species (S).

To make a simple relationship between L (trophic linkages) and S (species richness) an equation was used in the form of $L = \alpha S^{\beta}$ (where α and β are constants). A log–log regression of the food webs of both the managed and unmanaged pond estimated the value of β.

Results

A food web statistics concerning macrozoobenthos, zooplankton, phytoplankton, detritus and a detritivore fish of managed and unmanaged pond are summarized in Tables 7.1 and 7.2. In the managed pond the range of the number of species (S) in the 50 different food webs is from 8 to 19 and the maximum and minimum length of food chains (cl_{max} and cl_{min}) are 4 and 1 respectively, whereas in the unmanaged pond, the range of the number of species (S) in the 36 different food webs is from 6 to 20, and the maximum and minimum length of food chains (cl_{max} and cl_{min}) are 3 and 1 respectively (Tables 7.3 and 7.4). The mean value of chain length in the managed and unmanaged pond is 3.22 and 2.28 respectively. From Tables 7.1 and 7.2 it has been found that the ranges of total linkages are in between 7 and 52 in the managed pond and in between 1 and 79 in the unmanaged pond. The ranges of link fractions between top-basal (T–B), intermediate-basal (I–B), top-intermediate (T–I), intermediate-intermediate (I–I) are 0.019–1.0, 0.303–0.796, 0.105–0.333, 0.048–0.2581 respectively in the managed pond and those are 0.0556–1.0, 0.1136–0.75, 0.0227–0.2727, 0.188–0.0862 respectively in the unmanaged pond. Tables 7.1 and 7.2 summarized the species fractions of top (TF), intermediate (IF) and basal (BF) of the managed and unmanaged pond. The top, intermediate and basal species fraction varies from 0.048 to 0.6769, 0.167 to 0.7142, 0.211 to 0.875 respectively in the managed pond and from 0.091 to 0.6111, 0.0625 to 0.4211, 0.211 to 0.8571 respectively in the unmanaged pond. Figure 7.7 shows that in both the ponds the predator/prey ratio increases with the increase of species richness (S). The mean of predator/prey ratio is 0.74 in the managed pond and that is 1.015 in the unmanaged pond (Table 7.5).

A log-log regression of the 50 food webs in the managed pond estimate $\beta = 1.93$ ($\alpha = 0.184$; $\varepsilon = 0.93$) and in the unmanaged pond with 36 food webs $\beta = 2.67$ ($\alpha = 0.027$; $\varepsilon = 1.67$) (Table 7.5). Figure 7.5 shows that in both the ponds there is an exponential growth curve between L and S following the equation $L = 0.184\ S^{1.93}$ for the managed pond and $L = 0.027\ S^{2.67}$ for the unmanaged pond. The linkage density (d) is calculated here by using the formula ($d = L/S$) and plotted it against S in both the ponds. Figure 7.6 shows that linkage density (d) increases with the increasing of S following the linear regression equation, $d = 0.265 + 0.135\ S$ in the managed pond and $d = -0.564 + 0.207\ S$ in the unmanaged pond. Linkage density shows highly significant positive correlation with S in both the ponds ($r = 0.84$, $t = 10.73$, $p < 0.001$ in the managed pond $r = 0.91$, $t = 12.80$, $p < 0.0001$ in the unmanaged pond). In the managed pond, the fraction of basal species (BF) decreases with the increasing of S (slope $= -0.025$; SE of slope $= 0.101$; $R^2 = 0.37$, $p < 0.001$), whereas, the fraction of top species (TF) (slope $= 0.003$; SE of slope $= 0.16$; $R^2 = 0.003$) and intermediate species (IF) (slope $= 0.026$; SE of slope $= 0.145$; $R^2 = 0.24$; $p < 0.001$) increase with the S. The plots of these relationships for 50 webs are given in Figure 7.1. Same patterns are also found in the unmanaged pond (*i.e.* slope $= 0.037$; SE of slope $= 0.08$; $R^2 = 0.8$; $p < 0.0001$) for basal fraction (BF), slope $= 0.018$; SE of slope $= 0.135$, $R^2 = 0.31$, $p < 0.001$ for intermediate fraction (IF) and the relationships for 36 webs are given in Figure 7.2.

Table 7.1: Benthic Food Web Statistics of the Managed Pond

	1985						1986	
Date	*Oct.*		*Nov.*		*Dec.*		*Jan.*	
	10	*30*	*15*	*29*	*12*	*27*	*10*	*27*
Total Link (L)	7	7	34	52	44	33	21	28
Total Species (S)	8	8	15	18	18	15	12	15
Link Fraction								
TB	1.0	1.0	0.2647	0.019	0.2045	0.2730	0.429	0.286
TI	0	0	0.147	0.231	0.3182	0.333	0.190	0.286
IB	0	0	0.5882	0.5	0.3864	0.303	0.380	0.429
II	0	0	0	0.192	0.091	0.091	0	0
Species Fraction								
TF	0.125	0.125	0.333	0.333	0.389	0.533	0.417	0.267
IF	0	0	0.333	0.389	0.333	0.20	0.167	0.4
BF	0.875	0.875	0.333	0.278	0.278	0.267	0.417	0.333
Linkage density (d)	0.875	0.875	2.27	2.89	2.44	2.2	1.75	1.87
Predator/Prey	0.143	0.143	1.0	1.08	1.18	1.57	1.0	0.91

Contd...

Table 7.1–Contd...

	1986									
	Feb.		*Mar.*		*Apr.*		*May*		*June*	
	10	*25*	*10*	*24*	*8*	*21*	*5*	*20*	*5*	*20*
L	24	22	21	27	25	30	50	57	45	38
S	12	11	12	12	13	15	18	19	17	15
TB	0.25	0.045	0.048	0.037	0.04	0.033	0.02	0.175	0.022	0.026
TI	0.292	0.273	0.238	0.222	0.2	0.233	0.16	0.14	0.18	0.184
IB	0.458	0.591	0.667	0.667	0.68	0.633	0.64	0.667	0.578	0.579
II	0	0.091	0.048	0.074	0.08	0.1	0.18	0.175	0.222	0.211
TF	0.167	0.091	0.083	0.048	0.077	0.0667	0.0556	0.0526	0.0588	0.0667
IF	0.417	0.545	0.417	0.5	0.538	0.667	0.667	0.632	0.706	0.667
BF	0.417	0.364	0.5	0.417	0.384	0.267	0.2778	0.316	0.235	0.267
d	2.0	2.0	1.75	2.25	1.92	2.0	2.78	3.0	2.65	2.53
P/P	0.7	0.7	0.55	0.60	0.67	0.79	0.76	0.72	0.81	0.79

Contd...

Table 7.1–Contd...

1986

	July		*Aug.*		*Sep.*		*Oct.*		*Nov.*	
	7	*22*	*5*	*20*	*5*	*20*	*6*	*30*	*15*	*30*
L	33	30	36	49	41	49	36	43	32	23
S	14	13	15	19	18	17	16	15	13	13
TB	0.0303	0.0333	0.028	0.0204	0.0244	0.0204	0.0278	0.0233	0.0313	0.0434
TI	0.2121	0.267	0.222	0.1429	0.195	0.143	0.1667	0.1163	0.1563	0.3043
IB	0.576	0.467	0.556	0.7143	0.683	0.633	0.7222	0.6744	0.656	0.565
II	0.182	0.233	0.194	0.1224	0.0976	0.204	0.083	0.186	0.1563	0.0869
TF	0.0714	0.0769	0.0667	0.0526	0.0556	0.0589	0.0625	0.0667	0.0769	0.6769
IF	0.5714	0.692	0.6	0.6842	0.6667	0.647	0.563	0.6	0.5385	0.692
BF	0.3571	0.2307	0.333	0.211	0.2778	0.294	0.375	0.33	0.385	0.2307
d	2.36	2.31	2.4	2.58	2.28	2.88	2.25	2.87	2.46	1.77
P/P	0.69	0.83	0.71	0.82	0.76	0.75	0.67	0.72	0.67	1.48

Contd...

Table 7.1–Contd...

	1986		1987							
	Dec.		*Jan.*		*Feb.*		*Mar.*		*Apr.*	
	15	*30*	*13*	*28*	*8*	*23*	*8*	*23*	*6*	*21*
L	21	20	35	30	14	26	24	26	19	37
S	11	11	16	15	09	14	14	14	10	15
TB	0.0476	0.05	0.029	0.0333	0.0714	0.0385	0.0416	0.0385	0.053	0.027
TI	0.143	0.2	0.229	0.267	0.2143	0.269	0.2916	0.269	0.105	0.1892
IB	0.714	0.65	0.686	0.667	0.6428	0.6154	0.625	0.6154	0.737	0.5946
II	0.095	0.1	0.057	0.0333	0.0714	0.0769	0.0416	0.0769	0.105	0.1892
TF	0.091	0.091	0.625	0.0667	0.1111	0.0714	0.0714	0.0714	0.1	0.0667
IF	0.455	0.5454	0.625	0.6	0.444	0.6428	0.5714	0.6428	0.4	0.533
BF	0.455	0.3636	0.3125	0.33	0.444	0.2857	0.3571	0.2857	0.5	0.4
d	1.91	1.82	2.19	2.00	1.56	1.86	1.71	1.86	1.90	2.47
P/P	0.60	0.70	0.73	0.72	0.63	0.77	0.69	0.77	0.56	0.64

Contd...

Table 7.1–Contd...

1987

	May		*June*		*July*		*Aug.*		*Sep.*		*Oct.*	
	7	*22*	*6*	*20*	*10*	*26*	*8*	*23*	*5*	*18*	*3*	*18*
L	26	18	13	22	33	41	54	45	52	55	34	49
S	13	10	8	12	14	16	18	17	19	18	16	18
TB	0.0385	0.0556	0.0769	0.0455	0.0303	0.0244	0.0185	0.0222	0.0192	0.0182	0.0294	0.0204
TI	0.1923	0.1667	0.154	0.227	0.2424	0.195	0.1296	0.111	0.1154	0.1273	0.206	0.1837
IB	0.654	0.6667	0.6154	0.636	0.485	0.585	0.685	0.777	0.7885	0.673	0.6765	0.592
II	0.1154	0.111	0.154	0.091	0.2424	0.195	0.167	0.0888	0.0769	0.182	0.088	0.2041
TF	0.077	0.10	0.125	0.0833	0.0714	0.0625	0.0667	0.0588	0.0526	0.0667	0.0625	0.0667
IF	0.6154	0.50	0.50	0.5833	0.7142	0.625	0.6111	0.529	0.5263	0.6111	0.625	0.6666
BF	0.3077	0.40	0.375	0.333	0.2143	0.3125	0.333	0.412	0.421	0.333	0.3125	0.2777
d	2.0	1.80	1.63	1.83	2.36	2.56	3.00	2.65	2.74	3.06	2.13	2.72
P/P	0.75	0.67	0.71	0.73	0.85	0.73	0.72	0.62	0.61	0.72	0.73	0.78

Table 7.2: Benthic Food Web Statistics of the Unmanaged Pond

	1985				1986	
Date	*Nov.*		*Dec.*		*Jan.*	
	16	*30*	*14*	*28*	*12*	*28*
Total Link (L)	44	40	46	35	42	73
Total Speies (S)	16	18	17	16	16	20
Link Fraction						
TB	0.8636	0.675	0.6956	0.6286	0.5952	0.6849
TI	0.0227	0.10	0.0652	0.1143	0.1666	0.0959
IB	0.1136	0.175	0.1956	0.2	0.2381	0.2192
II	0	0.05	0.0435	0.0571	0	0
Species Fraction						
TF	0.5625	0.6111	0.5294	0.5625	0.5	0.5
IF	0.0625	0.1667	0.1765	0.1875	0.25	0.2
BF	0.375	0.222	0.2941	0.25	0.25	0.3
Linkage Density (d)	2.75	2.22	2.71	2.19	2.63	3.65
Predator/Prey	1.43	2.00	1.50	1.71	1.50	1.40

Contd...

Table 7.2–Contd...

1986

	Feb.		*Mar.*		*Apr.*		*May*		*Jun.*	
	11	*26*	*11*	*25*	*9*	*22*	*6*	*21*	*6*	*21*
L	16	20	16	14	15	16	16	13	17	19
S	47	79	53	12	37	52	48	34	49	49
TB	0.8085	0.5443	0.6981	0.6875	0.7838	0.8269	0.7708	0.5588	0.5702	0.5306
TI	0.0638	0.0759	0.0566	0.0938	0.0811	0.0385	0.0625	0.0588	0.0612	0.1429
IB	0.1277	0.3418	0.2264	0.1875	0.1351	0.1346	0.1667	0.3529	0.3673	0.3061
II	0	0.0379	0.0188	0.0313	0	0	0	0.0294	0.0612	0.0204
PF	0.5	0.35	0.4375	0.5714	0.533	0.5	0.4375	0.3077	0.294	0.316
IF	0.125	0.3	0.1875	0.2143	0.2	0.125	0.1875	0.2308	0.3529	0.4211
BF	0.375	0.35	0.375	0.2143	0.267	0.375	0.375	0.4615	0.3529	0.211
d	2.94	3.95	3.31	2.29	2.47	3.25	3.0	2.62	2.88	2.58
P/P	1.25	1.00	1.43	1.83	1.57	1.25	1.11	0.78	0.92	1.08

Contd...

Table 7.2–Contd...

1986

	July		*Aug.*		*Sep.*		*Oct.*		*Nov.*	
	8	*21*	*6*	*21*	*6*	*21*	*7*	*29*	*16*	*29*
L	18	14	17	19	17	9	11	11	10	6
S	52	36	58	55	53	11	18	18	16	5
TB	0.6923	0.5833	0.6379	0.4727	0.5849	0.091	0.0556	0.0556	0.0625	1.0
TI	0.0962	0.0833	0.0517	0.0909	0.0566	0.2727	0.1667	0.1667	0.125	0
IB	0.2115	0.3055	0.2241	0.3818	0.2830	0.6364	0.7222	0.7222	0.75	0
II	0	0.0277	0.0862	0.0545	0.0377	0	0.0556	0.0556	0.0625	0
TF	0.333	0.3571	0.4118	0.316	0.4118	0.111	0.091	0.091	0.10	0.333
IF	0.278	0.2857	0.235	0.4211	0.294	0.333	0.364	0.364	0.30	0
BF	0.389	0.3571	0.3529	0.211	0.294	0.5555	0.545	0.545	0.60	0.6666
d	2.89	2.57	3.41	2.89	3.12	1.22	1.64	1.64	1.60	0.83
P/P	0.92	1.00	1.10	1.08	1.20	0.50	0.50	0.50	0.44	0.50

Contd...

Table 7.2–Contd...

	1986		1987							
	Dec.		*Jan.*		*Feb.*		*Mar.*		*Apr.*	
	16	*31*	*15*	*30*	*14*	*27*	*9*	*24*	*7*	*22*
L	7	7	15	13	12	7	8	7	16	11
S	5	5	42	37	30	5	6	1	44	25
TB	1.0	1.0	0.7381	1.0	1.0	1.0	1.0	1.0	0.5682	1.0
TI	0	0	0.0714	0	0	0	0	0	0.0909	0
IB	0	0	0.1905	0	0	0	0	0	0.3182	0
II	0	0	0	0	0	0	0	0	0.0625	0
TF	0.2857	0.2857	0.4	0.538	0.5833	0.2857	0.25	0.1429	0.3125	0.5454
IF	0	0	0.2	0	0	0	0	0	0.3125	0
BF	0.7143	0.7143	0.4	0.4615	0.4166	0.7143	0.75	0.8571	0.375	0.4545
d	0.714	0.714	2.80	2.85	2.50	0.714	0.75	0.143	2.75	2.27
P/P	0.40	0.40	1.00	1.17	1.40	0.40	0.33	0.17	0.91	1.20

Table 7.3: Chain Length (Minimum and Maximum) of Benthic Food Web in the Managed Pond

Date	*C.I. Min.*	*C.I. Max.*	*S*
10-10-85	1	1	8
30-10-85	1	1	8
15-11-85	1	2	15
29-11-85	1	3	18
12-12-85	1	3	17
27-12-85	1	3	16
10-01-86	1	2	12
27-01-86	1	2	15
10-02-86	1	2	12
25-02-86	1	3	11
10-03-86	1	3	12
24-03-86	1	3	12
08-04-86	1	3	13
21-04-86	1	3	15
05-05-86	1	4	18
20-05-86	1	4	19
05-06-86	1	4	17
20-06-86	1	4	15
07-07-86	1	4	14
22-07-86	1	4	13
05-08-86	1	4	15
20-08-86	1	3	19
05-09-86	1	3	18
20-09-86	1	4	17
06-10-86	1	3	16
30-10-86	1	4	15
15-11-86	1	4	13
30-11-86	1	3	13
15-12-86	1	3	11
30-12-86	1	3	11
13-01-87	1	4	16

Contd...

Table 7.3–Contd...

Date	C.I. Min.	C.I. Max.	S
28-01-87	1	4	15
08-02-87	1	3	9
23-02-87	1	3	14
08-03-87	1	3	14
23-03-87	1	3	14
06-04-87	1	3	10
21-04-87	1	4	15
07-05-87	1	3	13
22-05-87	1	3	11
06-06-87	1	3	8
20-06-87	1	3	12
10-07-87	1	4	14
26-07-87	1	4	16
08-08-87	1	4	18
23-08-87	1	3	17
05-09-87	1	3	19
18-09-87	1	4	18
03-10-87	1	3	16
18-10-87	1	4	18

CI Max. = Maximum Chain Length; CI Min. = Minimum Chain Length.

Table 7.4: Chain Length (Minimum and Maximum) of Benthic Food Web in the Unmanaged Pond

Date	C.I. Min.	C.I. Max.	S
16-11-85	1	2	18
30-11-85	1	3	18
14-12-85	1	3	16
28-12-85	1	3	16
12-01-86	1	2	16
28-01-86	1	2	20
11-02-86	1	2	16

Contd...

Table 7.4–Contd...

Date	*C.I. Min.*	*C.I. Max.*	*S*
26-02-86	1	3	20
11-03-86	1	3	16
25-03-86	1	3	14
09-04-86	1	2	15
22-04-86	1	2	16
06-05-86	1	2	16
21-05-86	1	3	13
06-06-86	1	3	17
21-06-86	1	3	19
08-07-86	1	2	18
21-07-86	1	3	14
06-08-86	1	3	17
21-08-86	1	3	19
06-09-86	1	3	17
21-09-86	1	2	9
07-10-86	1	3	11
29-10-86	1	3	11
16-11-86	1	3	10
29-11-86	1	1	6
16-12-86	1	1	7
31-12-86	1	1	7
15-01-87	1	2	15
30-01-87	1	1	13
14-02-87	1	1	12
27-02-87	1	1	7
9-03-87	1	1	8
24-03-87	1	1	7
07-04-87	1	3	16
22-04-87	1	2	11

CI Max. = Maximum Chain Length; CI Min. = Minimum Chain Length.

Table 7.5: Statistical Values of Zoobenthic Food Webs in the Managed and Unmanaged Pond

	Managed Pond	*Unmanaged Pond*
Range of S	8–19	6–20
Cl Max.	4	3
Cl Min.	1	1
ε	0.93	1.67
k	0.18	0.027
Mean of Predator/Prey	0.74	1.015
Mean of Chain Length	3.22	2.28
Slope of TF	0.003	0.018
Slope of IF	0.026	0.018
Slope of BF	– 0.025	– 0.037
Slope of TB	– 0.031	– 0.014
Slope of IB	0.019	0.008
Slope of TI	0.003	0.004
Slope of II	0.009	0.002

TF = Top species fraction;

IF = Intermediate species fraction;

BF = Basal species fraction;

TB = Link between top species and basal species;

IB = Link between intermediate species and basal species;

TI = Link between top species and intermediate species;

II = Link between intermediate species and intermediate species;

Cl Max. = Maximum Chain Length;

Cl Min. = Minimum Chain Length;

ε = Exponent;

k = Constant.

The link fractions between top to basal species (T–B), top to intermediate species (T–I), intermediate to intermediate (I–I) and intermediate to basal (I–B) species of the managed and unmanaged

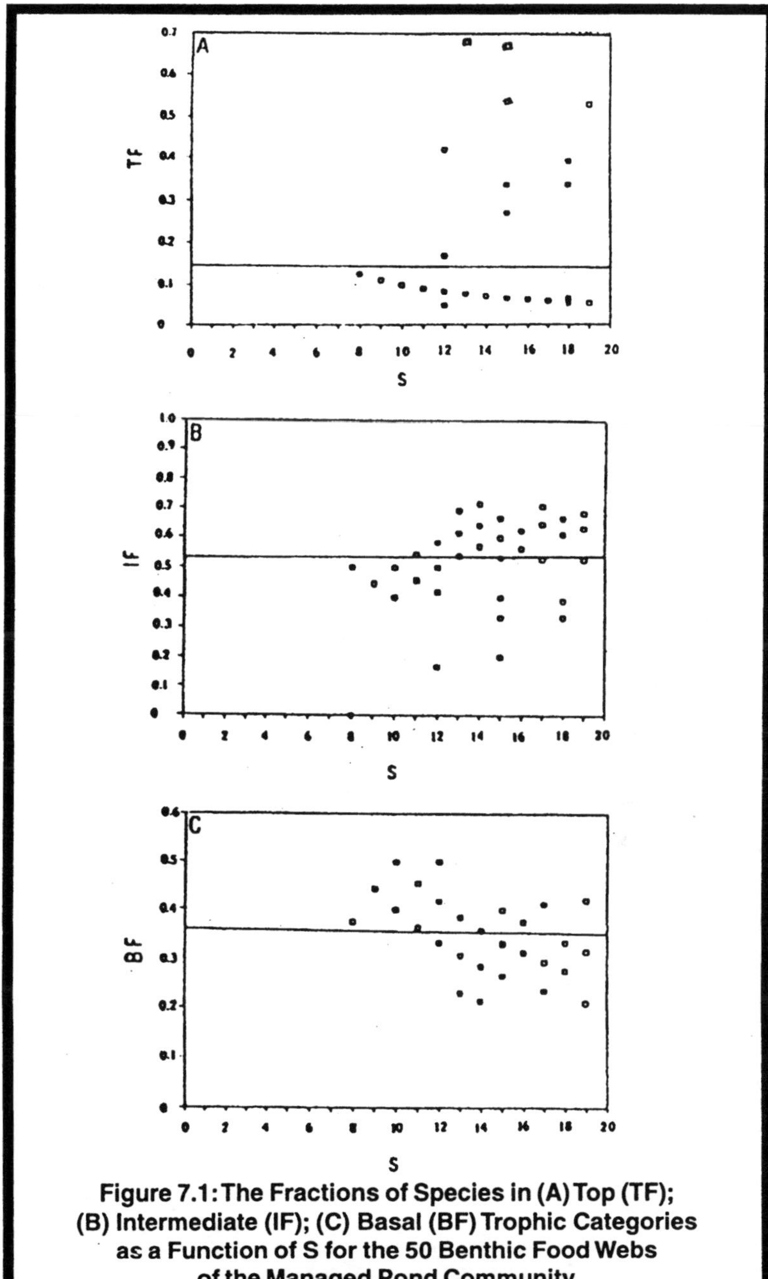

Figure 7.1: The Fractions of Species in (A) Top (TF); (B) Intermediate (IF); (C) Basal (BF) Trophic Categories as a Function of S for the 50 Benthic Food Webs of the Managed Pond Community

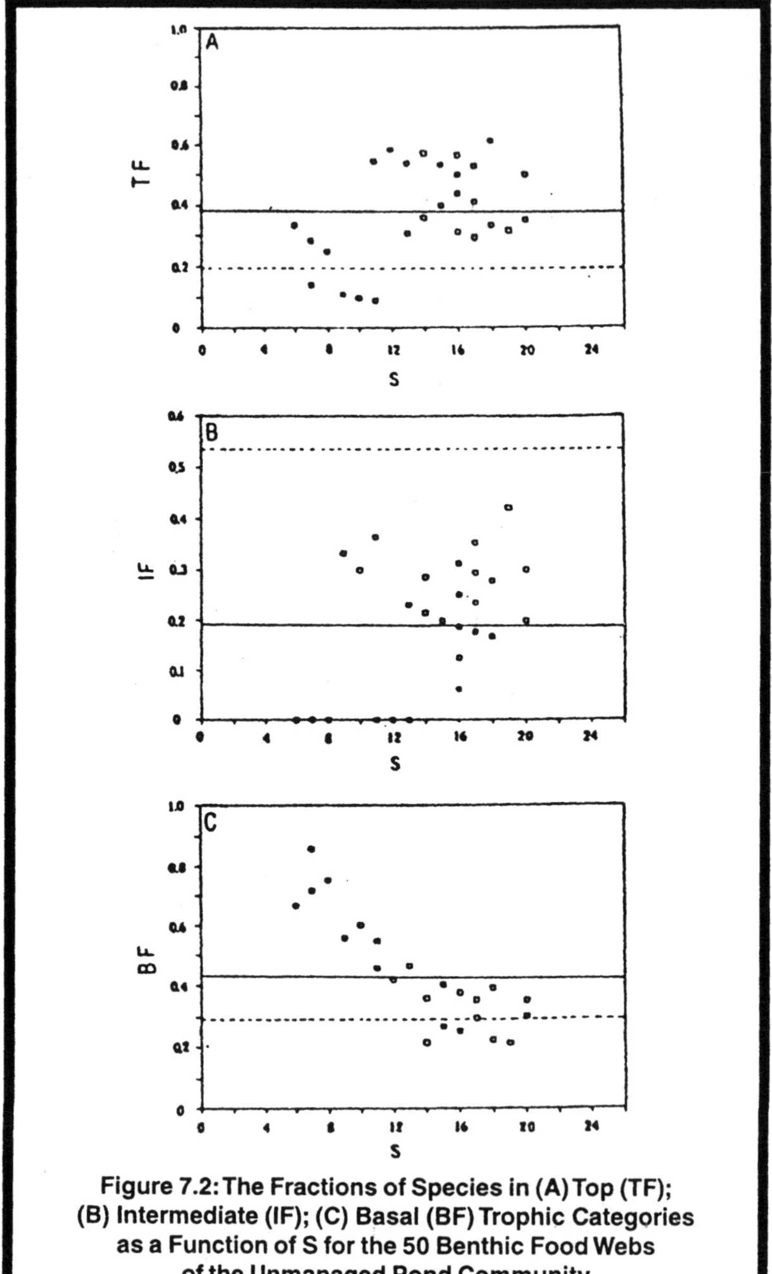

Figure 7.2: The Fractions of Species in (A) Top (TF); (B) Intermediate (IF); (C) Basal (BF) Trophic Categories as a Function of S for the 50 Benthic Food Webs of the Unmanaged Pond Community

are summarised in Tables 7.1 and 7.2. These linkage criteria are plotted against the food web size (S). Top to basal species linkage (T–B) criterion decreases with the increase of S, where as the other linkages (T–I, I–B, I–I) criteria increase with the increase of S in both the managed and unmanged pond (Figures 7.3 and 7.4).

In the webs of the managed pond, the fractions of top, intermediate and basal species are not scale invariant with means of 0.146, 0.528 and 0.358 respectively and those of unmanaged pond are 0.38, 0.19, 0.43 respectively. The ranges of these fractions are affected by linkage criteria as well as by the number of species in the web. Food web structure for different samplings of both the managed and unmanaged pond are designed (Figures 7.8 and 7.9) and some of the webs of different patterns are shown (Figures 7.10 and 7.11).

Discussion

Natural systems are dynamic and dietary links continuously change with time, environmental fluctuations and physiology of organisms. This implies that, focus on dynamic rather than static properties of food web would constitute a primary desideratum (Paine, 1988).

In the present study, trophic links (L) are directly calculated instead of estimating connectance (C) values as these linkages (L) are more precise than connectance measurement (Briand, 1985; Martinez, 1991). Several authors contend that L is generally underrepresented in webs with large S because only the most important links are included in large webs (Cohen and Briand, 1984; Cohen and Newman, 1988; Paine, 1988; Martinez, 1991). A simple relationship in the form of $L = \alpha S^{\beta}$ has been calculated and it has been shown that L increases exponentially with S (Figure 7.5). The value of true exponent (β) is equal to 1.85, which is much closer to two than to one in the managed pond (50 seasonal) webs. This results supports the constant connectance hypothesis (Martinez, 1992), which states that the links in a web increase approximately as the square of S, but not the link-species scaling law (Cohen and Newman, 1985). In the unmanaged pond, b is equal to 2.67 with 36 seasonal webs containing 7–20 species which is closer to the Martinez's (1992) analysis of 12 high resolution webs ($\beta = 2.18$). The low value of β according to link-species scaling law is due to "artistic convenience" (Paine, 1988), *i.e.* L is underestimated due to the difficulty of graphically recording L in large webs. However, in this

Figure 7.3

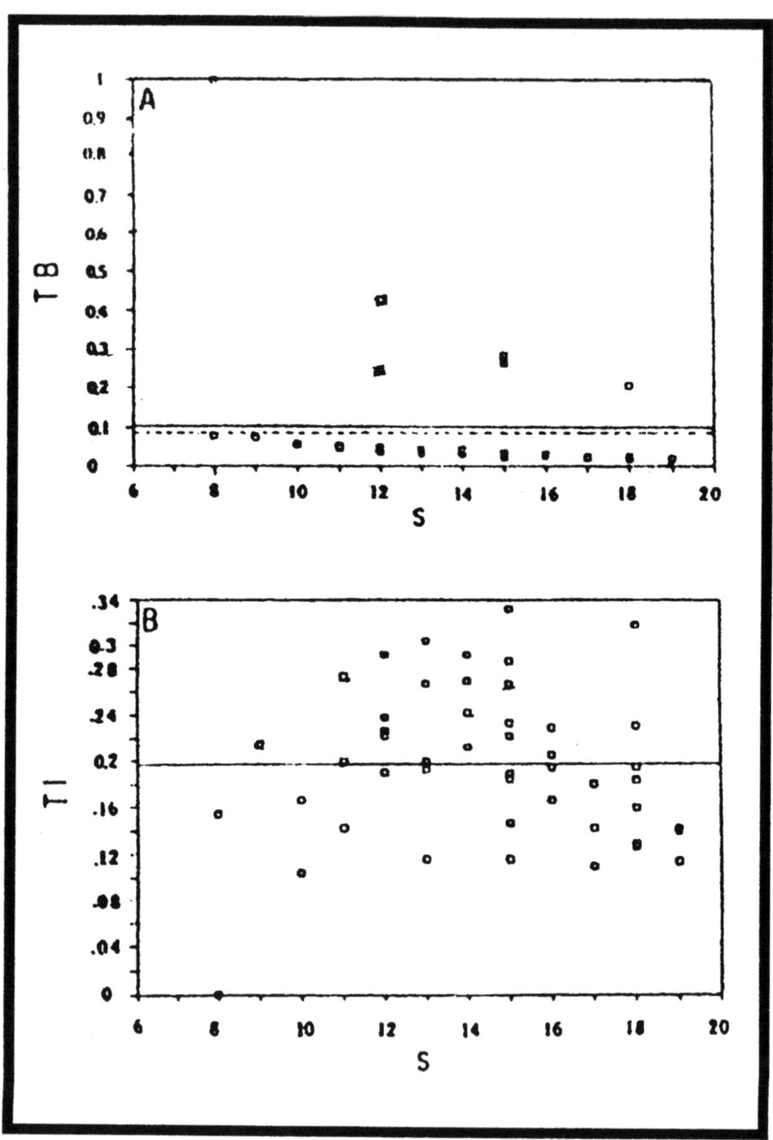

Contd...

Figure 7.3–Contd...

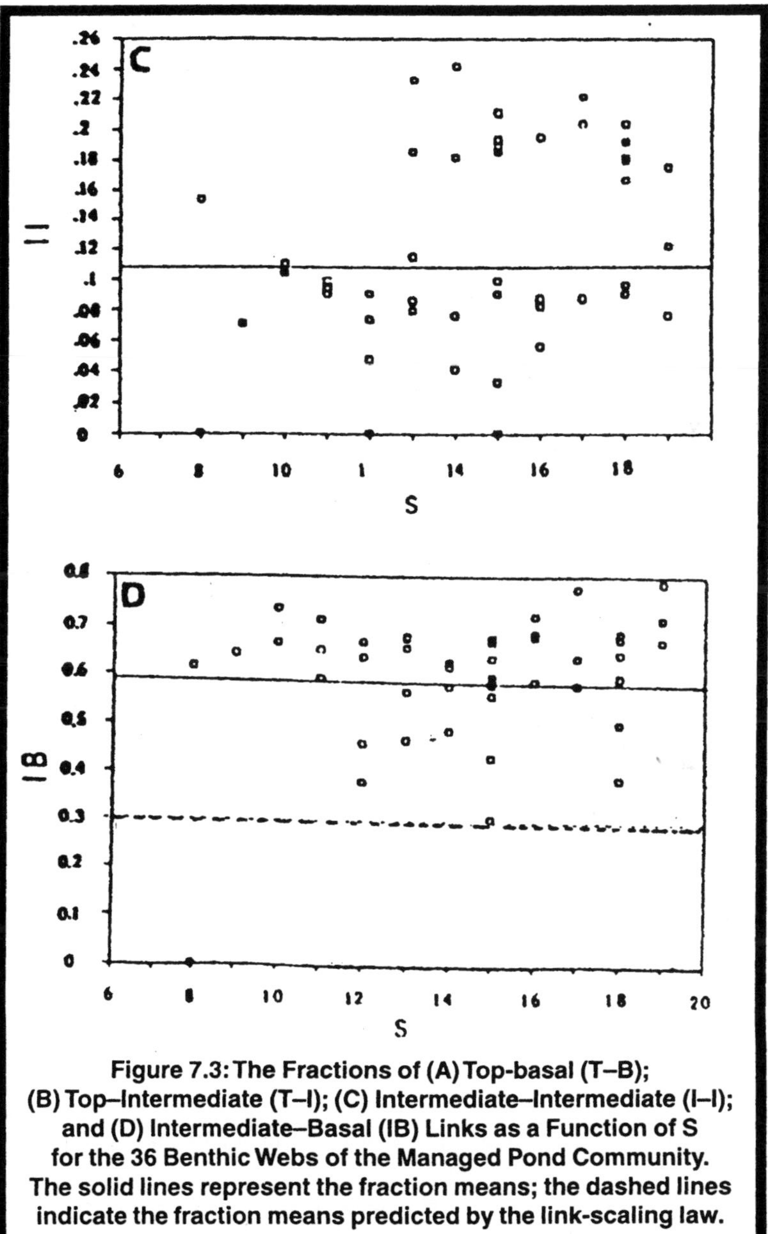

Figure 7.3: The Fractions of (A) Top-basal (T–B); (B) Top–Intermediate (T–I); (C) Intermediate–Intermediate (I–I); and (D) Intermediate–Basal (IB) Links as a Function of S for the 36 Benthic Webs of the Managed Pond Community. The solid lines represent the fraction means; the dashed lines indicate the fraction means predicted by the link-scaling law.

Figure 7.4

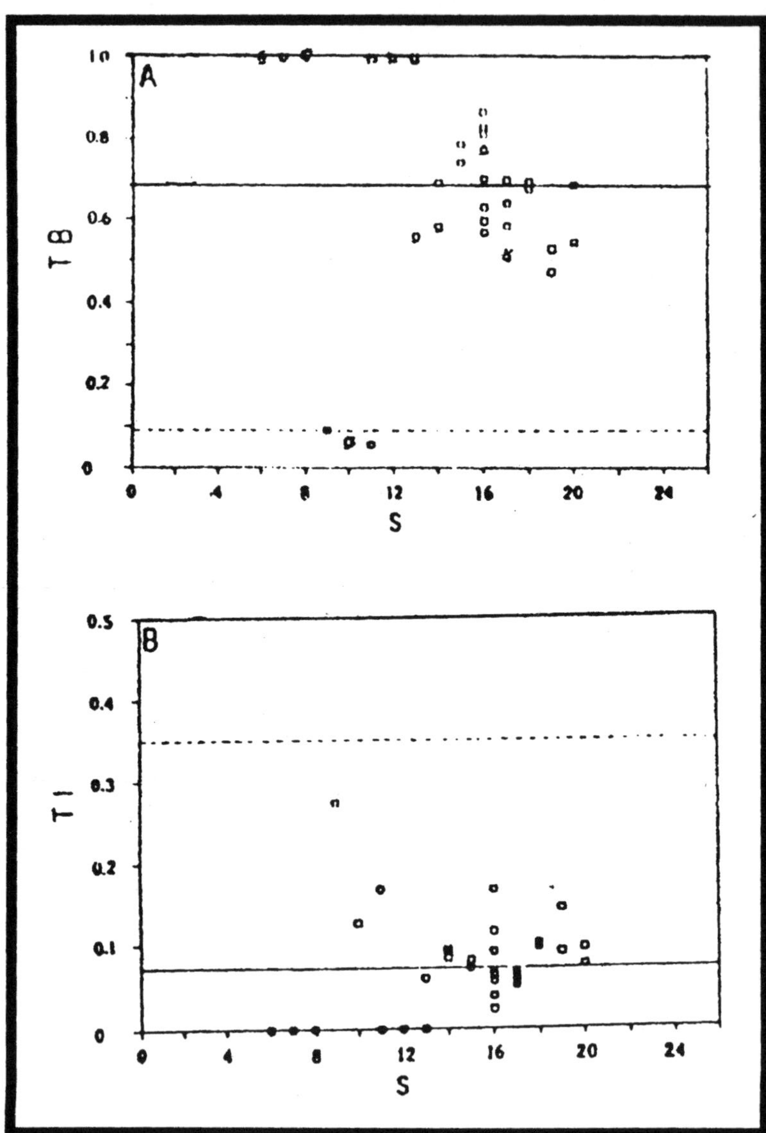

Contd...

Figure 7.4–Contd...

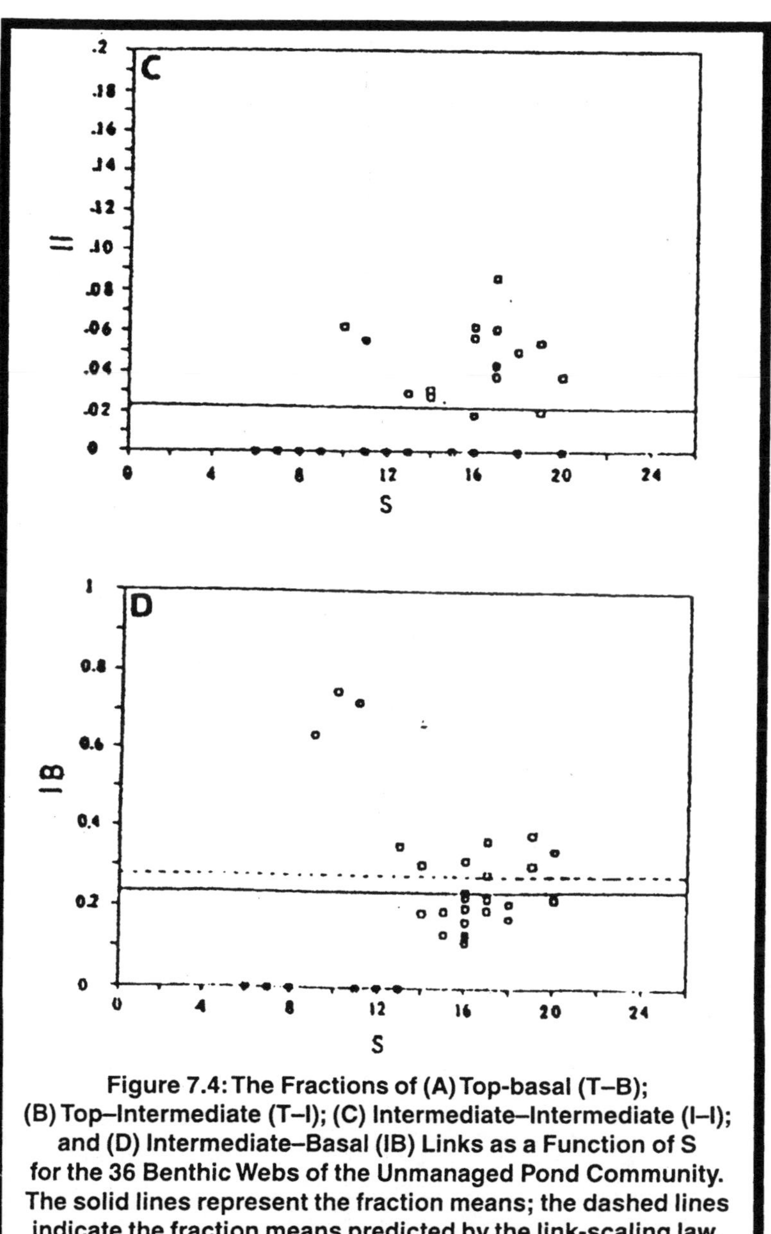

Figure 7.4: The Fractions of (A) Top-basal (T–B); (B) Top–Intermediate (T–I); (C) Intermediate–Intermediate (I–I); and (D) Intermediate–Basal (IB) Links as a Function of S for the 36 Benthic Webs of the Unmanaged Pond Community. The solid lines represent the fraction means; the dashed lines indicate the fraction means predicted by the link-scaling law.

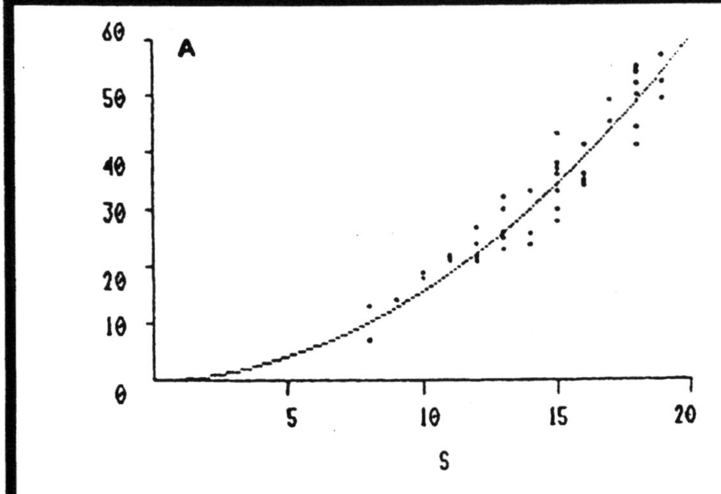

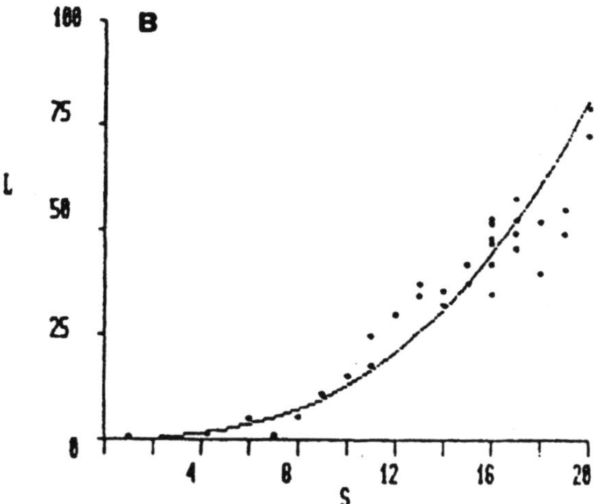

Figure 7.5: The Links (L) as a Function of S for the (A) 50 and (B) 36 Benthic Food Webs of the Managed and Unmanaged Pond Community Respectively. The solid lines are fitted to the data points the exponential growth curves represent the equation ($L = 0.184\ S^{1.93}$ for the managed pond and $L = 0.027\ S^{2.67}$).

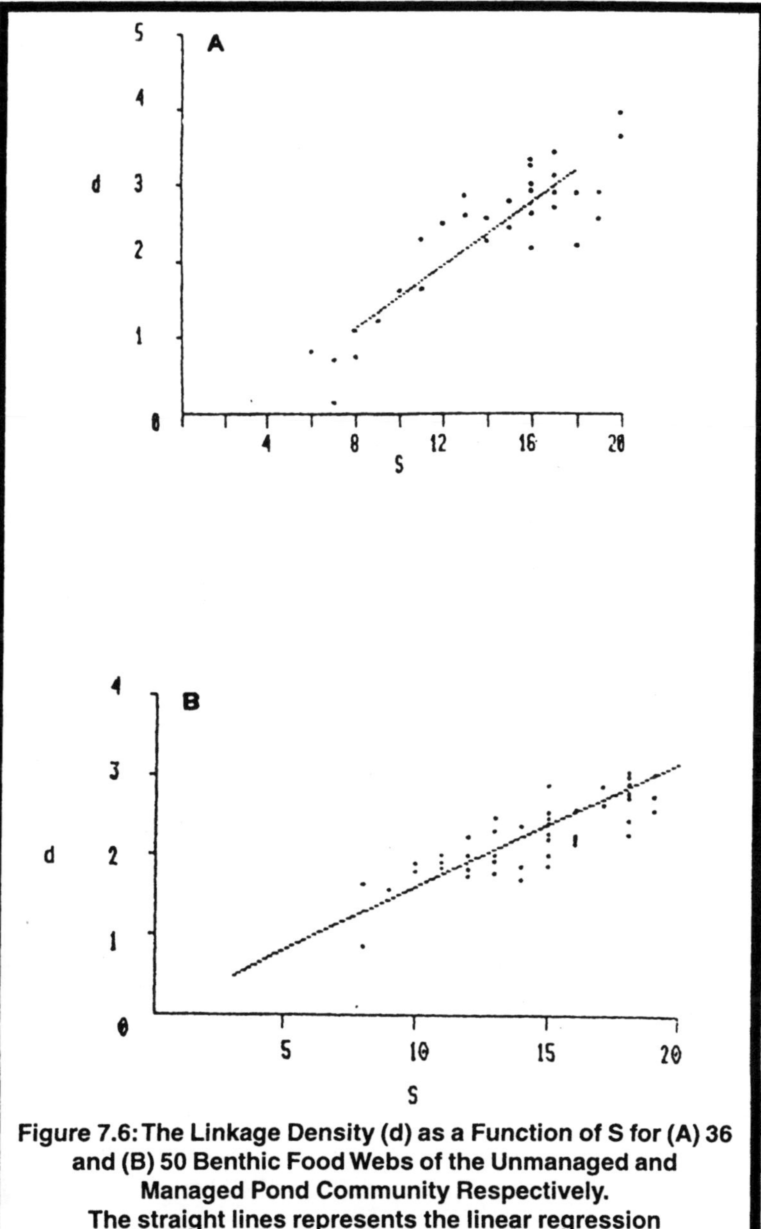

Figure 7.6: The Linkage Density (d) as a Function of S for (A) 36 and (B) 50 Benthic Food Webs of the Unmanaged and Managed Pond Community Respectively. The straight lines represents the linear regression (d = 0.564 + 0.207 S for the unmanaged pond and d = 0.265 + 0.135 S for the managed pond.

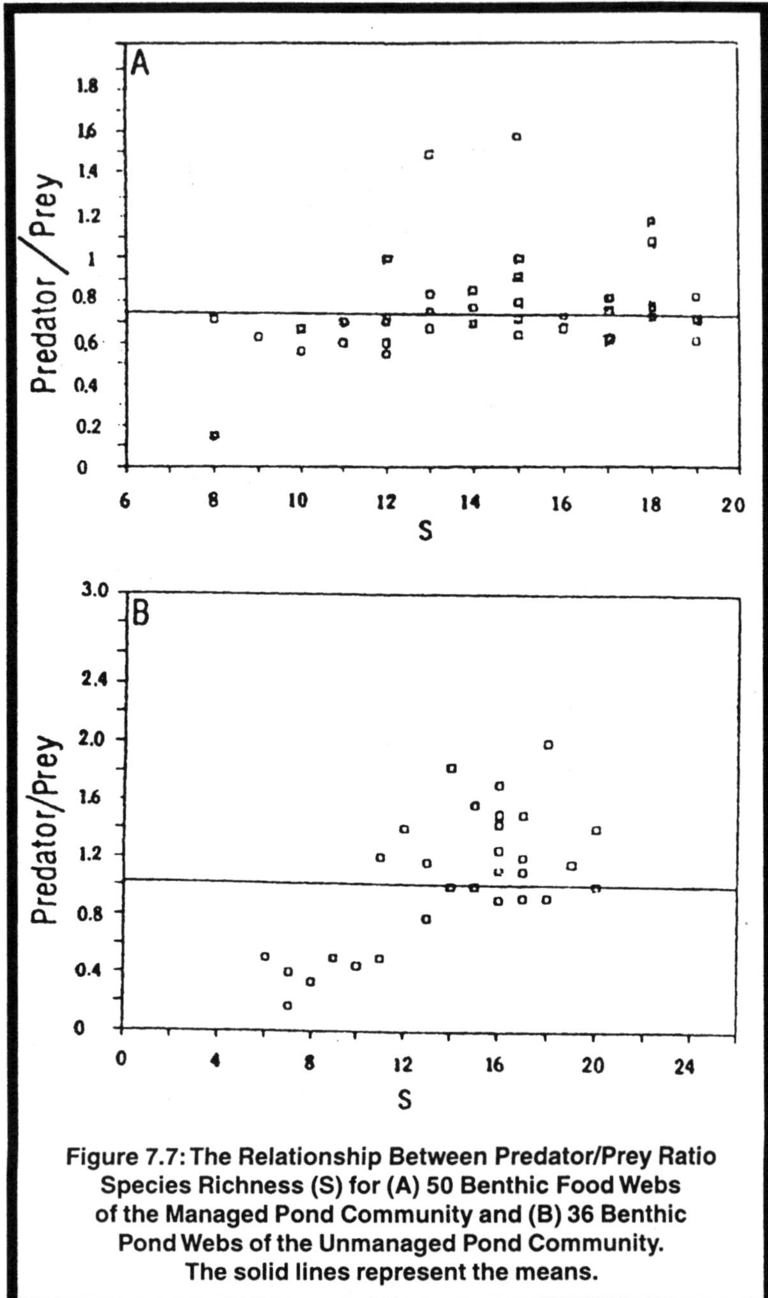

Figure 7.7: The Relationship Between Predator/Prey Ratio Species Richness (S) for (A) 50 Benthic Food Webs of the Managed Pond Community and (B) 36 Benthic Pond Webs of the Unmanaged Pond Community. The solid lines represent the means.

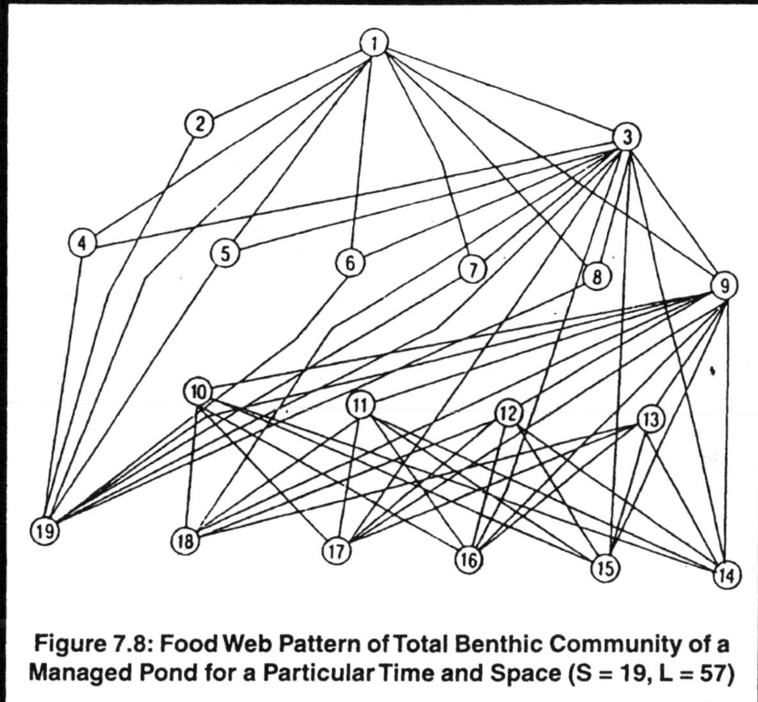

Figure 7.8: Food Web Pattern of Total Benthic Community of a Managed Pond for a Particular Time and Space (S = 19, L = 57)

1: *Cirrhinus mrigala*; 2: *Cullcoides* sp.; 3: *Tanypus* sp.; 4: *Branchiura sowerbyl*; 5: *Branchiodrilus semperi*; 6: *Limnodrillus hoffmoisteri*; 7: *Dero dorsalis*; 8: *Aulophorus furactus*; 9: *Chironomus* sp.; 10: *Brachionus rubens*; 11: *B. angularis*; 12: *B. calyoiflorus*; 13: *B. quadridentatus*; 14: *Chlamydomonas* sp.; 15: *Chlorella* sp.; 16: *Clusterius* sp.; 17: *Scenedesmus* sp.; 18: *Gyrosigma* sp.; 19: *Detritus*.

study, the webs were sufficiently small to allow counting of directed linkage between pairs of species. Hence the webs were described graphically (Figure 7.5).

In the present study linkage density (d) are estimated instead of connectance (C) estimation which shows a linear regression with the species richness (S) following the equation $d = 0.265 + 0.135\ S$ in case of managed pond and $d = -0.564 + 0.207\ S$ in case of unmanaged pond (Figure 7.6). This results discard the link-species scaling hypothesis (Briand and Cohen, 1984), but support the findings of

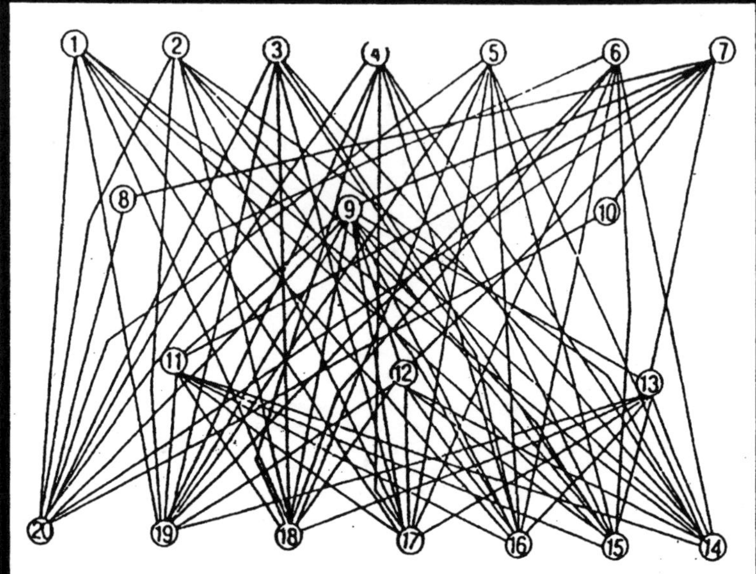

Figure 7.9: Food Web Pattern of Total Benthic Community of a Unmanaged Pond for a Particular Time and Space (S = 19, L = 57)

1: *Bellamya bengalensis*; 2: *Bellamya* sp.; 3: *Thiara granifera*; 4: *Digoniostoma cerameopoma*; 5: *Lymnea luteola*; 6: *Lamellidens marginalis*; 7: *Cirrhinus mrigala*; 8: *Branchiura sowerbyi*; 9: *Chironomus* sp.; 10: *Culicoides* sp.; 11: *Brachionus angularis*; 12: *B. falcatus*; 13: *B. quadridentatus*; 14: *Chlamydomonas* sp.; 15: *Closterium* sp.; 16: *Diffingia* sp.; 17: *Microcystis* sp.; 18: *Oedogonium* sp.; 19: *Volvox* sp; 20: Detritus.

Winemiller, 1989; Polis, 1991; Martinez, 1992; and Havens, 1992. The presence of filter feeders contribute to an increase in the linkage density, for they consume all organisms of suitable size (Deb, 1995).

The data from the 50 species benthic food webs of managed pond and 36 species food webs of unmanaged pond (Figures 7.1 and 7.2) do not support the species scaling law (Briand and Cohen, 1984; Cohen and Briand, 1984; Cohen and Newman, 1985).

In both the ponds, the fractions of basal species (BF) decreases with the increasing S, whereas fractions of top species (TF) and

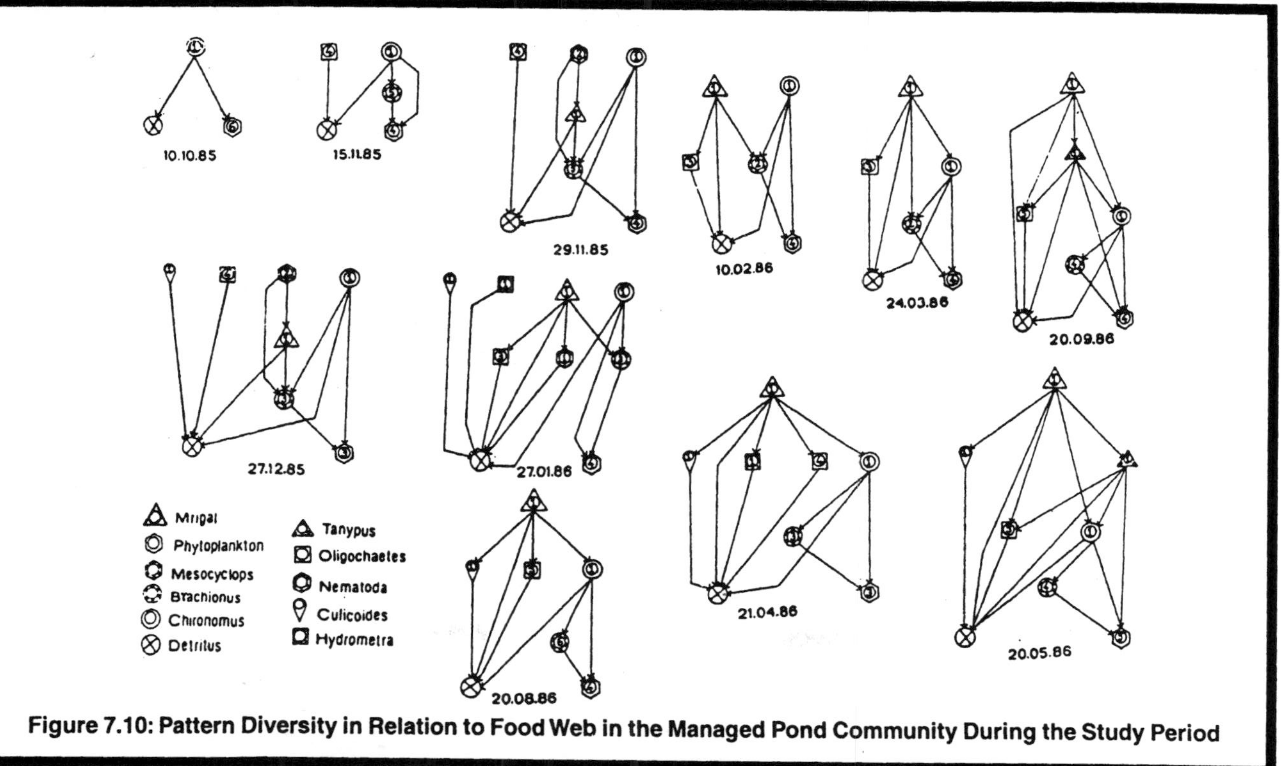

Figure 7.10: Pattern Diversity in Relation to Food Web in the Managed Pond Community During the Study Period

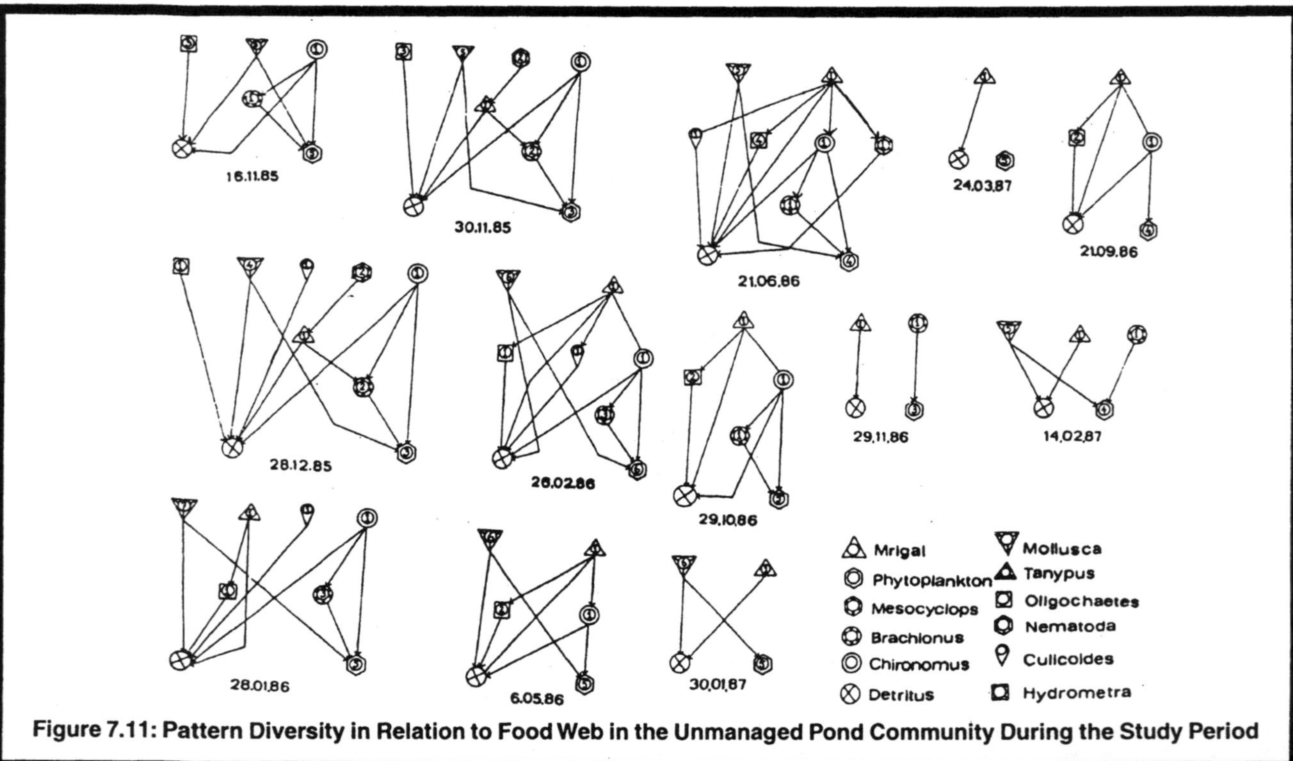

Figure 7.11: Pattern Diversity in Relation to Food Web in the Unmanaged Pond Community During the Study Period

intermediate species (IF) increases with the increasing S (Figures 7.1 and 7.2). The present results indicate that Martinez's (1991) view for benthic food web is also correct in that only a fraction of the species are top predators. The results of the present study (Figures 7.3 and 7.4) partly support the link-scaling law (Cohen and Briand, 1984; Briand and Cohen, 1984; Cohen and Newman, 1985). The link fractions T–I, I–I in the managed pond are scale variant with means of 0.195, 0.114, 0.588 respectively, and T–B is scale-invariant with mean of 0.103. The fractions of I–I link is increased in this study, while the fractions of T–B link is reduced about to zero. The changes of the link fractions with S are statistically significant which is contrary to the results of Martinez (1991) and Havens (1992). The I–I and I–B fractions in Havens' data have strong correlations with S ($p < 0.001$), whereas those in the pond webs herein do not. The means of I–B, I–T, I–I links are higher than those of Havens' data whereas the mean of I–B link is lower.

The results of chain-Iength analysis (Tables 7.3 and 7.4) is comparable with the mean chain length of 3.19 (Jeffries and Lawton, 1985) and 3.77 (Warren, 1989) for freshwater, webs and corroborates with the findings of Winemiller (1990) and Martinez (1991). Previous studies (Sugihara *et al.*, 1989) report that chain length is invariant and robust to aggregation of trophically redundant species. This conclusion appears to be an artifact of incomplete analysis (Martinez, 1991). In the present study, chain length seems to increase significantly with webs size S ($r = 0.51$; $t = 4.11$; $p < 0.001$ for managed pond; and $r = 0.62$, $t = 4.61$, $p < 0.001$ for unmanaged pond). This finding signifies the incompleteness of Cohen and Briand's as well as Havens' data. The mean of the predator/prey fractions could be derived from the mean values of the species fractions (Cohen and Briand, 1984).

$$(TF + IF)/(IF + BF) = (I\text{–}BF)/(I\text{–}TF)$$

The calculation yields the mean of predator/prey ratio as 0.74 and 1.015 for the managed and unmanaged pond respectively (Table 7.5). The ratio > 1 indicates that there are more top species than basal species as in case of unmanaged pond, and < 1 indicates that the situation is reverse as in case of managed pond. This results corroborate with the Martinez's (1991) analysis.

The predator/prey ratio is variant with the S in the present study (Figure 7.7) which conforms the analysis of Schoener (1989), but differs with the Cohen a Briand's (1984) claim.

The results of the present study invalidate the link-species scaling laws and support the findings of Winemiller (1989), Polis (1991), Martinez (1992) and Havens (1992). The value of exponent β differs from the values given by Sugihara and Havens. The species-scaling law is thus not supported in this study and disagrees with the results of Briand and Cohen (1987), Sugihara *et al.* (1989), Martinez (1991) and Havens (1992). The difference of the results seems ultimately to lie in the predilection of details. The elaborate description of the benthic web variable and a moderate emphasis on the dynamic nature of interactions in this study have increased the number of links between the species. It may also be said that there is a basic difference in the food web patterns between benthic and pelagic levels. A conclusion may only be drawn after a thorough analysis of different food webs *i.e.* full descriptions of ecological interactions of all the species comprising a system in a real-life situation, though it is virtually impossible in the current state of knowledge (Schaffer, 1981; Schoener, 1989). In future, ecologists may reduce the shortcomings of the food-web analysis.

Acknowledgements

We thank the Head of the Department of Zoology, University of Calcutta for permitting necessary laboratory facilities for the contribution of the present research work.

References

Briand, F. (1983a). Environmental control of food web structure. *Ecology*, **64**: 253-263.

Briand, F. (1983b). Biologeographic patterns in food web organisation. In: *Current Trends in Food Web Theory*, D.L. De Angelis, W.M. Post, G. Sughihara, eds. ORNL–5983. Oak Ridge, National Laboratory, Oak Ridge, Tennessee, USA, pp. 37-39.

Briand, F. (1985). Structural singularities of freshwater food webs. lnternationale Vereinigung fur theoretische und angewandte Limnologie, Verhandlungen, **22**: 3356-3364.

Briand, F. and Cohen, J.E. (1984). Community food webs have scale-invariant structure. *Nature*, **307**: 264-266.

Briand, F. and Cohen, J.E. (1987). Environmental correlates of food chain length. *Science*, **238**: 956-960.

Cohen, J.E. (1978). *Food Webs and Niche Space*. Princeton University Press, Princeton, New Jersey, USA.

Cohen, J.E. (1989). Just proportions in food webs. *Nature*, **341**: 104-105.

Cohen, J.E. and Briand, F. (1984). Trophic links of community food web. *Proceedings of National Academy of Sciences, USA*, **81:** 4105-4109.

Cohen, J.E. and Newman, C.M. (1985). A stochastic theory of community food webs I. Models and aggregated data. *Proceedings of Royal Society London*, **224B**: 449-461.

Cohen, J.E., Briand, F. and Newman, C.M. (1986). A stochastic theory of community food webs. III. Predicted and observed lengths of food chains. *Proceedings of Royal Society London*, **228B**: 317-353.

Cohe, J.E., Luczak, T., Newman, C.M. and Zhou, Z. (1990). Stochastic structure and nonlinear dynamics of food webs: Qualitative stability in a Lotka-Volterra cascade model. *Proceedings Royal Society London*, **240B**: 601-627.

Deb, D. (1995). Scale-dependence of food web structures: tropical ponds as paradigm. *Oikos*, **72**: 245-262.

Hartig, J.H., Jude, D.J., and Evans, M.S. (1962). Cyclopoid predation in Lake Michigan fish larvae. *Canadian Journal of Fisheries Aquatic Sciences*, **39**: 1563-1568.

Havens, K. (1992). Scale and structure in natural food webs. *Science*, **257**: 1107-1109.

Homechaudhury, S., Pandit, T., Poddar, S. and Banerjee, S. (1986). Effects of mahua oil cake on the blood cells and blood values of an air breathing catfish *Heteropneustes fossilis* and a carp *Cyprinis carpio*. *Proceedings Indian Academy of Sciences*, **96**: 617-622.

Jeffries, M.J. and Lawton, J.H. (1985). Predator-prey ratios in communities of freshwater invertebrates: The role of enemy free space. *Freshwater Biology*, **15**: 105-122.

Jhingran, V.G. (1983). *Fish and Fisheries of India*. Hindustan Publishing Corporation (India), Delhi.

Lawton, J.H. (1989). Food webs. In: *Ecological Concepts*, J.M. Cherrett, ed. Blackwell Scientific, Oxford, England, pp. 43-78.

Martinez, N.D. (1991). Artifacts or attributes? Effects of resolution on the Little Rock Lake food web. *Ecological Monograph*, **61**: 367-392.

Martinez, N.D. (1992). Constant connectance in community food webs. *American Naturalist*, pp. 1217-1218.

May, R.M. (1983a). The structure of food webs. *Nature*, **301**: 566-568.

May, R.M. (1983b). Food web structures: some thoughts and some problems. In: *Current Trends in Food Web Theory*, D.L. Angelis, W.M. Post, G. Sugihara, eds. ORNL–5983, Oak Ridge National Laboratory, Oak Ridge, Tennessee, USA, pp. 127-129.

Paine, R.T. (1988). Food webs: Road maps of interactions or grist for theoretical development? *Ecology*, **69**: 1648-1654.

Pimm, S.L. (1982). *Food Webs*. Chapman and Hall, London.

Polis, G. (1991). Complex desert food webs: An empirical critique of food web theory. *American Naturalist*, **138**: 123-155.

Schaffer, W.M. (1981). Ecological abstraction: the consequences of reduced dimensionality in ecological models. *Ecological Monograph*, **51**: 383-401.

Schoener, T.W. (1989). Food webs from the small to the large. *Ecology*, **70**: 1559-1589.

Schoenly, K. and Cohen, J.E. (1991). Temporal Variation in food web structure: 16 empirical cases. *Ecological Monograph*, **61**: 267-298.

Sinha, V.R.P. (1979). Contribution of supplementary feed in increasing fish production through composite fish culture in finfish nutrition and fish feed technology. Vol. I. H. Heinemann, Berlin.

Sprules, W.G. and Bowerman, J.E. (1988). Omnivory and food chain length in zooplankton food webs. *Ecology*, **69**: 418-426.

Strong, D.R. (editor) (1988). Food web theory: A ladder for picking strawberries? (Special feature). *Ecology*, **69**: 1647-1676.

Sugihara, G., Schoenly, K. and Trombla, A. (1989). Scale invariance in food web properties. *Science*, **245**: 48-52.

Warren, P.H. (1989). Spatial and temporal variation in the structure of a freshwater food web. *Oikos*, **55**: 299-311.

Winemiller, K.O. (1989). Must connectance decrease with species richness? *American Naturalist*, **134**: 960-968.

Winemiller, K.O. (1990). Spatial and temporal variation in tropical fish trophic networks. *Ecological Monograph*, **60**: 331-367.

8

Study of Algae Tolerating Organic Pollution of Sonvad Dam and Devbhane Dam of Maharashtra

☆ *S.N. Nandan & D.S. Jain*

Introduction

In Maharashtra, Gunale and Balakrishnan (1981) has used algae as biomonitors of eutrophication in the study of Pavana, Mula and Mutha rivers flowing through Pune city. Jagdale *et al.* (1984) studied the pollution of Godavari river water at Nanded. Pandey *et al.* (1992) showed species diversity of plankton in fish pond at Pune. Hosetti *et al.* (1994) observed water quality in Jayanthi Nalla and Panchaganga at Kolhapur. Nandan and Ansari Ziya (1999) observed pollution tolerant genera and species of Mausam river of Malegaon city (Maharashtra).

In present investigation algal communities which were used as indicators of organic pollution were reported from all stations of two dams.

Materials and Methods

The Sonvad project dam is located near the village Dongargaon of Taluka Shindkheda Distt. Dhule, on small river of Bhat. Geographically Sonvad project dam is located at latitude 21°.4' North and longitude 74°.50' East. Three stations of dam *viz.*, SD–I, SD–II and SB–III stations were selected for the collection of water and algal samples.

The Devbhane dam is situated near village Devebhane of Taluka Dhule. It lies at 21°.2' North latitude and 74°.48' East longitude. Three stations of dam *viz.*, DD–I, DD–II and DB–III stations were selected for the collection of water and algal samples.

For qualitative study of algal samples were preserved in 4 per cent formalin for further study and identification of algae, line drawings of different forms of algae were made by Camera lucida. The algae were identified by relevant monographs and recent available literature. The pollution tolerant genera and species were recorded from all stations of 2 dams. Algal pollution indices of Palmer (1969) based on genus were used in rating water samples for high organic pollution.

Twenty most frequent genera were taken into account. A pollution index factor was assigned to each genus by determining the relative number of total points scored by each alga. For rating of water samples as high or low organically polluted, observations were made according to Palmer (1969).

The following numerical values for the individual zones have been followed:

1. 0–10: Suggests lack of organic pollution.
2. 10–15: Indicate moderate pollution.
3. 15–20: Indicate probable high organic pollution.
4. 20 or more: Confirmed high organic pollution.
5. 44: Theoretical maximum.

The pollution index was calculated for all stations of both the dams.

Results

Pollution tolerant genera and species of 4 groups of algae from 3 stations of Sonvad dam and 3 stations of Devbhane dam were observed and recorded in Tables 8.1 and 8.2. The most pollution tolerant species of *Scenedesmus, Oscillatoria, Chlorella, Ankistrodesmus* and *Navicula* are listed in Table 8.3. The pollution tolerant genera and species are listed in order of decreasing emphasis according to Palmer (1969) of 3 stations of Sonvad dam and 3 stations of Devbhane dam (Tables 8.1 and 8.2).

Table 8.1: Pollution Tolerant Genera of Algae from 3 Stations of Sonvad Project Dam and 3 Stations of Devbhane Dam in Order of Decreasing Emphasis (Palmer, 1969)

Sl.No.	*Genus*	**Group*	*Total Point*	*Sonvad Dam*			*Devbhane Dam*		
				SD–I	*SD–II*	*SB–III*	*DD–I*	*DD–II*	*DB–III*
1.	*Euglena*	E	172	+	+	+	+	+	+
2.	*Oscillatoria*	B	161	+	+	+	+	+	+
3.	*Scenedesmus*	G	112	+	+	+	+	+	+
4.	*Chlorella*	G	103	–	+	+	+	+	–
5.	*Nitzschia*	D	98	–	–	+	+	+	+
6.	*Navicula*	D	92	+	+	+	–	+	+
7.	*Synedra*	D	58	+	–	+	–	+	+
8.	*Ankistrodesmus*	G	57	+	–	–	+	–	+
9.	*Phacus*	E	57	+	+	+	+	–	+
10.	*Phormidium*	B	52	+	+	+	+	+	+
11.	*Gomphonema*	D	48	+	+	+	–	–	+
12.	*Cyclotella*	D	47	+	–	+	+	+	+
13.	*Clostedum*	G	45	+	+	+	+	+	+
14.	*Spirogyra*	G	37	+	–	+	–	–	+
15.	*Anabaena*	B	36	–	+	–	+	–	+
16.	*Pediastrum*	G	35	–	–	+	+	+	+
17.	*Arthrospira*	B	35	–	–	–	–	–	+
18.	*Fragilalia*	D	33	–	–	+	–	+	–
19.	*Lyngbya*	B	28	+	–	+	–	–	–
20.	*Oocystis*	G	28	+	–	–	–	–	+
21.	*Spirulina*	B	25	+	+	+	+	+	+
22.	*Cymbella*	D	24	+	–	–	+	+	+
23.	*Coelastrum*	G	24	–	–	–	+	–	+
24.	*Achnanthes*	D	19	+	–	+	+	–	–
25.	*Pinnularia*	D	18	+	–	+	–	–	+
26.	*Chlorococcum*	G	17	–	–	–	+	–	–
27.	*Cocconies*	D	17	+	–	–	–	–	–
28.	*Cosmarium*	G	17	+	+	+	+	+	+
29	*Selanastrum*	G	15	–	–	–	+	–	–
30.	*Crucigenia*	G	14	+	–	+	–	–	–

* Groups: E: Euglenoids, B: Blue-green algae, G: Green algae, D: Diatoms.

Table 8.2: Pollution Tolerant Species of Algae from 3 Stations of Sonvad Project Dam and 3 Stations of Devbhane Dam in Order of Decreasing Emphasis (Palmer, 1969)

Sl.No.	Algal Species	*Group	Total Point	Sonvad Dam			Devbhane Dam		
				SD–I	SD–II	SB–III	DD–I	DD–II	DB–III
1.	*Euglena viridis*	E	93	–	–	–	+	–	–
2.	*Scenedesmus quadricuda*	G	41	–	+	+	+	–	+
3.	*Oscillatolia tenuis*	B	40	+	–	–	–	–	+
4.	*Synedra ulna*	D	33	+	–	–	+	–	+
5.	*Ankistrodesmus falcatus*	G	32	+	–	–	+	–	+
6.	*Chlorella valgaris*	G	29	–	+	+	+	+	–
7.	*Cyclotella meneghiniana*	D	27	–	–	+	–	–	–
8.	*Navicula cryptocaphala*	D	25	–	–	+	–	+	–
9.	*Oscillatoria splendida*	B	19	–	–	–	–	+	–
10.	*Pediastrum boryanum*	G	18	–	–	+	–	–	–
11.	*Phacus pleuronectus*	E	16	–	+	–	–	–	–
12.	*Navicula Viridula*	D	16	–	+	–	–	–	–
13.	*Phormidium autumnate*	B	15	+	–	–	–	–	–
14.	*Cocconeis placentula*	D	14	+	–	–	–	–	–
15.	*Scenedesmus dimorphus*	G	13	+	+	+	–	–	+
16.	*Navicula cuspidate*	D	12	+	–	+	–	–	–
17.	*Closterium leibleini*	G	10	+	+	–	–	+	–
18.	*Oscillatoria brevis*	B	11	–	–	–	+	–	–
19.	*Euglena proxima*	E	10	–	–	–	+	–	–
20.	*Tetraedron muticum*	D	10	–	+	–	–	–	–

* Groups: E: Euglenoids, B: Blue-green algae, G: Green algae, D: Diatoms.

Table 8.3: The Most Pollution Tolerant Species from Stations of Sonvad Dam and 3 Stations of Devbhane Dam

Sl.No.	Pollution Tolerant Species	Total Point	Sonvad Dam			Devbhane Dam		
			SD–I	SD–II	SB–III	DD–I	DD–II	DB–III
(I)	Scenedesmus							
1.	*Scenedesmus quadricuda* (Chodat) G.M. Smith	41	–	+	+	+	+	–
2.	*Scenedesmus dimorphus* (Turpin) Kuetz.	13	+	+	+	–	–	+
(II)	Oscillatoria							
1.	*Oscillatoria tenus* Ag. ex. Gomont	40	+	–	–	–	–	+
2.	*Oscillatoria brevis* (Kuetz.) Gomont	11	–	–	–	+	–	–
(III)	Chlorella							
1.	*Chlorella vulgaris* Beijerinck	29	–	+	+	+	+	–
(IV)	Ankistrodesmus							
1.	*Ankistrodesmus falcatus* (Corda) Ralfs.	32	+	–	–	+	–	+
(V)	Navicula							
1.	*Navicula cryptocephala* (Kuetz.) Grun.	25	–	–	+	–	+	–

30 pollution tolerant genera were recorded at study areas of both dams. Out of 30 pollution tolerant genera 26 genera were observed at 3 stations of Sonvad dam and 28 genera were observed at 3 stations of Devbhane dam (Table 8.1).

20 pollution tolerant species were observed at all 3 stations of Sonvad dam and Devbhane dams. Out of 20 pollution tolerant species 16 species were observed at Sonvad dam and 12 species were observed at Devbhane dam (Table 8.2).

By using Palmer's index of pollution for rating of water sample as high or low organically polluted at 3 stations at Sonvad dam and 3 stations of Devbhane dam were studied. The results are shown in Table 7.4, out of 20 genera 15, 13 and 19 were observed at SD–I, SD–II and SB–III Stations of Sonvad dam respectively where as 9, 16 and

19 genera were observed at DD–I, DD–II and DB–III stations of Devbhane dam respectively (Table 8.4).

Table 8.4: Pollution Index of Algal Genera (Palmer, 1969) at 3 Stations of Sonvad Project Dam and 3 Stations of Devbhane Dam

Sl.No.	*Group/Genera*	*Palmers Pollution Index Number*					
		SD–I	*SD–II*	*SB–III*	*DD–I*	*DD–II*	*DB–III*
(I)	Cyanophyceae						
1.	*Oscillatoria*	4	4	4	4	4	4
2.	*Phormidium*	1	1	1	1	1	1
3.	*Microcystis*	–	1	1	–	–	–
4.	*Lyngbya*	1	–	1	–	–	1
5.	*Spirulina*	1	1	1	1	–	1
(II)	Chlorophyceae						
6.	*Ankistrdesmus*	2	–	–	2	–	2
7.	*Chlorella*	–	3	3	3	3	–
8.	*Closterium*	1	1	1	1	1	1
9.	*Scenedesmus*	4	4	4	4	4	4
10.	*Cosmarium*	1	1	1	1	1	1
11.	*Spirogyra*	1	–	1	–	–	1
12.	*Pediastrum*	–	–	1	1	1	1
(III)	Bacillariophyceae						
13.	*Cyclotella*	1	1	1	1	1	1
14.	*Gomphonema*	1	1	1	–	–	1
15.	*Navicula*	3	3	3	–	3	3
16.	*Nitzschia*	–	–	3	3	3	3
17.	*Synedra*	2	–	2	–	2	2
18.	*Fragilaria*	–	–	1	–	1	1
(IV)	Euglenineae						
19.	*Euglena*	5	5	5	5	5	5
20.	*Phacus*	2	2	2	2	–	2
	Total Score	**30**	**28**	**37**	**29**	**30**	**35**

The degree of organic pollution was decreased from SB–III to SD–I and SD–II and from DB–III to DD–II and DD–I stations. The total score of each stations of both dams was greater than 20 indicating the confirmed high organic pollution. The station SB–III of Sonvad dam highly organically polluted as compared to the SD–I, SD–II stations, whereas in Devbhane dam station of DB–III was highly polluted as compared to the DD–II and DD–I. It was observed that the number of pollution tolerant genera were recorded more at Sonvad dam than the Devbhane dam.

Discussion

Palmer (1969) has reported that the ganera like *Scenedesmus, Oscillatoria, Microcystis, Navicula, Nitzschia* and *Euglena* were found organically polluted water supported by (Goel *et al.*, 1986; More, 1997; Mahajan, 2001). Similar genera were recorded in present study.

The epilethic and epiphytic algae may found excellent indicators of water pollution (Round, 1965). In present investigation occurrence of *Oscillatoria, Phormidium, Lynbya* as epilethic algae and certain diatoms like *Gomphonema, Cymbella* and *Navicula* as epiphytic were recorded from sites of study area (Table 8.1). Pearsall (1932) was the first to show clear correlation between Organic pollution with blue green algae and centric diatoms. He also reported the algal genera like *Chlorella, Cosmarium, Navicula, Nostoc, Anabaena* and *Spirulina* from polluted habitats. All these genera were reported from study areas in present study. These were used as indicators of organic pollution as agreed with the views of Pearsall (1932) and Palmer (1969).

In present study the dominance of *Ocsillatoria* with 19 species was observed and they are indicated as pollutant of biological origin as agreed with observations of earlier workers (Rai and Kumar, 1976; Singh, 1979; Nandan and Patel, 1986; Joy *et al.*, 1990; More, 1997). Cholnoky (1968) gave detailed account of dominance of species of diatoms being used as indicators of water quality. This was also supported by present study.

According to Palmer (1969) 30 pollution tolerant genera were recorded from both sites of dam (Table 8.1) and 20 species of pollution tolerant were also recorded from both sites of dams.

The total score of each station of both sites was greater than 20 indicating all the stations confirmed organically polluted. Among all stations of study area. SB–III of Sonvad dam of Bhat river showed the highest organic pollution. Thus in present study algal communities were used as reliable indicators of organic pollution conferming the earlier observations (Hosmani and Bharati, 1980; Nandan and Ansari Ziya, 1999; Mahajan, 2001).

Acknowledgement

Authors are geateful to Principal B.M. Patil for providing laboratory facilities.

References

Cholnoky, B.J. (1968). Die okologie der diatomeen in binery ewassera. *J. Cramex, Germany*, p. 699.

Gunale, V.R. and M.S. Balkrishnan (1981). Biomonitoring of eutrophication in the Pavana, Mula and Mutha rivers flowing through Poona. *Indian J. Environ. Hlth.*, **23(4)**: 316-322.

Goel *et al.* (1985). Studies on the limnology of a few freshwater bodies in south-western Maharashtra. *Indian J. Env. Prot.*, **5(1)**: 19-25.

Hosetti *et al.* (1994). Water quality in Jayanthi Nall and Panchaganga at Kolhapur. *Indian J. Environ. Hlth.*, **36(2)**: 124-127.

Hosmani, S.P. and S.G. Bharati (1980). Algae as indicators of organic pollution. *Phykos.* **19(1)**: 121-124.

Jagdale *et al.* (1984). Pollution of Godavari river water at Nanded. *Poll. Res.*, **3**: 83-84.

Joy *et al.* (1990). Effect of industrial discharges on the ecology of phytoplankton production in the river Periyar (India). *Wat. Res.*, **24(6)**: 787-796.

Mahajan, S.R. (2001). Ecological study of algae from certain lakes of Jalgaon district of North Maharashtra.

More, Y.S. (1997). *Limnological Study of Panzaa Dam and River*. Ph.D. Thesis, N.M.U., Jalgaon.

Nandan, S.N. and Ansari Ziya (1999). Ecological study of algae from Mausam river flowing through Malegaon city (Maharashtra), *BRI's JAST*, **11(1)**: 31-38.

Nandan, S.N. and R.J. Patel (1986). Assessment of water quality of Vishwamitri river by algal analysis. *India J. Environ. Hlth.*, **29(2)**: 160-161.

Pandy *et al.* (1992). Species diversity of plankton in fish pond at Pune, Maharashtra. *Proc. of Acad. of Envi. Biol.*, pp. 167-172.

Palmer, C.M. (1969). A composite rating of algae tolerating organic pollution, *J. Phycol.*, **5**: 78-82.

Pearsell, W.H. (1932). Phytoplankton in English lakes–II. *J. Ecol.*, **22**: 241-262.

Rai, L.C. and H.D. Kumar (1976). Nutrient uptake by *Chlorella vulgaris* and *Anacystis nidulans* isolated from fertilizer factory effluent. *Ind. J. Ecol.*, **3**: 63-69.

Round, F.E. (1965). *The Biology of the Algae.* Edward Arnold Pub. London, p. 269.

Singh, Y. (1979). *Ecological Studies on River Algae with Special Reference to Bacillariophyceae*. Ph.D. Thesis, Lucknow Univ.

9

Nrusinghnath Spring Water Quality in District of Baragarh, Orissa

☆ *S.K. Samal, B. Pradhan & T.N. Tiwari*

Introduction

Orissa is a maritime state spreading over an area of 1,55,707 sq km and lying between 81°27′ and 87°30′ East longitude and 17°31′ and 22°27′ North latitude on the East coast of India. The coastline along the Bay of Bengal is about 480 km long. Orissa is studded with natural resources like mountains, rivers, springs, valuable forest heritages with high genetic diversity, historical and archaeological monuments, etc. Springs are places where a flow of water rises to the surface through natural rock opening under hydraulic pressure from the depth. The water of springs will be either in plenty or scarce, depending on the area and thickness of the aquifer or the volume of water. The water qualities of the springs mostly depend on geological and topological condition (Sharma, 2001) and intensity of anthropogenic activity. The source of the Nrusinghnath spring is probably a pool of water formed under the cap of laterite covering the hill through which rain water percolates. In the present investigation, an attempt has been made to assess the water quality and carry out correlation and regression analysis of the famous Nrusinghnath spring water, which is of great religious and tourist importance. The correlations among water quality parameters of water, if they do exist, give important indirect information regarding the mineral contents, geology and hydrochemistry of sub-soil rocks, aquifers and ground water of the concerned locality (Tiwari *et al.*, 1989). Water quality index enables large amount of water quality data to be reduced to a single numerical value in an objective, rapid and reproducible manner (Dwivedi *et al.*, 2000). The spring water is generally used for drinking, bathing and religious purposes. But in the downstream, people use the spring water for their domestic and irrigation purposes. Thus spring water acts as a lifeline for the local people.

Materials and Methods

Bargarh is located on 21°20′ North latitude and 83°371 East longitude. It is situated at a distance of 50 km from Sambalpur city and 370 km from the State capital Bhubaneswar. This town is well connected by road. It is at a distance of 1375 km away from Mumbai and 575 km away from Kolkata. This town is also connected with railway and name of the rail station is Bargarh Road. Bargarh is a part of the peninsular shield. One of the oldest and most stable land masses in the world. By and large, it is underlain by Precambrian rocks. The Precambrian rocks are represented by pelitic, pasammitic and pasamopelitic with subordinate psephitic and calcareous rocks, metamorphosed in several grades. The Pre-Cambrians are over three billion years old. They are scarcely recorded in any form other than different types of litho genesis, tectonic cycles, a combination of tectonics magmatism and metamorphism serve as the main basis for limiting the time boundary between two rocks assemblages, recognized in space. The Pre-Cambrian rocks are subdivided into three super groups:

1. Pre-Rihean (older than 1400 million years) *i.e.* Gabdhamardan Hill ranges of Eastern Ghat Groups,
2. Riphean (600–1400 million years), and
3. Vendian (younger than 600 million years).

Rocks of Khondalite suit garnet-sillimanite genesis, graphitic genesis, garnetiferrous quartzite and calc-silicate granulites groups are found in Bargarh district. There is one of the outstanding Vaishnava Temples of West Orissa located at the foot of Gandhamardhana hill in the district of Bargarh. Nrusinghnath, the largest seat of Nrusingha worship, lies in the foothills of Gandhamardhana hill range, located 120 km from Bargarh. The most important part of Nrusinghnath is the stone inscription found on the wall of the temple. The famous Gandharmadhan hills are popularly known as "Ayurvedists Paradise". The hill, as myths say was carried by Hanuman after he failed to locate the *Bisalya Karani*, a medicinal plant. It is said that a part of the hill had broken while it was being carried by Hanuman. From the temple, one will find the long Barapahad hills running south-west for a distance of nearly 48 km. This hill forms a barrier to communication with the rest of the subdivision. The Borasambar tract lies to the south-west of the Bargarh plain. It is bounded by hills on the north.

Nrusinghnath is a place great religious and tourist importance in Padampur subdivision of Baragarh district on the northern flank of Gandhamardana hills, situated about 32 km southwest of Padampur. A perennial stream, flowing close by, accumulates itself into five sacred *kundas*, higher upon the hill, towards the southwest, and near the first *kunda* are five colossal rock figures identified as *Panchpandavas*.

5 different samples were collected from the spring during the year 2001 and fifteen physico-chemical parameters were analysed. The samples were brought to the laboratory within 12 hours and stored at 4 to 15°C. The physico-chemical preservation and analysis were carried out according to standard methods given by APHA, AWWA and WPCF (1985); Trivedi and Goel (1986) and Manivaskaram (1986). The analysis was completed within 48 hours to avoid any variation in the result. The parameters Temperature, pH, EC, TDS, DO and Turbidity were measured using a portable water analysis kit (model–191 E) and with the standard methods supplied by the manufacturer. Sodium and Potassium were measured by flame photometer (Toshniwal-model no.–Chemito 1000). TH, Ca^{+2}, Mg^{+2}, Cl^- and Alkanity were determined by titrimetric methods and argentrometric method, respectively, while SO_4^{-2} and Fe^{+2} were measured spectrophotometrically. The statistics like max-min, mean, variation, standard deviation and standard errors, correlation and regression equation were evaluated by Trivedi and Goel (1986).

Results and Discussion

The standard IS, ICMR, ISI and WHO values of different parameters are shown in Tables 9.1 and 9.2 shows the classification of water quality index for drinking water.

Table 9.1: Standards for Drinking Water*

Parameter	*WHO*		*ICMR*		*IS*		*ISI*	
	PL	*EL*	*PL*	*EL*	*PL*	*EL*	*PL*	*EL*
pH	7–8.5	6.5–9.2	7–8.5	6.5–9.2	6.5–8.5	6.5–9.2	7–8.5	6.5–9.2
EC			300					
Turbidity	5	25	5	25	10	25		
DO			3	6				
TA			120				200	600
TH	200	600	300	600	300	600	300	600
TDS	500	1500	500	1500			500	1500
Cl^-	500		250				600	
Ca^{+2}	75	200	75	200	75	200	250	1000
Mg^{+2}	50	150	50	150	30	100	30	100
SO_4^{-2}	200	400	200	400	300	600	200	400

* All parameters are expressed in mg/l except EC and pH.

PL = Permissible limit; EL = Excessive limit.

Table 9.2: Classification of Water Quality Index Values

W.Q.I. Values (in mg/l)	*Water Quality Rating*
0–25	Excellent
26–50	Good
51–75	Poor
76–100	Very poor
Above 100	Unfit for drinking

The analysis results of different physico-chemical parameters of the spring during the year 2001 are shown in Table 9.3. The observed values of parameters, when compared with standard values, indicate that almost all the parameters are within the prescribed upper limit. The mean, variation, standard deviation and standard error values show typical representative, inferring values, degree of spread and reliability of observed and parameters respectively. The observed average values of Temp., EC, pH, TSS, TDS, Cl^- and SO_4^{-2} were 23°C, 267 µmho, 7.47, 13, 185, 8 and 5.84 ppm respectively. The average values of Turbidity, DO and Fe^{+2} may be due to the deposits of iron ore in Gandhamardhan hill range.

Table 9.3: Statistical Analysis of Physico-chemical Parameters of Nrusinghnath Spring in the Year 2001

Parameter	*Min.–Max.*	*Mean*	*Variation*	*S.D.*	*S.E.*
Temp	18–27	23	11.2	3.347	1.367
Turbidity	29–41	35.667	20.667	4.546	1.856
EC	246–290	267	325.6	18.045	7.367
pH	7.1–7.8	7.467	0.067	0.258	0.1054
DO	7.0–7.7	7.3	0.068	0.26	0.106
TSS	10–15	13	3.2	1.789	0.73
TDS	150–230	185	1000	31.623	12.909
Alkalinity	30–36	33	4.8	2.19	0.895
TH	21–30	25	12.8	3.578	1.46
Ca^{+2}	11–17	14	4.8	2.19	0.895
Ma^{+2}	10-14	11	3.2	1.789	0.73
Na^{+}	2–6	4	2	1.414	0.577
Cl^{-}	6–11	8	3.2	1.789	0.73
SO_4^{-2}	5.3–6.7	5.834	0.283	0.532	0.217
Fe^{+2}	0.679–0.834	0.751	0.004	0.058	0.024

* All parameters are expressed in mg/; except EC and pH;

S.D.: Standard Deviation; S.E.: Standard Error

Water quality of the spring was compared with standard values and it was observed that the water quality index value 61.094 indicates that the water quality of the spring comes under "poor" quality category for drinking purposes.

The systematic calculation of correlation coefficient between each possible pair of the 15 water quality parameters are shown in Tables 9.4 and 9.5 shows the numerical results of regression equation, where the results are given only for those parameters, which have |r| m 0.9. The very high value of "r" shows a strong correlation

Table 9.4: Correlation Coefficient of Nrusinghnath Spring for the Year 2001

	p–1	*p–2*	*p–3*	*p–4*	*p–5*	*p–6*	*p–7*	*p–8*	*p–9*	*p–10*	*p–11*	*p–12*	*p–13*	*p–14*	*p–15*
p–1	1.000														
p–2	– 0.092	1.000													
p–3	0.917	0.044	1.000												
p–4	0.694	0.261	0.695	1.000											
p–5	– 0.985	– 0.000	– 0.914	– 0.743	1.000										
p–6	– 0.200	0.984	– 0.068	0.130	0.129	1.000									
p–7	0.907	– 0.271	0.925	0.527	– 0.897	– 0.389	1.000								
p–8	0.930	0.093	0.948	0.548	– 0.922	0.000	0.894	1.000							
p–9	– 0.109	0.944	– 0.091	0.247	0.000	0.919	– 0.346	0.000	1.000						
p–10	0.017	0.910	0.118	0.260	– 0.150	0.844	– 0.088	0.198	0.919	1.000					
p–11	– 0.382	0.843	– 0.319	0.035	0.245	0.816	– 0.491	– 0.258	0.917	0.893	1.000				
p–12	– 0.100	0.812	– 0.031	0.087	– 0.043	0.750	– 0.159	0.079	0.868	0.969	0.919	1.000			
p–13	0.869	– 0.443	0.824	0.476	– 0.857	– 0.562	0.972	0.791	– 0.459	– 0.219	– 0.561	– 0.250	1.000		
p–14	0.843	– 0.226	0.721	0.855	– 0.851	– 0.358	0.738	0.612	– 0.172	– 0.126	– 0.361	– 0.231	0.778	1.000	
p–15	0.081	0.964	0.198	0.438	– 0.149	0.943	– 0.156	0.221	0.888	0.816	0.698	0.663	– 0.33	– 0.041	1.000

P1–Temperature; P2–Turbidity; P3–EC; P4–pH; P5–DO; P6–TSS; P7–TDS; P8–Na^{+}; P9–Alkalinity; P10–TH; P11–Ca^{+2}; P12–Mg^{+2}; P13–Cl^{-}; P14–So_4^{-2}; and P15–Fe^{+2}.

between the parameters and the value of "r2" shows the per cent of variation in one parameter due to the other. Y is influenced by corresponding parameter X, but the rest per cent variation can be attributed to other reasons.

Table 9.5: Regression Analysis Equation Y = A + BX Among Various Parameters: | r | ≥ 0.9

X	*Y*	*r*	*A*	*B*
p–3	P–1	0.917	153.232	4.946
p–5	P–1	– 0.985	5.699	0.069
p–5	P–3	– 0.914	10.825	– 0.013
p–5	P–8	– 0.922	40.5	– 5
p–6	P–2	0.984	– 0.806	0.387
p–7	P–1	0.907	– 12.143	8.571
p–7	P–3	0.925	– 247.973	1.622
p–8	P–1	0.930	– 5.036	0.393
p–8	P–3	0.948	– 15.845	0.074
p–9	P–2	0.944	16.777	0.455
p–9	P–6	0.919	18.375	1.125
p–10	P–2	0.910	– 0.542	0.716
p–10	P–9	0.919	– 24.500	1.500
p–11	P–9	0.917	– 16.250	0.917
p–12	P–10	0.969	– 1.109	0.484
p–12	P-11	0.919	0.500	0.750
p–13	P–7	0.972	– 2.175	0.055
p–15	P–2	0.964	0.310	0.012
p–15	P–6	0.943	0.351	0.031

P1–Temperature; P2–Turbidity; P3–EC; P4–pH; P5–DO; P6–TSS; P7–TDS; P8–Na^{+}; P9–Alkalinity; P10–TH; P11–Ca^{+2}; P12–Mg^{+2}; P13–Cl^{-}; P14–So_4^{-2}; and P15–Fe^{+2}.

Summary

Physico-chemical study of the Nrusinghnath spring water has been carried out to examine its suitability for drinking. We have also performed correlation and regression analysis among different parameters. The present study indicates that Nrusinghnath spring water quality is poor for drinking, but can be used for bathing purposes. Therefore, the Nrusinghnath spring water requires periodic monitoring, cleaning and making the public aware about the quality of water. The result of present investigation also gives the necessary information needed for spring water quality management of this region in district of Baragarh, Orissa.

Acknowledgement

Authors owe their thanks to the Director of N.I.T., Rourkela and the local people of that area for their active support in field work and giving necessary information about Nrusinghnath spring.

References

APHA, AWWA, WPCF (1985). *Standard Methods for the Examination of Water and Wastewater, 16th edition*. Washington, D.C.

Dwivedi, Smriti, Tiwari I.C. and Bhargava, D.S. (2000). Classification of water quality of river Ganga at Varanasi. *Indian J. Env. Poll.*, **20 (9)**: 688-697.

Manivaskaram (1986). *Physico-chemical Examination of Water Sewage and Indusirial Effluent*. Pragati Prakashan, Meerut.

Sharma, B.K. (2001). *Environmental Chemistry*. Goel Publishing House, Meerut.

Tiwari, T.N. and Manzoor Ali (1989). Groundwater of Nuzvid Town: Regression and Cluster Analysis of Water Quality Parameters. *Indian J. Env. Prot.*, **9(1)**: 13-18.

Trivedy, R.K. and Goel, P.K. (1986). *Chemical and Biological Methods for Water Pollution Studies*. Environmental Publications, Karad.

10

Water Quality of Hingni (Pangaon) Reservoir and Its Significance to Fisheries

✰ *V.B. Sakhare*

Introduction

Fish is an important agricultural commodity in market. The reservoir fisheries sector in India is viewed as a vital growing sector within food production because of its ability to provide inexpensive protein and employment to millions of people. Nowadays the extensive investments in alternative technologies and new management techniques have been proposed for improving the output from reservoir fisheries, many of those new techniques, especially those of that involve the introduction of Indian Major Carps along with some exotic carps, resulting in the harvest of multiple marketable species.

The Hingni (Pangaon) reservoir, one of the artificial water bodies in Barshi tahsil in Maharashtra is used for irrigation as well as fisheries. The reservoir was constructed during year 1976 in between villages Hingni and Pangaon on river Bhogawati. The average rainfall at reservoir area was recorded at 582 mm. The reservoir starts filling by the late July. The water level reaches its peak generally in the middle of August. During September and October it becomes somewhat stable. From the month of January, the level declines steadily. The reservoir water is used for irrigation and it has an opportunity to provide drinking water to about 15 villages of Solapur district. The reservoir water is also used for Bhogawati sugar factory at Vairag. The detail morphometric features of Hingni (Pangaon) reservoir are depicted in (Table 10.1). The reservoir is about 40 km away from Tuljapur town, on Tuljapur-Barshi state highway.

Physico-chemical studies of Hingni (Pangaon) reservoir were undertaken to enhance the limnological knowledge about the reservoir and its significance with fisheries. In present investigation various physico-chemical parameters like water temperature, pH, transparency, dissolved oxygen, free carbon dioxide, total alkalinity, total hardness, chlorides and total dissolved solids were analysed.

Table 10.1: Salient Features of Hingni (Pangaon) Reservoir

1.	Name of River	Bhogawati
2.	Year of construction	1976
3.	Average rainfall (mm)	582
4.	Length of dam (m)	1987.80
5.	Mean depth (m)	21.87
6.	Water spread area (ha)	1005.65
7.	Irrigation potential (ha)	5625
8.	Dead storage capacity (million m^3)	13.54
9.	Usable storage capacity (million m^3)	31.97
10.	Gross storage capacity (million m^3)	45.91
11.	Catchment area (km^2)	401.45
12.	Maximum flood (m)	504

Materials and Methods

The water samples were collected in 3-litre polythene can, which were thoroughly washed thrice with the water to be analysed. The parameters like total alkalinity, chlorides, and total dissolved solids were estimated by the methods followed by APHA (1985); Trivedy and Goel (1986). The other parameters like water temperature, pH and transparency were measures by mercuric thermometer, Hanna made pH meter and secchi disc respectively.

Results and Discussion

Physico-chemical characters for 12 months period (from August 2001 to July 2002) are, shown in the Table 10.2. Water temperature in the reservoir ranged from 19 to 26°C with an average of 23.5. Water temperature influences aquatic life and concentration of dissolved gases like CO_2, O_2 and chemical solutes. The minimum temperature was recorded in November while maximum in the month of April.

Table 10.2: Water Quality of Hingni (Pangaon) Reservoir

Sl.No.	*Water Parameters*	*Range*
1.	Water temperature (°C)	19–26
2.	pH	8.1–8.6
3.	Transparency (cm)	15–43
4.	Dissolved oxygen (mg/l)	3.7–9.4
5.	Free carbon dioxide (mg/l)	Nil–4.1
6.	Total alkalinity (mg/l)	49–69.8
7.	Total hardness (mg/l)	48–53.2
8.	Chlorides (mg/l)	14.3–59.2
9.	Total dissolved solids (mg/l)	39–43

Transparency ranged from 15 to 43 cm (av. 33 cm). Low values were recorded during September, while the high values in May. The low values in month of September might be due to low and moderate velocity of winds.

The pH of the reservoir water varied from 8.1 to 8.6 (av. 8.2) indicating alkaline, range throughout the study period. The pH in a water body normally depends on the prevailing biological activity. A high pH value of waters is usually associated with higher photosynthetic activity, which results in utilization of carbon dioxide from bicarbonates and formation of carbonates. According to Jhingran (1990) the pH range of Indian reservoirs is from 6.5 to 9.2 and the pH in between 6.0 to 8.5 indicates the medium productive reservoir. The reservoir under present investigation falls under medium productive category. It is due to its average pH value recorded at 8.2.

Total alkalinity is the sum of phenolphthalein and methyl orange alkalinity. During present investigation total alkalinity value ranged from 49 to 69.8 mg/l with an average value of 52.5 mg/l. The minimum value was recorded in March, while the maximum was during November.

Dissolved oxygen in the water is greatly affected by the biological activity, such as photosynthesis or from air due to turbulence of the water surface by wind action, which is dissolved into the water at its surface. The dissolved oxygen content of warm water fish habitat should not be less than 5 mg/l during at least 16 hours of any 24-hour period. It may be less than 5 mg/l for a period of not exceeding 8 hours within any 24-hour period and no time shall the dissolved oxygen content be less than 3 mg/l (Boyd, 1982; Kee and Wolf, 1963). It has been found that fish can survive at dissolved oxygen in the range of 1 to 5 mg/l, but its growth and reproduction are retarded (Das, 1996). Dissolved oxygen in Hingni (Pangaon) reservoir varied from 3.7 to 9.4 mg/l with maxima in October and minima in March. Thirumathal *et al.* (2002) showed that the dissolved oxygen varied between 5.5 to 7.2 mg/l in Amaravathy reservoir.

In an aquatic ecosystem sources of CO_2 are community respiration and decomposition, while it is consumed in the photosynthesis. The values of free carbon dioxide varied between Nil to 4.1 mg/l with an average value of 2.8 mg/l. The maximum concentration (4.1 mg/l) was recorded during August and it was totally absent during months of November.

Chlorides occur naturally in all types of waters. High concentration of chlorides is considered to be the indicator of pollution due to high organic wastes of animal or industrial origin. The present study found a range of 14.3 to 59.2 mg/l of chloride. The maximum concentrations of chlorides were recorded during April and minimum during summer (Singh, 1965; Verma, 1969 and Tripathi and Pandey, 1990).

Total hardness value was found maximum (53.2 mg/l) in April and minimum (48 mg/l) in October. According to Das and Das (1995) the fishes have been found to be susceptible to diseases in water with hardness below 20 mg/l. Epizootic Ulcerative Syndrome outbreak has been observed to be more frequent in waters with low hardness. The productive water bodies should have hardness above 20 mg/l. Optimal hardness for fish culture has been observed to be around 75 to 150 mg/l. The total hardness values recorded in Hingni (Pangaon) reservoir indicate that the reservoir is productive.

Total dissolved solids ranged from 39 to 43 mg/l with an average value of 41.9. The reservoir water showed maximum value in the month of May, while its minimum was during August.

References

APHA (American Public Health Association) (1985). *Standard Methods for the Examination of Water, Sewage and Industrial Wastes, 16th edition.* American Water Association and Water Pollution control federation. APHA, Washington, USA.

Boyd, C.E. (1982). *Water Quality Management for Pond Fish Culture.* Elsevier Publication.

Das, Manas Kr. and R.K. Das (1995). Fish diseases in India: A review. *Environment and Ecology*, pp. 533-541.

Das, R.K. (1996). Monitoring of water quality, its importance in disease control. *Proc. of National Workshop on Fish and Prawn Disease, Epizootics and Quarantine Adoption in India at Central Inland Fisheries Research Institute*, Barrackpore, pp. 51-54.

Jhingran, A.G. (1990). Recent advances in the reservoir fisheries management in India. In: *Reservoir Fisheries of Asia*, Sena De Silva, ed. Proceedings of the 2nd Asian Reservoir Fisheries Workshop held in Hangzhov Peoples Republic of China, 15-19 October, 1990.

McKee, J.E. and H.V. Wolf (1963). *Water Quality Criteria.* State of California, State Water Quality Control Board, Sacramento.

Singh, M. (1965). Phytoplankton periodicity in a small lake near Delhi: I. Seasonal fluctuations of the physico-chemical characteristics of the water. *Phykos.*, **4**: 61-68.

Thirumathal, K., A.A. Sivakumar, J. Chandrakantha and K.P. Suseela (2002). Physico-chemical studies of Amaravathy reservoir, Coimbatore district, Tamil Nadu. *J. Ecobiol.*, **14(1)**:13-17.

Tripathy, A.K. and S.N. Pandey (1990). *Water Pollution.* Ashish Publishing House, New Delhi.

Verma, M.N. (1969). Hydrobiological study of a tropical impoundment, Tekeapur reservoir, Gwalior, India with special reference to the breeding of Indian carp. *Hydrbiol.*, **34**: 358-368.

11

Physico-chemical and Biological Aspects of Pollution of Groundwater due to Sewage and Waste Disposal in Pingalgad Nala Stream of Parbhani City, Maharashtra

☆ *Md. Babar*

Introduction

Water quality surveillance is much felt in recent days due to contamination of rivers, lake water, and groundwater and even treated piped drinking water as a result of rapid industrialization and increased men's activities. Periodical monitoring of water quality in all the habitation is indeed cumbersome due to lack of laboratory facilities and it involves high cost in developing countries like India. Groundwater is a replenishable source and is also an economical resource. It has a few inherent advantages over the surface water. The wide distribution, negligible evaporation loss, low risk of pollution, fairly closeness at hand, more uniform character, relatively free from harmful bacteria and the capability to make it available at a low capital cost in a short time, are a few advantages. Despite these advantages the groundwater is found to be polluted because of the release of untreated sewage into nearby streams and rivers is a common practice in many areas.

The study area around Pingalgad nala stream in Parbhani district is located between latitude 19°10′ N and 19°20′ N and longitude 76°40′ E and 77°15′ E (Figure 11.1). The tributary stream of Pingalgad nala is flowing through the heart of the Parbhani City. The total length of Pingalgad nala is 40.2 km and meets the Purna river at Neela village near Purna city in Parbhani District. Finally, at a distance of 15 km the Purna River meet the mighty Godavari River at Kanteshwar in Parbhani District. Nearly 1 Million litres per day of sewage waste generated in Parbhani city is released in the stream of Pingalgad nala (Chavan *et al.*, 2002), which pollute the area, round the nala it moves in the Purna River and finally discharged in the Godavari River. For the present study water samples from dug wells and bore wells situated around Pingalgad nala are considered for physico-chemical and biological aspects.

Geologically, the study area belongs to Deccan basalts of late Cretaceous to early Eocene age. Hydrogeology of the area revealed that the types of basalt such as compact (massive) basalt and amygdaloidal basalt yield good amount of groundwater only when the joints, fractures and deep weathering occurs in the rock (Dhokarikar, 1991; Babar and Kaplay, 1999; Babar, 2000 and Chavan *et al.*, 2002). Many fracture lineaments have been reported in Parbhani District, which are better sites of high bore well yield (Murthy and Jayaram, 1996).

Materials and Methods

Groundwater samples from 20 open dug wells and bore wells and 2 stream water samples were collected in 1 litre plastic cans. The physico-chemical characteristics of water samples determined with reference to methods of APHA (1989) are pH, total hardness (TH), total dissolved solids (TDS), calcium (Ca), magnesium (Mg), sodium (Na), potassium (K), chlorides (Cl), carbonates (CO_3) and bicarbonates (HCO_3).

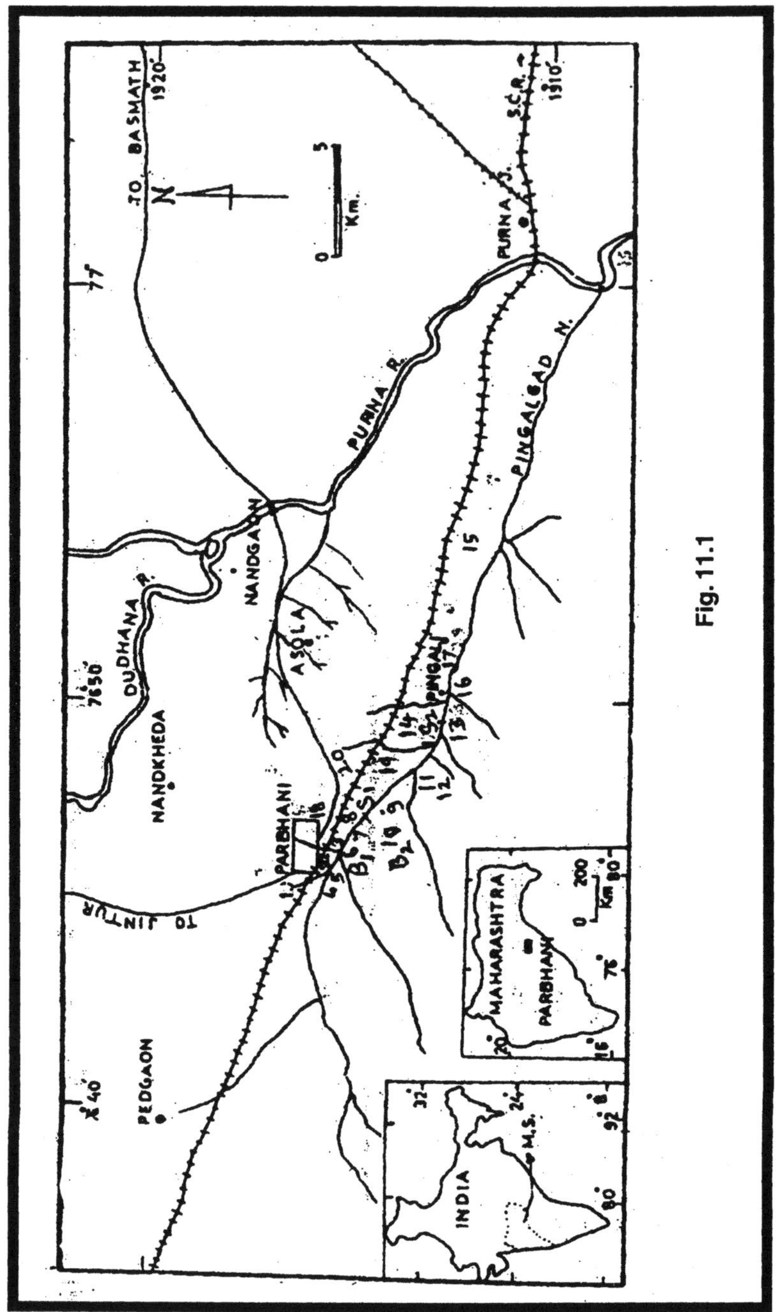
TO BASMATH
PURNA R.
PINGALGAD N.
DUDHANA R.
NANDGAON
ASOLA
NANDKHEDA
PARBHANI
PEDGAON
TO JINTUR
PURNA J.
S.C.R.
PIMPALA
Km.
MAHARASHTRA
INDIA
M.S.

Fig. 11.1

For zooplankton collection conical type Nylon plankton net of 150-175 was used. The plankton net was dipped in well water at different patches from surface up to 4 metre depth. The plankton samples were separately fixed and preserved by using 100 ml wide mouth bottles in 4 per cent formalin solution. The qualitative analysis was carried out to determine different species of phytoplanktons and zooplanktons by using compound microscope. The identification of plankton species is done by referring APHA (1989) and biological methods for field and laboratory investigations by (Trivedy and Goel, 1986).

The presence or absence of the Coliform bacteria in the water samples is detected by using PA Coliform Kit prepared by Himedia Laboratories Ltd., Mumbai. The kit contains sterile disposable bottle–100 ml capacity, dehydrated medium–3 X concentration and 1 ziplag bag. 100 ml of the water sample to be tested is transferred in the disposable bottle. The entire quantity of dehydrated medium (Lauril Triptose Broth with BCP/P A Broth) is added slowly in the water by swirling to dissolve the powder completely. After dissolution, the bottle is incubated for 14 to 48 hours at 30–35°C. The colour change of the medium is observed from reddish purple to yellow, indicating the presence of Coliform bacteria.

Result and Discussion

During the field work it was found that the bad smell of the water in the wells is very much like the stream water. By the oral discussion with farmers we came to know that the working in such water causes the problem of etching and skin disease. Physico-chemical characteristics of groundwater samples around Pingalgad Nala stream are given in Table 11.1. The pH of the water samples varies from 6.9 to 8.4, which indicates that all the water samples are alkaline.

Total hardness (TH) varies between 260 and 960 mg/l. The values of TH for sample numbers 3, 4, 5, 17 and 19 (Table 11.1) are higher (> 600 mg/l) as compared to maximum permissible limits of BIS (1991) and WHO (1992). TH values between 150 and 300 mg/l means the water is hard and the value greater than 300 means it is very hard (Todd, 1995). High concentration of TH in water may cause kidney stone and heart disease in human (Jain, 1996).

Table 11.1: Physico-chemical Parameters of Groundwater and Effluent Water

Sample No.	*pH*	*TH*	*TDS*	*Total Alkalinity*	*Ca*	*Mg*	*K*	*Na*	*Cl*	*SO_4*
1.	7.6	284	583	232.56	75.00	36.50	8.0	73.4	23.00	29.25
2.	7.8	352	529	275.00	92.52	52.85	11.3	122.0	20.80	26.25
3.	8.4	766	1892	470.00	156.91	102.16	23.4	102.2	135.00	424.35
4.	8.4	880	1744	380.00	153.10	72.24	13.7	220.0	192.00	403.25
5.	7.9	608	1595	335.00	139.78	56.41	9.7	120.0	208.00	450.00
6.	6.9	350	591	235.00	100.40	36.62	2.0	92.2	114.50	29.00
7.	7.9	427	743	493.00	148.29	54.23	4.6	145.6	154.00	35.90
8.	7.8	580	568	480.50	79.85	51.65	1.4	126.8	165.00	25.65
9.	7.9	960	1534	422.50	162.20	61.52	12.9	190.0	250.00	300.00
10.	7.9	420	811	227.50	120.24	42.84	1.7	63.6	80.00	31.35
11.	8.3	425	504	320.00	64.12	51.42	1.3	186.0	163.00	132.50
12.	7.7	340	849	220.00	92.18	73.84	4.2	41.4	130.90	200.00
13.	7.9	260	547	270.00	92.38	38.43	6.0	150.0	96.00	40.00
14.	8.1	382	628	301.55	101.34	43.25	7.2	58.2	75.05	123.25
15.	7.5	372	565	352.00	85.50	40.00	3.8	71.5	43.00	23.50

Contd...

Table 11.1—Contd...

Sample No.	*pH*	*TH*	*TDS*	*Total Alkalinity*	*Ca*	*Mg*	*K*	*Na*	*Cl*	*SO_4*
16.	8.4	568	1205	450.00	165.50	70.54	12.3	154.0	225.50	312.00
17.	8.2	753	1673	437.50	187.53	87.08	15.4	235.9	324.00	404.00
18.	7.8	458	732	245.00	88.65	65.00	6.2	93.6	185.00	155.50
19.	7.2	607	835	298.00	78.63	64.75	3.6	135.7	231.00	176.00
20.	7.9	393	546	212.00	67.42	46.26	4.6	105.7	178.50	94.00
B1	7.1	500	1895	380.00	118.25	41.32	22	36	47.00	87.00
B2	7.2	409	822	192.00	49.35	24.52	26	14	47.50	46.50
S1	7.1	286	543	208.50	42.50	18.50	12	21	175.00	287.00
S2	7.4	308	524	262.00	89.00	32.00	3.8	24.5	212.50	325.00

Except for pH all concentrations are in mg/l.

As regards Total Dissolved Solids (TDS), the BIS (1991) maximum permissible limit is 1500 mg/l for drinking water. The values of TDS in the study area ranged from 504 mg/l to 1895 mg/l. The values of TDS in the water samples from well numbers 3, 4, 5, 9, 17 and BI are found to be greater than the maximum permissible limit of BIS (1991). It could be due to the sewage and solid waste disposal in the natural streams. It is non-palatable and may induce unfavourable physiological reactions in the transient consumer. Further high TDS indicates that the water is highly mineralized, which is, in turn unsuitable apart from potability for industrial applications also. The concentrations of dissolved solids over 2000 mg/l produce a laxative affect (Dhembare *et al.*, 1998 and Jain, 1996).

The concentrations of Ca and Mg determined in the present area ranged from 42.50 mg/l to 187.53 mg/l for Ca and 18.50 mg/l to 102.16 mg/l for Mg (Table 11.1). The values of Ca and Mg are found to be well below the maximum permissible limits given by BIS (1991).

Natural water contains low Chlorides (Cl). The concentration of Cl in the water of the area ranged from 20.80 mg/l to 324.00 mg/l. The BIS (1991) maximum permissible limit for Cl in drinking water is 1500 mg/l. In all the well samples the values of Cl are well below this limit. The Sulphate (SO_4) concentration in natural water ranges from 0.0 to 1000 mg/l. As per NRC (1977) SO_4 cannot be removed from the range by any of the common treatments. The maximum permissible limit of SO_4 for drinking water given by BIS (1991) is 400 mg/l. The range of SO_4 concentrations in the groundwater in the present study area is from 23.50 to 450.00 mg/l. The values of SO_4 are found to be less than the maximum permissible limit of BIS (1991) except for the water sample numbers 3, 4, 5 and 17. The high concentration of SO_4 may induce diarrhoea similarly the laxative effect may occur at lower concentration if Mg is present in the water at an equivalent concentration.

Most of the water samples were consisting of zooplanktons like *Phyllodiaptomus, Mesocyclops, Cletocampus, Heliodiaptomus* and *Daphnia* (Table 11.2). *Phyllodiaptomus* is common in almost all the water samples while *Mesocyclops* also found all the water samples except sample numbers 1, 6, 12, 13, 15 and 20. It is well known that the *Cyclops* acts as an intermediate host tu carry the larval stages of dreadful roundworms.

Table 11.2: Zooplanktons and Coliform Bacteria Recorded from Water Samples

Well No.	*Location*	*Phyllo-di-aptomus*	*Moeso-cyclops*	*Cletoca-mpus (Male & female)*	*Heliodia-ptomus*	*Daphnia*	*Coli-form*
1.	Parbhani	+	–	–	–	–	–
2.	MAU Parbhani	+	+	+	+	+	–
3.	MAU C–block Parbhani	+	+	+	+	+	–
4.	Raipur	+	+	+	+	+	+
5.	Raipur	+	+	+	+	+	+
6.	Balsa	+	–	–	–	–	–
7.	Balsa	+	+	+	+	+	+
8.	Balsa	+	+	+	+	–	–
9.	Sayala	+	+	+	+	+	+
10.	Sayala	+	+	+	+	+	–
11.	Paralgavan	+	+	+	+	+	–
12.	Paralgavan	+	–	+	+	+	–
13.	Takalgavan	+	–	–	–	–	–
14.	Sendra	+	+	+	+	+	+
15.	Mirkhel	+	–	+	+	+	–
16.	Pingli	+	+	+	+	+	–
17.	Pingli	+	+	+	+	+	+
18.	Parbhani	+	+	+	+	+	–
19.	Sendra	+	+	+	+	+	–
20.	Khanapur	+	+	+	+	+	–
B1	Balsa	–	–	–	–	–	–
B2	Sayala	–	–	–	–	–	–
S1	Raipur	+	+	+	–	+	+
S2	Sendra	+	+	–	–	+	+

* MAU = Marathwada Agriculture University; + = Detected; –: Not detected

The occurrence of *Cyclops* may pose an alarming danger for possible occurrence of Dracunuclosis in the region (Chavan *et al.*, 2002). The zooplanktons like *Cletocampus, Heliodiaptomus* and

Daphnia are also found in most of wells except for water sample numbers 1, 6 and 13.

The study of bacteriological contamination of the area is carried out to find out the presence of coliform bacteria. Regarding the bacteriological contamination of the well water located near Pingalgad nala stream (well Nos. 3, 4, 5, 7, 9, 14, 17, S1 and S2) it is observed that the contamination is too high. The Faecal coliform growth in the groundwater and stream water is similar to that of sewage.

Conclusion

The present study indicates that the groundwater in the wells located along the Pingalgad nala are contaminated due to sewage and waste disposal in stream. The chemical concentrations of TH, TDS and SO_4 of well numbers 3, 4, 5, 9, 16, 17 and 19 are found to be greater than the maximum permissible limit given by BIS (1991). The bad smell of water and the problem of itching during working in the water in the field may cause problems of health hazard and crop damage.

The occurrence of heavy load of Cyclops and other zooplanktons in nearly all the wells indicates dangerous sign of pollution. The bacteriological investigation clearly reveals a quantitative and qualitative picture of pathogens, which transmit the harmful diseases to human being, through the groundwater available adjacent to sewage. The presence of pathogenic bacteria in such a high rate, may poses many risks to public health. Certain austerity measures should be taken immediately to prevent pollution of the stream and by the stream.

Such problems can be solved by taking the preventive measures like training of natural streams or nalas, proper disposal of refuse, monitoring of tap and valves of pipelines, bleaching of drinking water wells and awareness for keeping the clean environment. In order to prevent the pollution, indiscriminate dumping of waste and debris, as well as misuse of stream bank by slum dwellers and villagers has to be prevented. The sewage flow in the stream can be controlled by widening the existing sewage drain and by constructing parallel sewage drain. Seepage from the Pingalgad nala stream is to be restricted by effective concrete lining on the sides of the stream.

Acknowledgement

The author gratefully acknowledges the financial assistance given by U.G.C. Western Region Office, Pune towards minor research project. I also take an opportunity to thank Dr. S.P. Chavan Dept of Zoology and Dr. A.V. Manwar Deptt. of Microbiology, Dnyanopasak College, Parbhani, for their help during fieldwork and laboratory analysis work.

References

APHA (1989). *Standard Methods for the Examination of Water and Wastewater, 17th edition.* American Public Health Association, Washington D.C., pp. 1131-1138.

Babar, Md. (2000). Management of groundwater resources through watershed developments in Parbhani district of Maharashtra. In: *Resource Management and Contours of Development Reflections through Macro-Micro Narratives,* B.C. Barik, (ed.) Rawat Publications, Jaipur, pp. 350-364.

Babar, Md. and Kaplay, R.D. (1999). Groundwater quality around Pingalgad Nala, Parbhani District. *J. Ecology, Environment and Conservation,* **5(2)**: 141-142.

BIS (1991). *Indian Standard Specification for Drinking Water.* BS–10500.

Chavan, S.P., P.K. Joshi, S.B. Pawar and Md. Babar (2002). Qualitative analysis of planktons in wells around Pingalgad Nala in Parbhani District (Maharashtra) India. In: *Biotechnology in Agriculture, Industry and Environment,* A.M. Deshmukh, (ed.) Proc. Vol. of International Conf. of SAARC Countries, pp. 336-342.

Dhembare, A.J., G.M. Pondhe and C.R. Singh (1998). Groundwater characteristics and their significance with special reference to public health in Pravara area, Maharashtra. *Poll. Res.,* **17(1)**: 87-90.

Dhokrikar, B.G. (1991). *Groundwater Resources Development in Basaltic Rock Terrain of Maharashtra,* B.G. Dhokrikar, (ed.) Water Industry Publ., Pune, pp. 122-185.

Jain, P.K. (1996). Hydrogeochemistry and ground water quality of Singhari river basin, District Chhatarpur (MP). *Poll. Res.,* **15(4)**: 407-409.

Murthy, K.N. and Jayaram, K.M. (1996). Groundwater resources and development potential of Parbhani district, Maharashtra, C.G.W.B Report on Annual Action Programme (1995-96). 702/ DIS/96, Nagpur, pp. 1-39.

NCR (1977). *Drinking Water and Health.* National Research Councils, Vol. 4, National Academy Press, Washington, D.C.

Todd, D.K. (1995). *Groundwater Hydrology*, John Wiley and Sons, N.Y., p. 282.

Trivedi, R.K. and P.K. Goel (eds.) (1986). *Chemical and Biological Methods for Water Pollution Studies.* Environmental Publications, Karad, India, pp. 35-80.

WHO (1992). Revision of WHO guidelines for drinking water quality: Report of the final task group meeting at Geneva, Switzerland, pp. 21-25.

12

Relationship of the Rate of Growth of Plants with the Effect of Industrial Pollution at Kalu River in the Content of Minerals (Sodium, Calcium, Magnesium) in Its Vegetation

✰ *S.A. Salgare & R.N. Acharekar*

Introduction

The term 'Pollution' has a variety of meanings for different people. To some it means only the discharging of untreated wastewater, or sediments from urban development, to others introduction of pesticides and agricultural chemicals. To the sportsman, pollution may be water temperature too high for sports, fish to survive or massive fish kill resulting from toxic chemicals present in terms of algae and decaying plant growth. To those who work in water analysis and toxicological effect on the vegetation, pollution is all of this and more. This study might include physical measurement such as pH, temperature, conductivity in addition to various biochemical and chemical determinations. Included in this later category are nitrates, nitrites, phosphates, sulphates, chlorides, fluorides, major cations (sodium, potassium, calcium and magnesium) and last but not least trace elements.

Materials and Methods

Kalu is a small river which in its course upstream is very close to Titwala. This river after flowing westwards in its downstream receives a small river Bhatsai and later meets the Ulhas river near Kalyan, an industrial suburb of Mumbai. The combined river flows for a further distance of approximately 45 kms by receiving effluents from several industrial units. Ambivali a little further from Titwala also receives the Kalu river. The dumping of the major industries situated in that area (*i.e.* Ambivali) are of the Balakrishna paper mill, Rayon factory, Dye factory. Their effluents are dumped into this river. The Kalu river ultimately meets the Arabian sea approximately 50 kms north of Mumbai on the west coast of peninsular India.

Present investigation deals with the relationship of rate of growth of plants with the effect of industrial pollution on the content of minerals (sodium, calcium, magnesium) of the vegetation of the Kalu river (Table 12.1). To detect the toxic effect of polluted water of Kalu river, the following 3 weeds (*Celosia argentea, Corchorus capsularis, Corchorus olitorius*) were collected from the sites of Titwala, Ambivali and other clean areas of Mumbai (treated as control). Ambivali (considered to be highly polluted) where the number of industries are situated. The dumping of these industries is directly into the Kalu river polluting it. Titwala situated close to Ambivali is considered as less polluted since there are no industries in that area and being in the upstream of Kalu river there is dilution of the water and toxic effects are less felt.

Sodium and calcium were determined quantitatively from ash solutions as per Crosly (1977). EDTA method of Vogel (1978) was used for the estimation of magnesium. Data obtained is statistically analyzed applying 't' test. Mineral content was presented in mg/100 g dry weight. Percentage inhibition was also determined.

Table 12.1: Relationship of Rate of Growth of Plants with Effect of Industrial Pollution at Kalu River on the Content of Minerals of its Vegetation (Values given are mean ± SE of 10)

			Mineral Content in mg/100 g Dry Weight					
			Sodium		Calcium		Magnesium	
Species	H	Sites	Sodium	% DFC	Calcium	% DFC	Magnesium	% DFC
C. argentea	I	I	10.26±0.02	– 36.66	120.00±0.01	– 10.44	75.20±0.01	– 14.64
		II	08.29±0.01	– 48.82	104.00±0.01	– 22.38	62.24±0.02	– 37.13
	II	I	16.10±0.02	– 19.50	146.00±0.02	– 07.59	86.25±0.01	– 12.87
		II	12.20±0.01	– 39.00	112.26±10.01	– 29.07	72.26±10.01	– 27.01
		I	22.46±10.01	– 14.53	151.00±0.01	– 10.65	92.20±0.01	– 09.60
	III	II	18.26±0.02	– 30.51	126.00±0.01	– 25.44	78.21±0.02	– 23.32
C. capsularis	I	I	20.26±10.01	– 19.79	110.20±0.01	– 12.70	66.21±0.02	– 19.46
		II	15.21±0.02	– 57.98	100.0±0.02	– 20.78	58.29±0.01	– 29.09
	II	I	32.12±0.01	– 12.70	121.60±0.01	– 08.23	78.29±0.01	– 12.31
		II	20.18±0.02	– 44.25	113.00±0.02	– 14.72	63.26±10.01	– 29.15
			42.20±0.01	– 06.65	128.60±0.01	– 07.13	80.21±0.01	– 13.06
	III	I	24.20±0.01	– 46.47	117.20±1.02	– 15.23	72.26±0.02	– 21.67
		II	22.00±0.01	– 28.18	130.50±0.01	– 08.42	58.29±0.01	– 14.88

Contd...

Table 12.1–Contd...

Species	H	Sites	Mineral Content in mg/100 g Dry Weight					
			Sodium		Calcium		Magnesium	
			Sodium	*% DFC*	*Calcium*	*% DFC*	*Magnesium*	*% DFC*
C. olitorius	I	I	14.46±0.01	– 48.83	113.20±0.01	– 20.56	52.21±0.01	– 23.75
		II	24.65±0.01	– 18.53	142.20±0.01	–10.14	66.24±0.01	– 10.79
	II	I	20.20±0.01	– 33.24	122.16±0.02	– 22.81	58.26±0.01	– 21.54
	III	II	36.21±0.01	– 14.21	150.26±0.01	– 09.50	72.21±0.01	– 08.92
		II	24.21±0.01	– 42.64	12813±0.02	– 26.37	63.26±0.01	– 20.21

DFC: Difference from Control; H: No. of Harvests; P: mineral Content in Samples Collected from Polluted Area; Site I: Less Polluted Site; Site II: High Polluted Site.

Results and Discussion

In a natural ecosystem there is a balance between input and output, with hardly any accumulation or waste material. But rapid urbanization, industrialization and increasing human population are all responsible for the discharge of excessive waste into environment which adversely effect the life of organisms.

Industrial pollution at Kalu river inhibited the content of all the 3 minerals (sodium, calcium and magnesium) in all the 3 weeds (*Celosia argentea, Corchorus capsularis* and *Corchorus olitorius*) in all the 3 harvests collected from both the sites *i.e.* Ambivali as well as Titwala. It should be pointed out that very high percentage of inhibition in the content of all the 3 minerals in all the 3 weeds investigated was recorded in the collections made from Ambivali than Titwala (Table 12.1). Suwarna Gawde (1988) and Sudha Asthana (1988) also observed the similar results in case of plants treated with heavy metals. Andhyarujina (1988) recorded decrease in mineral content of the weeds growing along the bed of polluted Ulhas river. Salgare and Andhyarujina (1990) stated that industrial pollution at Ulhas river caused inhibition in inorganic content of its vegetation. Recently Salgare and Shobha Andhyarujina (1990, 96) recorded an inhibition in the mineral contents of the vegetation of Patalganga caused by the industrial pollution.

It should be pointed out that as such no definite conclusions could be drawn as far as the relationship of rate of growth of plants with the effect of industrial pollution on the content of minerals of the vegetation of the Kalu river is concerned (Table 12.1).

References

Andhyarujina, K.B. (1988). *Ecotoxicological Studies of Ulhas River–I.* Ph.D. Thesis, Univ., Bombay.

Asthana, Sudha (1988). *Effect of Industrial Pollution on Physiological and Biochemical Aspects of Crops–I.* Ph.D. Thesis, Univ. Bombay.

Crosly, N.T. (1977). Determination of metals in foods. *The Analyst,* **102**: 225-269.

Gawde, Suwarna (1988). *Effect of Industrial Pollution of Crop Physiology–I.* Ph.D. Thesis, Univ. Bombay.

Salgare, S.A. and K.B. Andhyarujina (1990). Effect of polluted water of Ulhas river on the inorganic content of its vegetation–II. *J. Phytol. Res.*, **3**: 39-44.

Salgare, S.A. and Shobha Andhyarujina (1990). Effect of polluted water of Patalganga on the mineral contents of its bank vegetation–I. *New Agriculturist*, **2**: 9-14.

Salgare, S.A. and Shobha Andhyarujina (1996). Effect of polluted water of Patalganga on the mineral contents of its bank vegetation–II. In: *Dimensions of Safe Environment*, R.N. Trivedi and M. Roy, (eds.) Anmol Publications, New Delhi, Chapter IV, pp. 85-94.

13

Physico-chemical Characteristics of a Small Percolation Tank

☆ *Yogesh Shastri*

Introduction

Considerable hydrobiological studies have been carried out on man made lakes in India (Abbasi *et al.*, 1996; Madan Mohan Rao *et al.*, 1996 and Bahura 1998). Few studies have been conducted on Physico-chemical aspects from Malegaon region of Maharashtra. Therefore present work was undertaken. The present study was undertaken to ascertain water quality status of small percolation tank, near Ravalgaon (Malegaon). The water of this tank is used for irrigation. The tank play an important role in maintaining the water table.

Materials and Methods

The study site *i.e.* percolation tank is situated 20 kms away from Malegaon city, near Ravalgaon. Malegaon, a taluka headquarter, is the second, largest city located at 20°32′ N and 74°35′ East in the Nasik District of Maharashtra State. Maximum depth of water is 2 meters in rainy season. Three sampling stations C_1, C_2 and C_3 were established for sampling purpose. Water samples were collected at monthly intervals in acid washed 2-litre plastic bottles from three stations. Water samples were analysed for number of physico-chemical parameters as per the procedures given in standard methods (APHA, 1985). Except temperature and pH rest of the parameters are expressed in mg/l.

Results and Discussion

Temperature

The temperature of tank water varied from 17°C to 26°C, showing higher temperature 26°C in May and 17°C in July at all the three stations of tank which was due to seasonal impact. In summer water temperature was generally higher compared to winter months. Generally water temperature was corresponding with air temperature. The samples were collected from shallower zones in which fluctuation in air temperature had a direct impact on temperature of water. Welch (1952) has observed that shallow water reacts more quickly to changes in atmospheric changes.

Table 13.1: Water Quality of Ravalgaon Percolation Tank

Parameters	*Stations*		
	C_1	C_2	C_3
Temperature	17°–26°C	17°–26°C	17°–26°C
pH	6.80–9.5	6.28–8.7	6.21–9.16
DO	5.6–24.96	4.4–25.77	4.4–32.21
Free CO_2	00–96.8	00–90.2	00–110
Calcium	4.08–44.08	2.40–34.46	9.6–49.6
Hardness	130–270	140–270	144–290
Total Alkalinity	Pa = 00–05	Pa = 00–05	Pa = 00–05
	Ta = 30–50	Ta = 30–60	Ta = 30–50
Phosphate	00–0.28	0.02–0.26	00–0.22
Nitrate	00–0.35	0.01–0.23	00–0.34

All values in mg/l; Pa: Phenolphthalein alkalinity; Ta: Total alkalinity.

pH

pH is one of the important abiotic factors that serve as an index for the pollution. In present study variation in pH between 6.21 to 9.5 was observed. Minimum pH 6.21 was observed in the month of

June at station C_3, while maximum of 9.5 in the month of August at station C_1. Abbasi *et al.* (1996) recorded the range of 6.04 to 7.5 in the case of Kuttadi Lake. Tripathi and Pandey (1990) also observed maximum pH value during rainy season in pond waters of Kanpur. Low pH in the month of June, may be due to temperature condition. The pH value ranges between 6.21 to 9.5 which is not safe for aquatic life, irrigation and domestic use.

Dissolved Oxygen (DO)

DO is one of the most important parameters to indicate water quality and its relation to the distribution and abundance of various algal species. Presence of DO in water may be due to direct diffusion from air or photosynthetic activity to autotrophs or because of both. Minimum of 4.4 mg/l was recorded in the month of February at station C_2, while maximum of 32.21 mg/l was recorded in the month of June at station C_3. Masood Ahmad and Krishnamurthy (1990) showed a positive correlation between temperature, duration of sun light and soluble gases like DO. In present study such correlation between DO and water temperature was evident.

Free CO_2

Free CO_2 was totally absent at all the three stations from March to June. The maximum value of free CO_2 level *i.e.* 110 mg/l was recorded at Station C_3 in the month of December. Water showed free CO_2 in morning samples due to biotic community respiration. During day time due to photosynthesis water was generally CO_2 free (Sahu and Behera, 1995).

Present observation suggest that in winter biomass of primary producers is higher that exerts profound influence on the two cases (O_2 and CO_2).

Calcium

Calcium is very important element influencing the flora of ecosystem which plays potential role in metabolism and growth. It has an effect on pH and carbonate of the system. The range of calcium content varied from 2.40–49.69 mg/l. The maximum and minimum calcium concentration was recorded in the month of February but at different stations *i.e.* C_3 and C_2 respectively. These values are well within the limits fixed for calcium (*i.e.* 75 mg/l). According to Ohle (1934) any value above 25 mg/l indicate calcium rich water. As per this definition the water of this tank is rich in calcium.

Total Hardness

The total hardness values in present study was varied within a range of 130–290 mg/l. The highest value of hardness 290 mg/l was recorded in the month of September at Station C_3 and minimum value of 130 mg/l was recorded in the month of December at station C_1. WHO (1984) permissible limit for total hardness is 150 mg/l and ISI (1983) limit is 300 mg/l. Todd (1995) suggested that the values between 150 to 300 mg/l of total hardness means water is hard and value greater than 300 mg/l means water is very hard. High concentration of hardness may cause problem of heart disease and kidney stones (Jain, 1996).

Alkalinity

Alkalinity plays an important role in controlling enzyme activity. Water alkalinity is a measure of acid present in water and of the cations balanced against them. Phenolphthalein alkalinity showed, a range from 00–5 mg/l. Total alkalinity showed variation between 30–60 mg/l.

Phosphate

Phosphate concentration was low in the water. The maximum of 0.28 mg/l was at station C_1 in November. The minimum of 00 mg/l was at Stations C_1 and C_3 in month of March. Zero value of phosphate in present study may be due to low productivity of percolation tank.

Nitrate

In present study water samples showed concentration of this nutrient *i.e.* 00–34 mg/l. The zero concentration of nitrate was recorded for Moosi and Vaigai rivers (Sabata and Nayar, 1955). The lower value clearly indicate oligotrophic nature of present water body.

On the basis of above physico-chemical parameters, it may be concluded that water of this tank is calcium rich and hard, this may cause problem of heart disease and kidney stones. The low value of nitrate and phosphate suggest that tank is oligotrophic. In present study positive correlation between temperature and DO was evident.

Water of this tank is unfit for human consumption but can be used for irrigation purpose.

Acknowledgement

Author is thankful to General Secretary M.G. Vidyamandir, Malegaon Camp and Principal S.P.H. Mahila College, Malegaon-Camp, for permission and providing necessary laboratory facilities.

References

Abbasi, S.A., K.S. Bhatiya, A.V.M. Kunhi and R.S. Soni (1996). Studies on the limnology of the Kuttadi Lake (North Kerala). *Eco. Env. Cons.*, **2**: 17-27.

Ahmad, Masood and R. Krishnamurthy (1990). Hydrobiological studies of Wohar reservoir, Aurangabad (Maharashtra State). *Indian J. Environ. Biol.*, **11(3)**: 335-343.

APHA (1985). *Standard Methods for the Examination of Water and Wastewater, 18th ed.* American Public Health Association, Washington, D.C.

Bahura, C. K. (1998). A study of physico-chemical characteristics of a highly eutrophic temple tank, Bikaner. *J. Aqua. Biol.*, **13**: 47-51.

ISI (1983). *Indian Standard Specification for Drinking Water*, ISI 10500.

Jain, P.K. (1996). Hydrochemistry and ground water quality of Singhari river basin, Distt. Chattarpur (M.P.). *Poll. Res.*, **15(4)**: 407-409.

Ohle, W. (1934). Chemical and physi kalishe under such innenhor seen. *Arch. Hydrobiol.*, **26**: 389.

Rao, Madan Mohan, A.V. Narendra Rao and S.K. Mahmood (1996). Assessment of water quality and pollution of Narsingi pond. *Eco. Env. Cons.*, **2**: 45-49.

Sabata, B.C. and M.P. Nayar (1995). *River Pollution in India: A case Study of Ganga River.* APH Publishing Corp., New Delhi.

Sahu, B.K. and S.K. Behera (1995). Studies on same physico-chemical characteristics of the Ganga river water (Rishikesh–Kanpur) within twenty four hour during winter, 1992. *Eco. Env. Cons.*, **1(1–4)**: 35-38.

Todd, D.K. (1995). *Groundwater Hydrology*. John Wiley and Sons, New York.

Tripathi, A.K. and S.N. Pandey (1990). *Water Pollution*. Ashish Publishing House, New Delhi.

Welch, P.S. (1952). *Limnology, 2nd Edition*. McGraw-Hill Book Co., New York.

WHO (1984). International standards for drinking water. *World Health Organisation Tech. Report*.

14

Soil Mapping in Datia District with Reference to Remote Sensing Technique

☆ *Satish Kumar Chakravarty & D.R. Tiwari*

Introduction

The utility of Landsat data for small soil mapping studies using various techniques of remote sensing applications has been demonstrated by several workers (Karale *et al.*, 1978 and Venkataratnam, 1981). They revealed that space borne data afford greater accuracy, economy and efficiency than the conventional method at reconnaissance level mapping. A fair degree of accuracy and saving in time of mapping have been reported by several workers, particularly, result reported by Hiwing *et al.* (1974), Mirajkar and Srinivasan (1975) proved that visual interpretation of landsat data is very promising in small scale soil mapping upto 1 : 50,000 scale.

The potential landsat thematic mapper data for soil mapping has been demonstrated by Rao *et al.* (1988). The interpretation technique involving conjunctive use of false colour composites of different band combinations at 1 : 25.000 and 1 : 50,000 scales, supported by limited ground checks revealed that mapping upto abstraction level of sub-grounds and associations of sub-groups was possible at 1 : 25,000 scale the 1 : 50,000 scale FCC's afforded segregation of families and their associations. Studies on soil mapping in part Tawa catchment, M.P. using IRS L1SS–II data showed better segregation of soil classes (Sinha *et al.*, 1989).

National Bureau of Soil (NBSS) and Land Use Planning (LUP) has employed the visual interpretation technique of Remote Sensing Landsat MSS FCC on 1 : 250,000 Scale for soil resource mapping with three approach (Quoted by Gawande, 1990).

Types of Soil in the Study Area

Soil maps prepared by NBSS and LUP and Development Alternatives, New Delhi a Non-Government Organization on 1 : 250,000 scale for Datia district has been used for the present study. Each map unit roughly conform to a natural geomorphic unit. Each unit consists of one or more major soils and some minor soils. The soils occurring in one unit may occur in other units but in a different pattern. Within a boundary of a soil map unit, occurrence of soil series which are not members of the association permitted upto the extent of 20 per cent (Soil of Datia District, Bundelkhand Region, Madhya Pradesh, NBSS and LUP Nagpur, 1985).

According to soil report prepared by development Alternatives, New Delhi (1992), the soils of Datia district are found to be related to physiography (Table 14.1 and Figure 14.1).

Description of the Soil

Soil description highlighting their characteristics and qualities and presented here based on report prepared by NBSS and LUP and Development Alternatives, New Delhi.

Entic Chromusterts (Black Soils)

1. These are dark greyish brown to greyish brown, silty clay to clay textured, very deep soils locally called as black soils or Kabar/Mar soils. These soils are friable, sticky and plastic with moderate, medium sub-angular blocky structure. These have high water holding capacity and high cation exchange capacity. These respond very favourably to irrigation and fertilisation. These are suitable for all climatically adopted crops of the area.

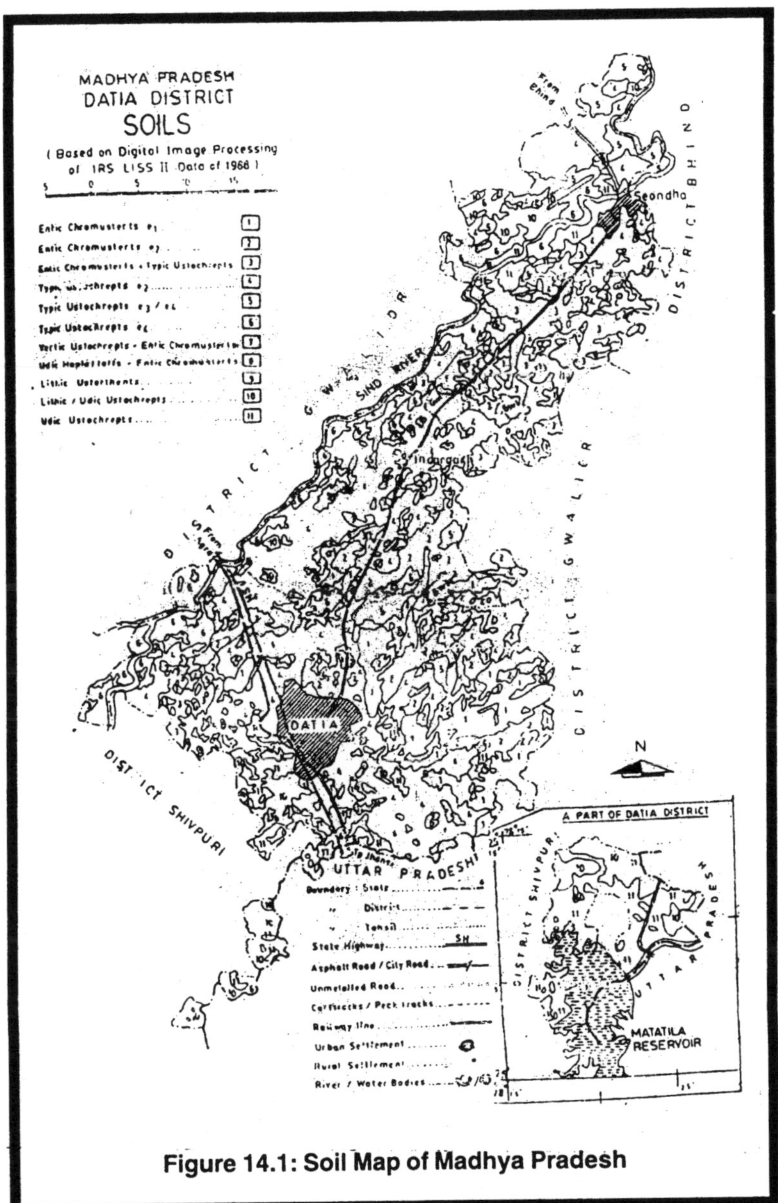

Figure 14.1: Soil Map of Madhya Pradesh

Table 14.1: Relationship Among Geology, Physiography, Soils and Soils Series

Geology	*Physiography/Geomorphic Unit*	*Soil Series*	*Soils*
Quartzite Granite	Hills and Hill Slopes	Bishramganj	Lithic Ustorthents Lithic Ustorchrepts Lithic Ustorchrepts
Granite/ Granite Gneiss	Pediment Inselberg complex, Burried Pediment	Datia Dumra	Lithic Ustorchrepts
Granite	Burried Pediplain Burried Pediment	Darsura Sundra	Vertic Ustorchrepts Entic Chromusterts Typic Ustochrepts e2 Entic Chromusterts Saline
Younger Alluvium	Younger Alluvial plain Dissected Alluvial plain (deep ravines), V highly dissected Alluvial plain (V deep ravines)	Morha Badankala Badankala	Typic Ustipsamment Typic Ustochrepts Typic Ustochrepts e3/e4 Typic Ustochrepts e4
Older Alluvium	Older Alluvial Plain	Marha Badankala Jamra	Udic Haplustalfs Entic Chromusterts

2. These soils moderate to moderately low permeability and imperfect drainage. These pose the problems of water logging and salinity under irrigation, if proper drainage improvement measures are not implemented. These also susceptible to erosion. Based on the variations in land slopes and erosion, two phases are encountered in the area. These together, cover 55792.65 or 27.4 per cent of the district area.

3. Locally, these soils have developed salinity in low lying basinal areas with impeded drainage. Such saline soils are, locally called "Usar", and account for 0.1 per cent of the district area. The Marha·Jamra and Sundra series consist of these soils. These soils of this association occur on nearly level to gently sloping land. The landscape consists of extensive alluvial plain formed by basaltic outwash.

Lithic Ustorthents (Red Hilly Soils)

These are reddish yellow to brown, sandy loam textured, stony, shallow to very shallow soils of hills and hill slopes. Locally known as red hilly soils or 'Rakar' soils, these are loose very friable, non plastic and non-sticky with rapid permeability well to excessive drainage and low cation exchanged capacity. The water holding capacity is very low. These soils are unfit for agriculture. These cover 3432.6 ha. or 1.7 per cent of the total area. The Bisaramganj series consists of this soil occurring on gently sloping to steep lands. The landscape consists of sandstone hills of Gwalior series and denudational hills of granite.

Udic Ustochrepts (Red Earth)

These are greyish-brown, moderately deep to very deep, sandy loam to sandy clay loam textured soils developed from granite and granite gneisses on upper pediplain. These are loose, friable, slightly sticky and slightly plastic. These have low to medium cation exchange capacity, low available water holding capacity and slightly acidic reaction. These are well drained with rapid permeability. These are suitable for agriculture but require intensive soil conservation measures to control erosion and soil loss. Depth variability of these soils are found to be at closer intervals. These cover an area of 17057.30 ha or 8.4 per cent of the total area. The Badankala series consists of this soil.

Typic Ustochrepts (Alluvial Soils)

These are yellowish brown to dark yellowish brown deep to very deep, sandy clay loam to silty loam textured, highly calcareous soils of alluvial plain and ravinous lands. Locally these are known as "Parwa" soils. These soils are loose, slightly sticky and slightly plastic. These have moderate to low water holding and cation exchange capacities and are moderately well drained with moderate permeability. These are non-comprehensive with very high hazard of erosion. The soils, located on alluvial plain and table lands are suitable to a variety of climatically adopted crops under irrigation, with more than two crops of favourable slopes. Choice of crops, under rainfed agriculture is determined by the moisture conditions of the soils related to physiography and slope. On the basis of variability in slope gradients and erosion, three phases have been mapped under this group of soils. They together cover 95,795.65 ha. or 47 per cent of the district area. Gullied land Badankal Misc. series consists of this soil.

Udic Haplustalf (Old Alluvial soils)

These are pale brown to dark greyish brown deep to very deep soils developed from granite and granitic gneiss on pediplains. These soils are characterised by sandy clay loam surface soil underlain by clay horizon for the latter giving rise to poor permeability. These are soft very friable, slightly sticky and plastic soils with weak, medium, crumb granular structure. The water holding capacity and cation exchange capacity of these soils are moderately high. These have slightly acidic reaction. These soils are moderately well drained with moderate to low permeability and susceptible to erosion. These are suitable for all climatically adopted crops of the area including rice. These show high response to nitrogenous and phosphatic fertilizers. These occupy only small areas of 937.70 ha. in the district accounting from 0.5 per cent. Dumra series consists of this soil.

Typic Ustipsamments (Reverine Sands)

These are light coloured, moderately deep to deep non-cohensive, structureless loamy sand to sand with slight to moderate calcareousness, with only limited extent (0.4 per cent of the district area). Locally these soils are encountered along the remnants of old levees and in the dry stream beds. Because of very low cation exchange and water holding capacities, these soils have very limited agricultural use. With heavy inputs of manners and fertilizers, these can be used for cucurbits under irrigation. Very high infiltration rates and excessive drainage restrict the choice of irrigation to drip/pitcher systems.

Vertic Ustochrepts (Dark Grayish Brown Soils)

These are dark greyish brown, deep to very deep, moderately fine textured, weakly cracking soils with moderate erosion encountered on the fringe of pediplain. The soils are moderately well drained with low permeability. The water holding capacity and cation exchange of these soils are moderately high. The soils are under cultivation. The Darsura series consists of the soils.

Lithic Ustochrepts (Rocky Soils)

These are brown to strong brown, moderately coarse textured, stony and rocky soils of hill slopes with erosion. The soils moderately deep with weak structural development. These are well to excessively drained with moderate to rapid permeability. The water holding capacity and cation exchange are low. The soils are mostly under bushy forests with patches of cultivation. Datia series consists of the soil.

References

Anon. (1972). Rev. Res Work (1959-1969) Vol-II Soil Conservation Deptt. D.V.C. Hazaribagh Quoted by Gautam and Narayan in Wastelands of India. 1988, 59 p.

Central Soil Water Conservation and Training Research Centre, Datia.

Census, Data 1991, (Obtained from NIC), Bhopal.

Dhruvanarayana, V.V. (1986). Soil and Water Conservation Research in India. *Indian J. Soil Conserv.*, **14(3)** 22-31.

Lillies and Keifer. *Interpretation of Remote Sensing*.

Soil report prepared by Development Alternatives, New Delhi.

15

Toxicity of Some of the Heavy Metals on Freshwater Crab *Barytelphusa cunicularis*

☆ *B.S. Yadav*

Introduction

It is a matter of great proud that, there is growing importance and awareness for environmental conservation and protection. At last we have realized the fact that we should no longer disturb the finally adjusted ecological balance between environmental factors and the physiological and behavioural adaptation of an organism. Otherwise our own existence on this earth would become insecure. Our society has witnessed and realized in the past few decades that, large scale pollution in a variety of ways of our environment is an evil and dangerous bi-products of tremendous technological advancement in various fields.

In the recent decade, metals in addition to pesticides are known to pollute the environment through indiscriminate discharge of un-treated and partially treated industrial effluents into water resources. They also cause detrimental effects to the living resources of river, estuarine, coastal and oceanic life of food chain importance. Many metals have been extensively used in agricultural operation in recent years. The run-off water from agricultural land contaminates the aquatic system with number of metals used in pesticide preparation.

Heavy metals discharged into the aquatic environment as by-products of industrial processing of ores and metals drainage of residues of mining and leaching of metals from garbage and solid waste dumps. As a result the organism tend to accumulate these metals in excess, leading to severe metabolic disturbances which may ultimately results in the gradual elimination of the effected species.

Materials and Methods

In the present experiment fresh water crab *Barytelphusa cunicularis* were procured from "*Girana Dam*" near Malegaon, Nasik District, Maharashtra. In the laboratory crabs were maintained in plastic troughs containing tap water. Crabs were fed with pieces of earth worms and water was changed every alternate day. Prior to the commencement of the experiment crabs were acclimatized to laboratory condition for at least 3 to 5 days. During experimentation crabs of approximately same size were used. Crabs were not fed during experiment.

The stock solution of Mercuric chloride, Mercuric sulphate and Copper sulphate were prepared by dissolving 1.0 g in 100 ml of tap water and from the stock solution different concentrations were prepared. Water parameters were measured throughout the experiment (pH–7.4; temp. –28°C; D.O.–4.6 ml/litre and hardness–210 mg/litre).

Crabs were divided into several batches and each batch comprising 10 crabs. Each batch of crabs were kept in plastic troughs containing two litres of water. The crabs were exposed separately to different concentrations of Mercuric chloride, Mercuric sulphate and Copper sulphate and mortality rate was recorded after 24, 48, 72 and 96 hours of exposure. The experiment was repeated thrice and the date were plotted for calculating the LC_{50} values using the un-weighed regression method of probit analysis. (Finney, 1971).

Results

The results of the present investigation have been summarised in the Table 15.1. The percentage mortality increased progressively upto 96 hours, in all the concentration of Mercuric chloride, Mercuric sulphate and Copper sulphate, The LC_{50} values decreased with increasing exposure period showed an inverse relation.

Table 15.1: Toxicity of Heavy Metals on Freshwater Crab *Barytelphusa cunicularis*

Exposure Period	*Pollutants (LC_{50} Values in ppm)*		
	Mercuric Chloride	*Mercuric Sulphate*	*Copper Sulphate*
24 hours	2.2±0.0426	2.8+0.0264	3.6±0.0296
48 hours	1.6±0.0084	2.2±0.0048	2.94±0.0094
72 hours	1.2±0.0062	1.8±0.0634	2.4±0.0156
96 hours	0.92±0.0034	1.2±0.0042	1.62±0.0234

The LC_{50} values of Cadmium chloride, Cadmium sulphate and Copper sulphate for 24, 48, 72 and 96 hours are shown in Table 15.1. The value clearly shows that, Copper sulphate is least toxic to crab *B. cunicularis* were as Cadmium chloride is most toxic. The toxicity of these heavy metals to the crabs is an increasing order from $CuSo_4 > CdSo_4 > CdCl_2$.

Discussion

The main water pollutants of the river system are metals, pesticides, detergents, chemicals, industrial and domestic waste including organic substances. One of the harmful effects of these pollutants in water system is on crabs. The present study obviously indicates that as the concentrate of Cadmium chloride, Cadmium sulphate and Copper sulphate in the media increased, the percentage survival of *B. cunicularis* decreased. When the physical and chemical conditions of the test solution were kept constant.

In the present experiment it was seen that, Mercuric chloride is more toxic than Mercuric sulphate and Copper sulphate. Ghate and Mulherkar (1978) observed the toxic effect of Copper sulphate to fresh water prawn *Macrobrachium* and *Caridina*. They further reported that cause of the death in crustaceans after metal exposure may be due to the damage of respiratory surface. Similar observations were made by Nagabhushanam *et al.* (1981) on fresh water prawn *Macrobrachium kistenensis* exposed to heavy metal pollutants.

Machale *et al.* (1989) reported the toxicity of Copper sulphate and Cadmium sulphate on the fresh water crab *B. cunicularis.* Further she stated that heavy metals may cause depletion of body metabolite reserves and ultimately lead to death of the animal. Sarojini *et al.* (1990) reported the effect of Cadmium chloride on histology and bio-chemical contents of hepatopancreas of the fresh water crab *B. guerini.* Joshi *et al.* (1981) on fresh water prawn *Penaeus mergenensis* after acute exposure of mercury, reported that high susceptibility of prawn to mercury exposure may be due to impairment in osmoregulatory and oxygen consumption mechanism. Bodke (1983) studied the toxicity of different pesticide to fresh water crab *B. guerini* and observed that the order of toxicity was Mercuric chloride Sevimol, DDT and Copper sulphate. Shaikh (1996) studied the toxic effect of heavy metals on fresh water crab *B. guerini* and stated that Mercuric chloride is more toxic than Mercuric sulphate. Similar observations were made by Mule *et al.* (1987) on Bivolve mollusc–*Lamellidens marginalis* after Mercuric chloride exposure.

The presented data reveal that, all the heavy metals Mercuric chloride, Mercuric sulphate and Copper sulphate poses a threat to the environmental health. Hence it is imperative to rigorously screen all heavy metals from the effluents to prevent the toxic impact on aquatic animals, especially, which are of commercial importance.

Toxicity of Mercuric chloride, Mercuric sulphate and Copper sulphate to fresh water crab *B. cunicularis* were assessed, the present results reveals that the crabs are more sensitive to Mercuric chloride and less sensitive to Copper sulphate. It was also observed that the toxicity of heavy metals was inversely proportional to the time of exposure. Greater the exposure period lower the LC_{50} values. The lethal concentrations of heavy metals cause sluggish behaviour and uncoordinated movements in crabs.

Acknowledgement

Author is thankful to Principal S.C. Hale of Arts, Science and Commerce College, Nampur for providing laboratory facilities and also thankful to Shri D.G. Jadhav, Lecturer in Botany for help during the work.

References

Bodke, M.K. and R. Nagabhushanam (1983). Effect of carbamate pesticide on osmo-regulation of the fresh water crab *B.*

cunicularis. Proc. All. Ind. Symp. Physiol. Response of Animals to Pollutants, pp. 108-113.

Finney, D.J. (1971). *Probit Analysis*, 3rd ed., Cambridge University Pressm London.

Ghate, H.U. and Leela Mulherkar (1978). Histological changes in the gills of the fresh water prawn exposed to copper sulphate. *Ind. J. Exp. Biol.*, **77**: 838-840.

Joshi, P.K. and R. Nagabhushanam (1981). Acute toxicity of mercury to the juvenile of penaeid prawn *Penaeus merguinensis. J. of Env. Biol.*, **2**: 3-6.

Machale, P.R., A.K. Khan, R. Sarojini and R. Nagabhushanam (1989). Copper and cadmium induced changes in blood sugar level of the crab *B. cunicularis. Uttar Pradesh J. of Zool.*, **9(1)**: 113-115.

Muley, D.V. and U.H. Mane (1987). Sublethal effects of Mercuric chloride on the tissue composition of a mollusc *Lamellidens marginalis. Biol. Bull. of Ind.*, **9(1)**: 31-40.

Nagbhushanam, R. and G.K. Kulkarni (1981). Fresh water prawn *Macrobrachium kistensis*: Effect of heavy metal pollutants. *Proc. Ind. Nat. Sci. Acad.*, **47B(3)**: 380-386.

Sarojini, R., P.R. Machale, A.K. Khan and R. Nagabhushanam (1990). Effect of Cadmium chloride on histology and bio-chemical contents of the hepato-pancreas of the fresh water crab *B. guirini. Env. Series*, **4**: 91-97.

Shaikh, I.S. (1996). *Toxic Effects of Heavy Metals on Some Physiological Aspects of Crab B. guirini*. Ph.D. Thesis submitted to Dr. Babasaheb Ambedkar Marathwada University, Aurangabad.

16

Aquatic Pollution Induced Changes in the Lymphoid Organs of the Fish from Hussain Sagar Lake, Hyderabad (A.P.)

✫ *K.Y. Chitra & N. Sree Ramkumar*

Introduction

Hussain Sagar lake, situated in between the twin cities of Hyderabad and Secunderabad, Andhra Pradesh is recognized as the most heavily polluted water body (Khan and Hussain, 1976; Siddiqui and Rama Rao). The lake was originally built for the purpose of drinking water supply, but in the course of time due to urbanization and human settlements all around the lake, it has become a victim of pollution. Sewage, industrial effluents, dhobighats etc are the main sources contributing to the ever increasing pollution of the lake.

Alterations of immune functions form an important aspect of environmentally induced chemical toxicity. The simplest way of studying such alterations involves an assessment of morphological and histological changes in the lymphoid organs such as kidney and spleen. These two tissues are especially important in the immunological defence mechanisms as well as hematopoises of fish (Ellis *et al.*, 1976). The free and the fixed macrophages of these two tissues are considered to comprise the reticuloendothelial system (RES), which is the system of phagocytic cells, responsible for the removal of effete cells and particulate matter from the circulation.

Materials and Methods

The three different species of fish were captured live from both Hussain Sagar lake and the two relatively non-polluted water bodies. Fish of almost the same size and weight were sacrificed and the tissues *i.e.*, kidney and spleen, were aseptically removed and were weighed prior to fixation in 10 per cent buffered formalin and embedded in paraffin wax using the routine techniques. The sections were cut at 6μ thickness and were stained with hematoxylin and eosin for microscopical examination. Morphological observations were also made.

Results and Discussion

The kidney of the control fish consists of both excretory and lymphoid or hemopoietic tissue, well vascularised glomeruli, renal tubules and interstitial tissue (Figure 16.1). The kidneys of the affected fish showed considerable pathological changes. They were found ruptured, shrunken in size and found to be pale in colour. The interstitial hemopoietic tissue was especially affected, with normal tissue replaced by extensive granulormas. Occasionally in some of the tissue sections only a small area of normal hematopoietic tissue remained. Alterations such as vacuolation and degeneration in the tubular cells with varying degrees of cytoplasmic degeneration, shrinkage of glomeruli along with lymphocytic infiltration and macrophage invasion, decrease in the size of the renal tubular lumen and tubular necrosis was also noticed in the sections of the kidneys (Figure 16.2). Similar observations were reported by many workers in the laboratory animals exposed to different toxic agents (Tafanelli, 1972; Baker, 1969; Shimada, 1973). These pathological changes may be due to accumulation of pollutants that causes tubular degeneration leading to lymphocytic infiltration as a measure of resistance to the toxicants and tissue susceptibility. Histological

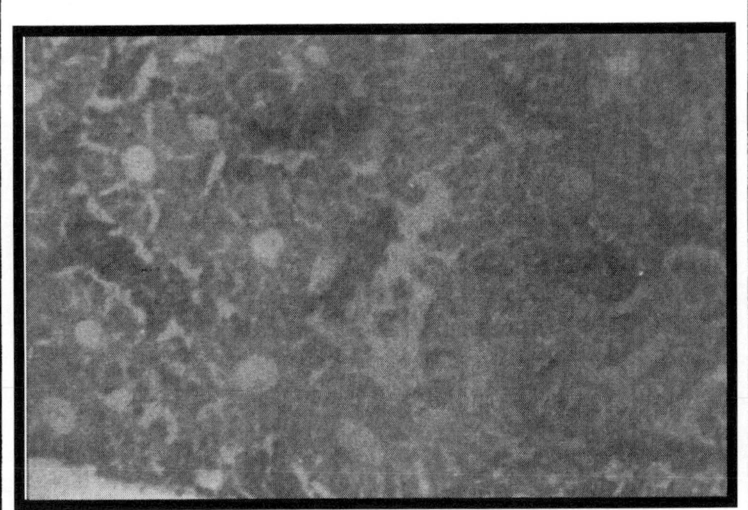

Figure 16.1: Selection of the Control Kidney Showing Hemopoietic Tissue, Glomerulus, Renal Tubules etc.

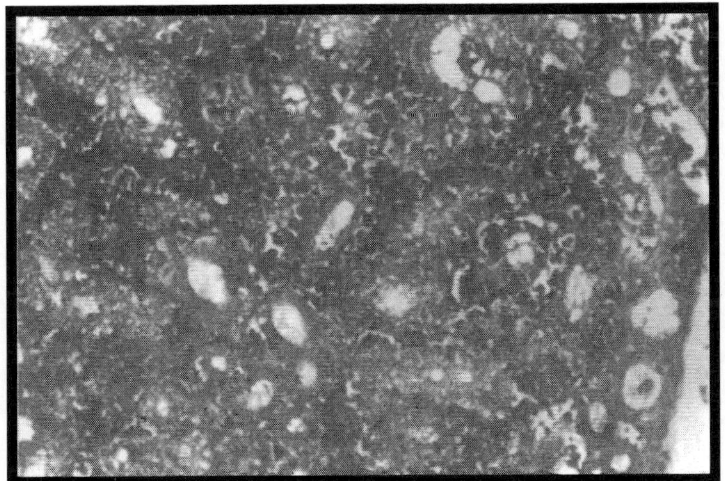

Figure 16.2: Selection of Polluted Water Fish, Showing Cytoplasmic Degeneration, Macrophage Invasion

changes in the glomeruli are seen in many disease processes, the incidence of which increases due to pollution. Because the renal tubular epithelium has, as its major function, the excretion of divalent ions, pollution with heavy metals such as mercury or cadmium is likely to affect these cells, as observed during the present investigation. Similar changes were reported in the southern flounder due to mercury poisoning.

The main elements of the control spleen are the ellipsoids, the white and the red pulp and the melanomacrophage centres (Figure 16.3). The ellipsoids are the thick walled filter capillaries which result from the division of the splenic arterioles. Each comprises of a thick basement membrane formed by a layer of sheathed compartments. These contain erythrocytes and phagocytic cells and are capable of trapping large quantities of particulate matter from the circulation. The spleens of the polluted water fish were small and pale with reduced hematopoietic tissue. The spleens of all the three species of fish examined were often significantly reduced in weight, shrunken, small in size, pale in appearance and atrophied. King (1962) also noted that the spleens of guppies exposed to DDT

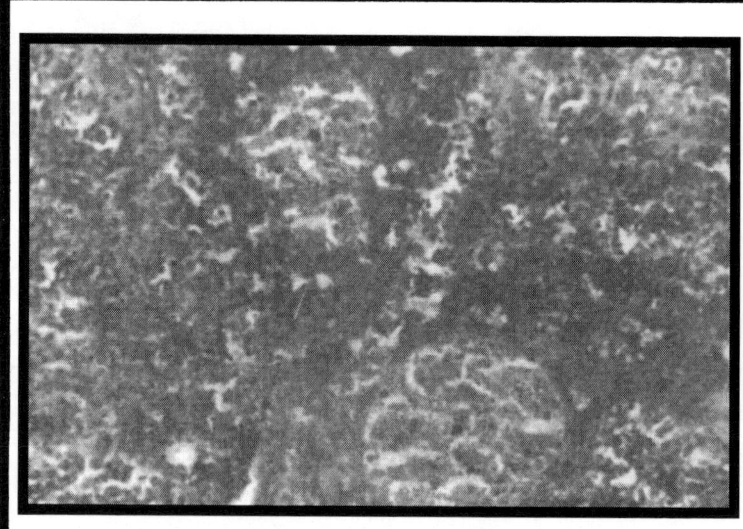

Figure 16.3: Selection of Control Spleen Showing White Pulp and Red Pulp

for one month were shrunken in size and paole in colour. Similarly, Walsh (1974) and Walsh and Ribelin (1975) reported increase in spleen weight in lake trout and coho salmon after 25 days of exposure to endosulfan.

The histological examination of the spleens of polluted water fish showed degeneration of red pulp and white pulp, vacuolation.and necrosis (Figure 16.4). Many sections of the spleen showed melano macrophage centres in the while pulp. The centres are surrounded by a collar of lymphoid cells with several pale staining ellipsoids in sections.

In the present study, it was observed that in the sections of kidneys and spleens of the polluted water fishes, the melano macrophage centres of the haemopoietic tissue of the lymphoid organs are usually ruptured and the pigment granules dispersed in toxaemic conditions.

The deleterious effects of pollutants on the histological structure of the target organs might result in immuno-deficiencies and

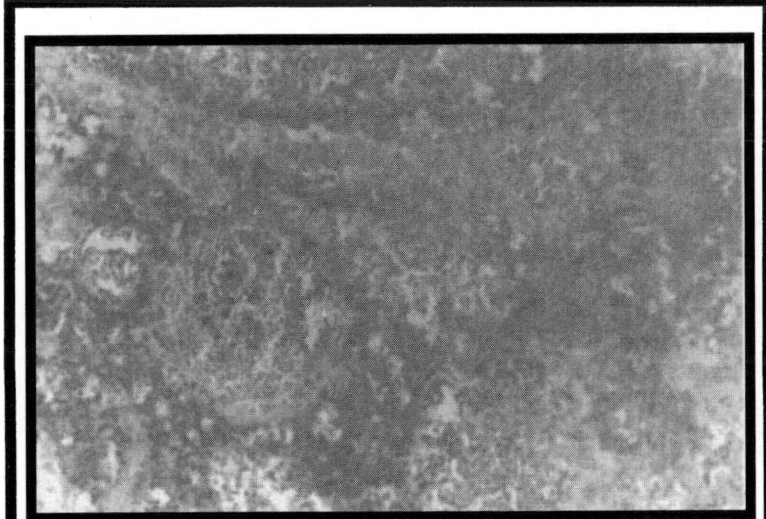

Figure 16.4: Selection of Spleen of Polluted Water Fish Showing (a) Degeneration of White and Red Pulp, Molonomacrophage Centres and Dispersion of the Pigments

susceptibilities of fish to infection and/or disease and this might be the cause for recurrent fish kills observed in the Hussain Sagar lake.

Acknowledgements

The authors are thankful to Prof. B. Raghavendra Rao and Prof. G.V. Rama Krishna for providing the laboratory facilities.

References

Baker, J.T.P. (1969). *J. Fish. Res. Bd. Con.*, **26**: 2785.

Ellis, A.E. Munroe, A.L.S. and R.J. J. Roberts (1976). *Fish. Biol.*, **8**: 67.

Khan, M.A. and Aziz Hussain (1976). *Indian J. Environ.*, **18(3)**: 227–232.

King, S.F. (1962). *U.S Fish Wild I. Serv. Spec. Sci. Rep. Fish.*, Washington D.C., **399**.

Shimada, J. (1973). English abstract in health aspects pest. *Abst. Bull.*, **6**: 32.

Siddiqui, S.Z. and K.V. Rama Rao (1991). *Poll. Res.*, **10(4)**: 191–198.

Tafanelli, R. (1972). Ph.D. Thesis, Oklahama State University.

Trump, B.F., Jones, R.T. and S. Sahaphong (1975). *The Pathology of Fishes*. Madison University Wisconsin Press, pp. 585–612.

Walsh, A.H. (1974). Ph.D. Thesis, University of Wisconsin, Madison.

Walsh, A.H. and W.E. Ribelin (1975). *The Pathology of Fishes*. University of Wisconsin Press, Madison, Chapter 22.

17

Integrated Studies on Land Resources Management for Western Part of Bhopal Using Remote Sensing and GIS

☆ *Satish Kumar Chakravarty & D.R. Tiwari*

Introduction

The optimum utilization and sustainable development of natural resources in general, land and water resources in particular on scientific lines without causing an imbalance to the natural geo-ecosystem is the major concern in present days. Remote Sensing is powerful spin off from space exploration and it has emerged as a useful tool for management of land characterization and conservation planning. Satellite Remote Sensing provides a reliable and accurate information on natural resources (land and water). Which is a pre-requisite planned and balanced development at watershed level (Ravindranan *et al.*, 1992). Remotely sensed data provides information for systematic analysis of various Lithological, Geomorphological, Soil and Land cover/Land use characteristics following synoptic and multi-spectral coverage of a terrain GIS.

Techniques having the capability of data storage, retrieval and manipulation, can play an important role in evolving suitable methods of arriving at alternate scenarios for natural resources (water and land), development and environmental planning. Remote sensing represents a powerful technology for providing input data for measurement mapping, monitoring and modelling with in the GIS content (Noviline *et al.*, 1993).

In the present study an attempt has been made to analyse terrain characteristics and generate maps of Soil, Slope, Geomorphology, Geology, Land cover/Land use and Land Resources Action Plan Information using SOI, Topographical, remotely sensed and field survey data.

Area of Study

The study covers Bhopal district lies in the Eastern edge of Malwa Plateau and is situated in the Central Part of the state between latitude 23°05′ and 23°55′ North and longitude 77°10′ and 77°39′ Eastern. The district is bounded by Rajgarh district in NW; Guna district in N; Vidisha, District in NE; Raisen district in E and SE, and Sehore district in the W and SW. The town Bhopal is Capital of M.P.

Climate and Rainfall

The district has three clearly distinguishable seasons which divide the year into three more or less equal parts. There are the summer, rainy and winter season corresponding to March, May, June–September and November–October, however witness a transition from the rainy to the cold weather.

The district receives rains from the Arabian Sea monsoon which commences by early June, July and August are the peak rainy months. The average rainfall is about 100 mm. December and January are the coldest months. The mercury touches 2°–4° when there is a cold wave. May is the hottest month when the average mean Max. and Min. temperature are 36°C and 24°C respectively.

Bhopal Plateau

This region is situated in the Southern Part of the district lying on the edge of Malwa Plateau Max. and Min. height of the region varies between 630 and 472 M (MSL). Most part of the plateau is with streams and thus the surface is rugged.

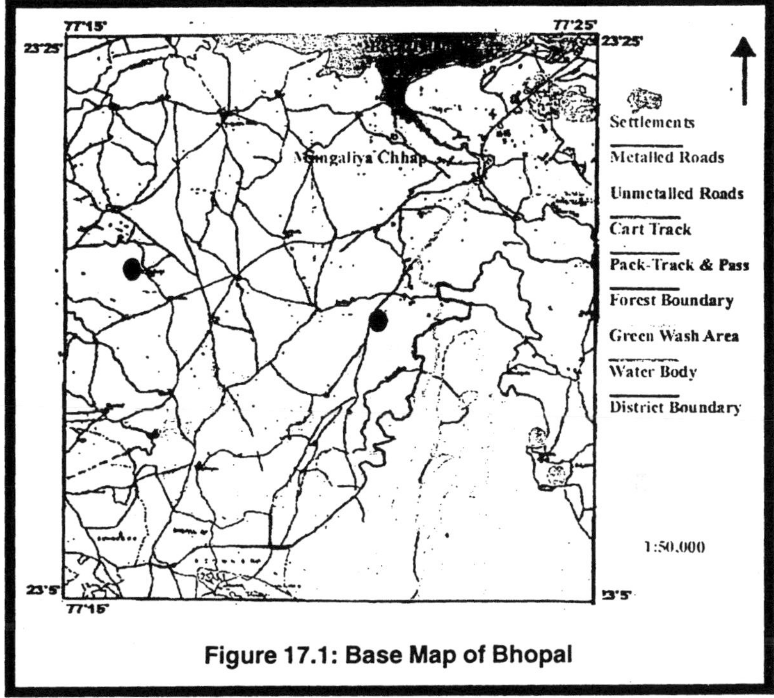

Figure 17.1: Base Map of Bhopal

The central part is at a higher altitude and the tract resembles the shape of Dome. It is the originating in plane of a number of streams draining in all directions. Notable among them are Halali, Betwa and Chamari in North, Kerwa and Kaliasot in South, Kolans in West and Ajner in East. The location of Bara Talab is an important physical feature of this region. A pocket of Bhopal Reserve Forest in located at the Southern part of the region.

Methodology

Data Used

Indian Remote Sensing Satellite IRS–ID, LISS–III digital data were used in the study. The false colour composite (FCC) was generated from green, red and near infrared (NIR) Spectral bands linear enhancement techniques were applied to improve visual information on Geomorphology, Land use/Land cover and linearnent pattern. Sulvey of India topograhical information on

1 : 50,000 scale and field soil survey data were used as ancillary information.

Data Preparation

The ground control points (GCP's) common to the map and image were identified and collected to register the image at sub-pixel accuracy using a 2nd order polynomial transformation. The digitally registered transformation. The digitally registered image was subsequently used to generated thematic information.

Base Map Preparation

The base map details for various themes should be uniform– Adequate ground features as available on the topomaps should be shown on the base map for facilitating ground referencing.

Major Drainage settlement with names, road network, railway lines, F.F. Boundary, Admin. Boundaries including block/tehsil boundary particulars with values, watershed boundary.

Drainage Pattern

The study area taken into consideration comes under Bhopal and Sehore district of Madhya Pradesh. The area is located at 77°15°25′ North and latitude 23°5′–23°25′ North longitude 77°15′ to 77°25′ East. There are two prominent water body *viz.*, Bara Talab in north and Chotta Talab in the north-east. Kolans river flows in the north-western side and meet into the Bara Talab.

Besides this towards east of Study are Kaliasot reservoir exist. The whole network comes under Betwa sub-Basin and a part of Ganga basin.

After careful observation and interpretation of the drainage map is prepared we come to the following conclusion which are discussed under the following heads:

Pattern

Dentritic drainage pattern is observed but in some places parallel pattern are also observed. Parallel pattern is predominant where the water is drained into the dendritic pattern in places homogeneous rocky strata.

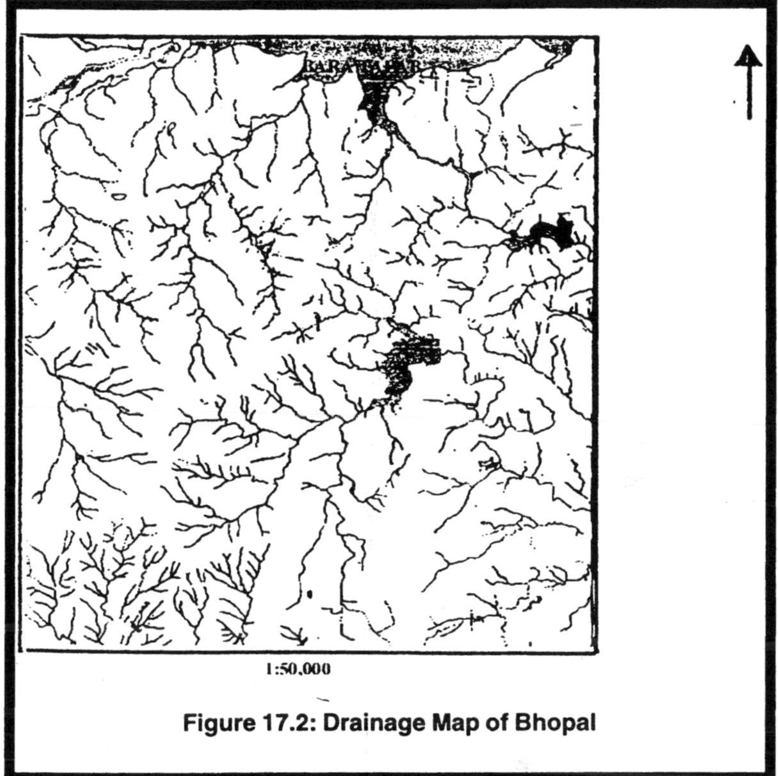

Figure 17.2: Drainage Map of Bhopal

Directions

The drainage forms a grid pattern and its convergence suggest that in most cases the flow is towards the permanent water bodies. In the North and NW the water flows towards the North direction and drains into the Kolans Nadi and the Bara Talab. Whereas in other parts the flow does not have definite patterned flows part of the study area in the west, SW and SE and Eastern areas. Another notable feature is the drainage in the central region where the parallel drains into the Kaliyasot reservoir.

Density

More or less the drainage is distributed evenly over entire area but we can observe high drainage density in the WS region whereas the SE region have the least network.

Order

Drainage order ranging from 1 to 6 are observed. Higher order drainage are seen in high density region (WS).

Runoff and Recharge Zone

Surface runoff as a result of rainfall is discharged into the permanent water bodies and in the North while in the other parts surface runoff is flown act of the study area except in the central part.

River Pattern

Both perennial and seasonal river are presents but seasonal rivers the out numbers the perennial ones.

Soil Map

Good understanding of soil with reference to their nature and distribution is essential to formulate any land based production system. Among agriculture soils, its suitability for different varies

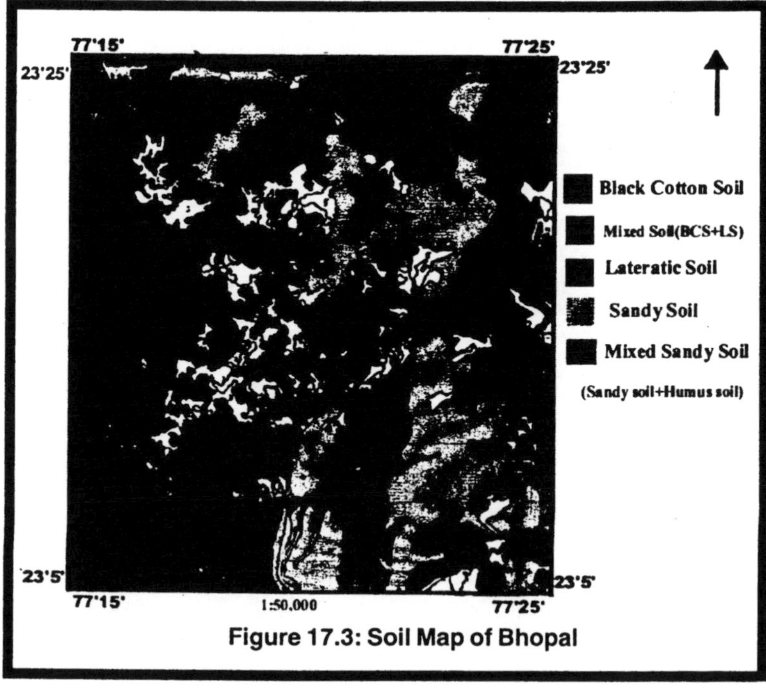

Figure 17.3: Soil Map of Bhopal

species. Soils are also known for their unique behavior under integrated and non-integrated conditions. Soil is a three dimensional body field work can be considerably reduced by Remote Sensing Technique use. In satellite image based soil mapping, emphasize on "Vision of Soil Domain". Thus a good understanding of its relationship with physiography and parent material is essential.

Classification

Soil Resource Map is be prepared on 1 : 50,000 scale. Soil mapping units will contain numerals as symbolized to represent soil series/association of soil series.

In integrated study phases like erosion, stones, slope etc. can be inferred from toposheet/other thematic maps.

Data Requirements

Satellite Imagery (geocoded standard FCC) on 1 : 50,000 scale of summer season having minimum land cover is used. At the same times rabi/kharif season data may be supplemented for additional informations to be extracted. 3 major components of soil mapping are:

1. Profile study
2. Laboratory analysis
3. Plotting soil boundaries.

Methodology

Overlays with base map details are super imposed on satellite imagery with the help of toposheet information, geology, geomorphology and by using the elements of the image interpretation, physiographic units are delineated. The details of soil survey procedures are given in soil survey manual IARI, 1971. Soil classification should be done as per latest version of USDA Soil Taxonomy. Subsequently classified soils are to be interpreted for different users following land capability and land irrigability classification.

Slope Map

Slope, aspect and altitude are important terrain parameters from land utilization point of view. Among the three, slope is very vital one for land irrigability and level capability assessment.

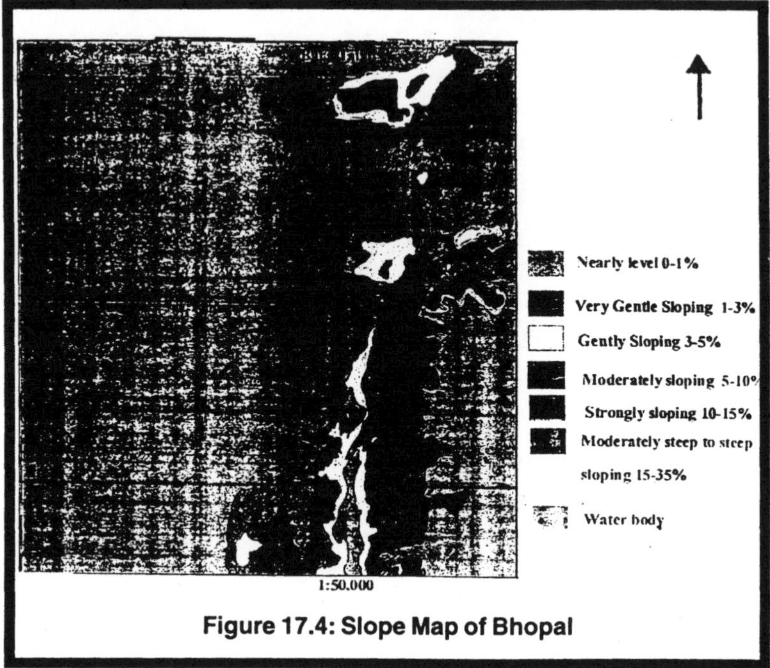

Figure 17.4: Slope Map of Bhopal

Classification

The slopes are broadly classified into seven groups as given in Table 17.1.

Table 17.1

Sl.No.	*Slope Category*	*Slope %*
1.	Nearly level	0–1
2.	Very gently sloping	1–3
3.	Gently sloping	3–5
4.	Moderately sloping	5–10
5.	Strongly sloping	10–15
6.	Moderately steep to steep sloping	15–35
7.	Very steep sloping	> 35

Steeper slopes of more than 35 per cent can be further sub-divided as per local need especially in hilly areas.

Methodology

Survey of India toposheets on 1 : 50,000 scale are to be used for deriving the information on slopes, aspect and altitude.

Toposheet on 1 : 50,000 scale give contours with 20 metre interval and its interval and its multiple *i.e.* 40 m, 60 m, 80 m etc. The vertical drop can be estimated/measured from the contour intervals and the horizontal distances in between the contours can be measured from maps by multiplying the map distance with the scale faster close spaced contours on the map have greater percentage slope as compared to sparse contours in the same space. Thus, density of contours on the map can be used for preparing the slope map that gives various groups/categories of slopes.

Finally each slope category should be numbered. The altitude contours should be drawn as thick lines. The final slope aspect map should thus show slope categories.

Land Use/Land Cover

Information on existing land use/land cover and pattern of their spatial distribution forms the basis for any developmental planning. The current land use has to be assessed for its suitability in the light of land potential before suggesting alternate land use practices.

Data Requirement

IRS geocoded FCC on 1 : 50,000 scale pertaining to two periods *i.e.* Kharif and Rabi seasons (preferably of the same agriculture year) are to be used for this purpose. Besides other collateral data as available in the form of maps, charts, census records, reports and especially topographical maps on 1 : 50,000 scale should be used.

Methodology

The land use/land cover maps have to be prepared adopting visual interpretation techniques in conjunction with collateral data such as topomaps and census record. The total number of classes would vary from area to area. The imagery has to be interpreted and ground checked for corrections. After field checking, the maps can be finalized.

The maps should be checked for quality by the quality assurance standardisation (QAS), *Team Before the same are finalised*. The quality

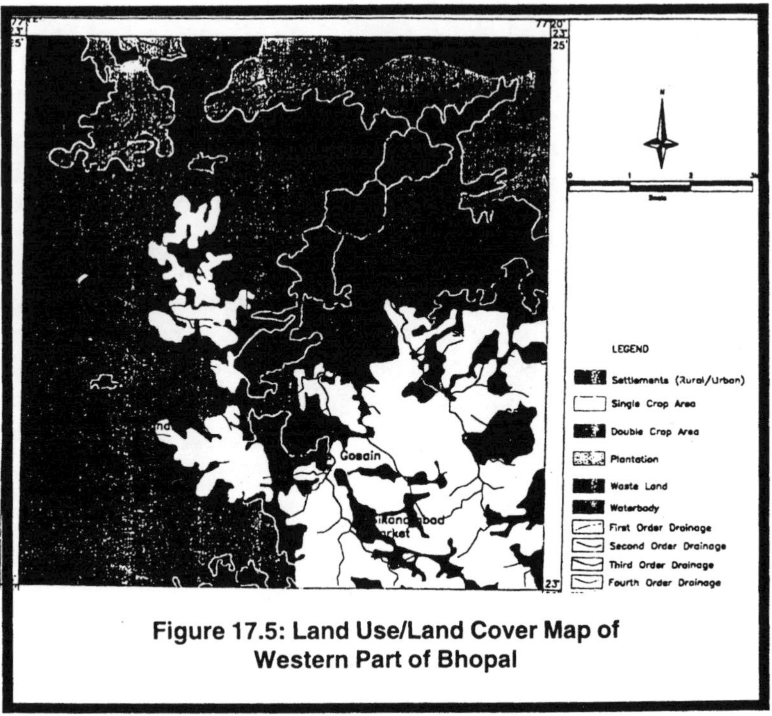

Figure 17.5: Land Use/Land Cover Map of Western Part of Bhopal

assurance check would be carried out two time *(i)* pre-field interpretation stage and *(ii)* field verification before finalizing the map.

Lithology

The lithology present in this area are Vindhyan sandstone in patelies in southern, south eastern and north eastern part. The rest of the area is made up of Deccan Traps (Basalts), basalt with soil cover in western and north western side.

The general geological succession of the area as established by GSI in Table 17.2.

Table 17.2

Group	*Period*	*Formation*	*Lithology*
Tertiary	Cretaceous/Eocene	Deccan Traps Vindhyan Sandstone	Basaltic Lava flows Sandstones

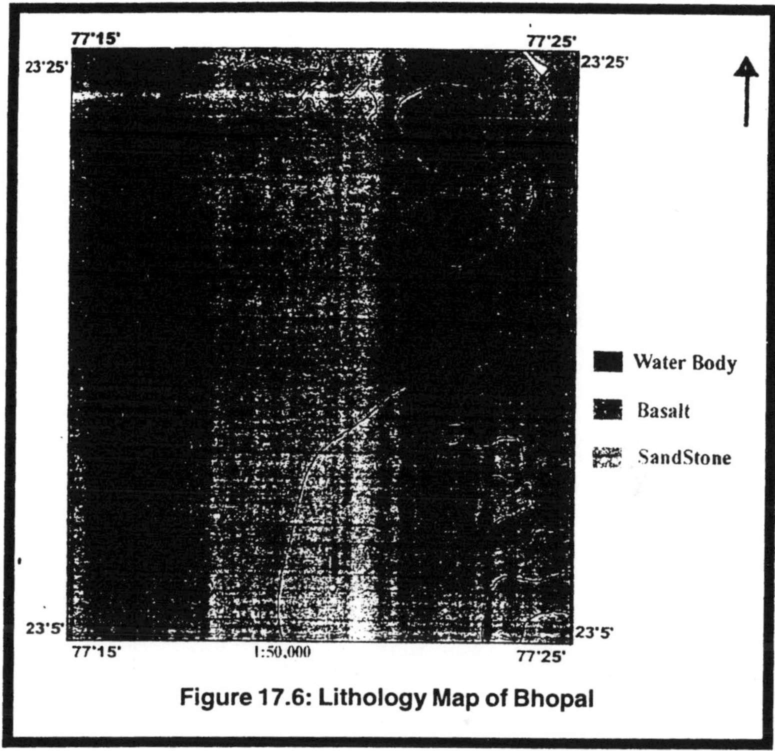

Figure 17.6: Lithology Map of Bhopal

Lineaments

A huge number of lineaments transverse the area. They are almost equally distributed all over the area. The majority of lineaments are trending is NE–SW directions. Some of the remaining lineaments are also trending in NW–SE directions.

Geomorphology

Geomorphologically the area can be divided into four parts *viz.,*

Buried Pediplain Sandstone

In southern part buried pediment sandstone is the coalescence of buried pediments where a thick overburden of weathered materials occur.

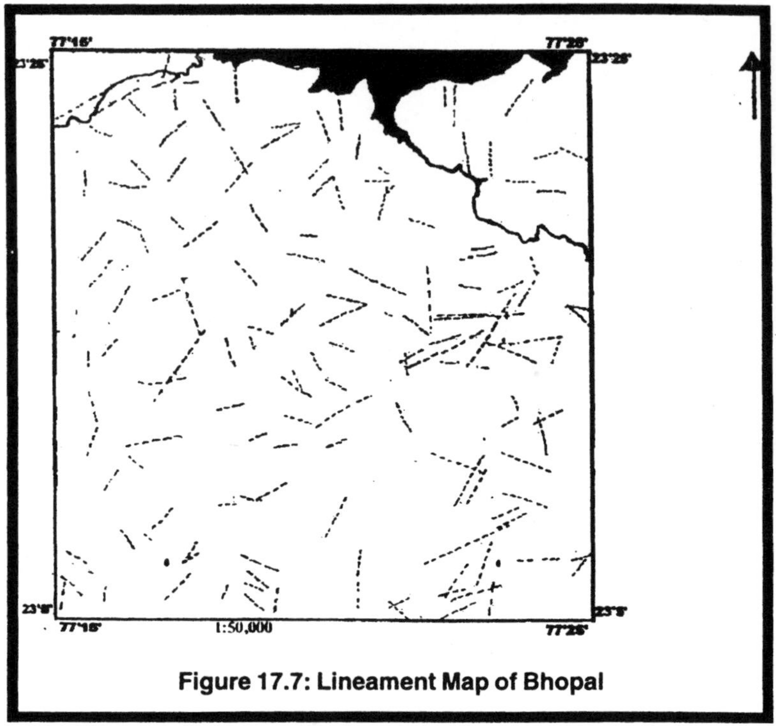

Figure 17.7: Lineament Map of Bhopal

Structural Hill Sandstone

In southern part which are linear to arcuate hills showing definite trend lines.

Denudational Hill Basalt

In the southern part formed by differential erosion and weathering, so that a more resistant formation of intrusion stand as mountains/hills.

Buried Pediplain Basalt

The rest of the area is Buried Pediment Basalt.

GIS Analysis

The drainage Contours, Geomorphology and Lineament were digitised in Autocad Map GIS as line vector layers. The covers were cleaned and topology, was built to generate the thematic layers of

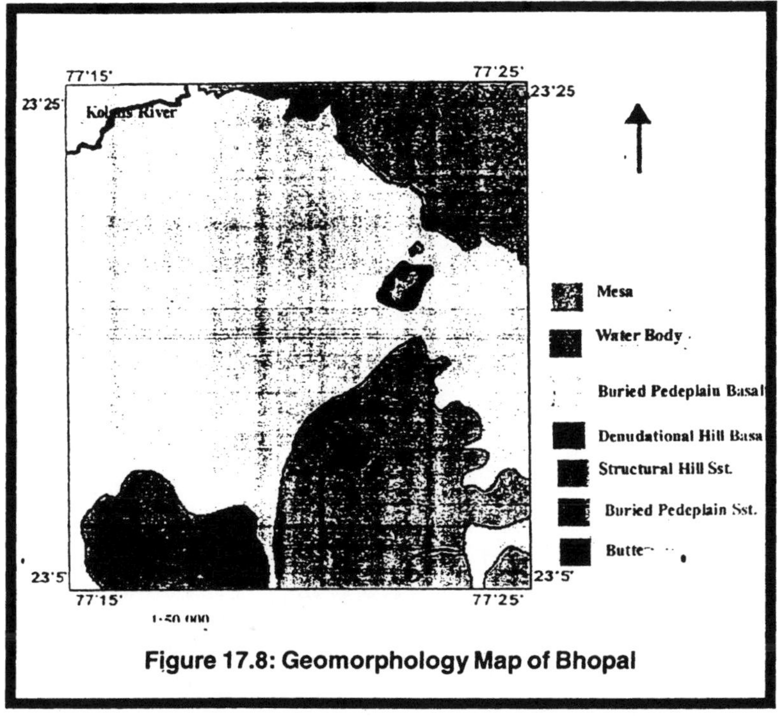

Figure 17.8: Geomorphology Map of Bhopal

Slope, Geomorphology, Soil depth and Lineaments. A buffer zones were generated for the Lineaments. The thematic information of Slope, Geomorphology and Lineament buffer zone were integrated using raster overlay (with union option) in GIS to delineate ground water prospect zones with respects Land Resources Management.

Results and Discussion

Environment and ecological system forms a complex factor in relation to support each other. Changes in these relational parameter is considerably slow in natural conditions, but with the fast development and due to some natural calamities these process variables can be changed abruptly. Thus it is necessary for sustainable development it is an essential and mandatory condition that their must be optimum balance kept all/major influencing parameters. To attain this goal a multi-dimensional study of the area is taking place to collect majority of information and then by their integration a best suited development plan can be extracted that can reach to provide optimum balance belt resources utilization and conservation.

Land Resource Management

Various theme maps are generated from SOI toposheet and satellite imagery. These covers Base map, Roads network and settlement map, Land use/Land cover map, Drainage Soil map, Slope map, Geological map, Geomorphological map, Lithological map, Lineament map. After integration of the theme "Development Land Resources Action Plan".

1. Land Resource Action Plan
   - (a) It was observed that soil and water conservation and efficient utilisation of these two seems an essential step throughout the area.
   - (b) The soil and water management practice if of prime importance in high slope as major influencing factor are marked separately from integration of land use/ land cover soil, slope, geological and drainage map. Area of priority was delieneated by giving weightage to all these themes. Various practices to be found suitable for different categories are listed as follows:

(c) Silvipasture with plantation of fast growing plants.

2. Silvipasture
3. Afforestation in forest blanks.
3. Double crop with suitable soil water conservation measures to retain good potential of the area.
5. Gap plantation in open forest and scrub forest area.
6. Pasture development in village nearby wasteland for animal growing purpose.
7. Soil and water protection measure in higher drainage density and degraded soil conditioned area, like forcing, earthen bunding, stop dams, check dams, drop structure, chute spillway, etc. should be adopted as per need of the area. Here only macro level guideline is given. For implementation their should be consideration of socio-economic viability and detailed engineering plan workout for the demarcated area is necessary.

Table 17.3: Land Resources Action Plan

Action	Landuse	Soil	GW/Excellent	Soil Capability	Slope Class	Remark
No action	Existing double crop	–	–	IIes	I, II	
Suitable soil conservation	Double crop	Good	Moderate	IIIes	II, III	
Double crop with suitable soil conservation	Existing Kharif	Good	Good Excellent	IIes	II	
	Existing Rabi	Good	Good Excellent	IIes	II	
Agrohorticulture	Existing Kharif	Either soil will be good or ground-water will be good		IIes, IIIes and IVes	III	
	Existing Rabi	Either soil will be good or ground-water will be good				
Horticulture	Existing fellows	Soil will be moderate/groundwater moderate		IIIes	II	
		Soil will be good/groundwater poor		IIes	II or III	
		Sites of check dam/stop dam			II, III and IV	

Contd...

Table 17.3–Contd...

Action	*Landuse*	*Soil*	*GW/Excellent*	*Soil Capability*	*Slope Class*	*Remark*
Agroforestry	Kharif or Rabi	Soil will be moderate or poor		Iles, Ives	II, III, IV	Graded bunding on higher slopes
Silvipasture with gulley plugging	Existing Gulley/Ravinous land			IIIes, Ives	II, III	
Silvipasture	Land with scrub			IIIes, Ives	II, IV, V	Soil conservation measures
Plantation of fast growing trees	Land without scrub		Iles, IIIes, Ives		II, III, IV	
Horticulture	Land with or without scrub		Either good to moderate groundwater	Iles	II	Surface water harvesting site is required
Afforestation	Kharif + Rabi + Double crop (encroachments) with good soil cover			Iles	II, III	Encroachments in notified areas
Gap filling	Open/degraded forest			IIIes, IVes, Ves	II, III, IV, V	Soil conservation inputs or higher slopes
Plantation of fast growing trees	Scrub forest/forest blank			IIIes, Ives	III, IV	
No action	Dense forest					

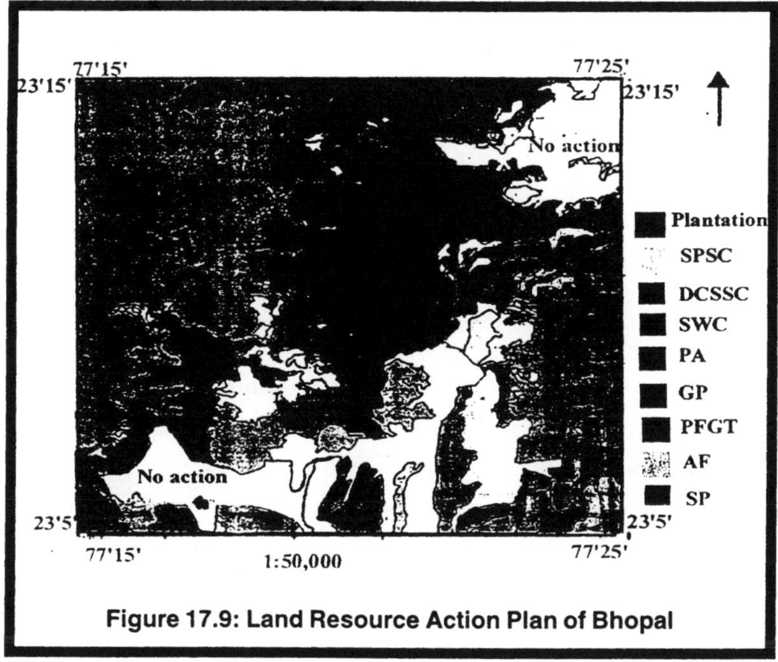

Figure 17.9: Land Resource Action Plan of Bhopal

Hence utilization and replenishment of resource as per its capability is justified and will avoid the exploitation of the available resources. But as we know that all development phases are dynamic, thus it is necessary to have a dynamic equilibrium belt demand and supply. Thus periodic and efficient assessment of resources is a vital part of sustainable development.

References

Agrawal, A.G. and Mishra, D. (1992). Hydrogeomorphological Assessment Satellite Remote Sensing Techniques, Photonirvachak. *Journal of the Indian Society of Remote Sensing.*

Anderson, J.R. *et al.* (1976). A landuse and landcover classification system for use with remote sensing data. *Professional Paper,* 964, USGS Reston, Verginia.

Burrough, P.A. (1990). *Principles of Geographic Systems for Land Resources Assessment* Jeffrey Star and John Estes, Geographic Information Systems: An Introduction, Prentice Hall, Engliwood Cliff, New Jersey.

Campbell, J.W. *Introduction of Remote Sensing.*

Census of Bhopal (1971).

Classification of Soil Data. Indian Institute of Soil Science, Bhopal.

Dhruvnarayana, V.V. (1986). *Soil and Water Conservation Research in India.*

Gazetteer of Bhopal.

Geological Study of Bhopal Data by Geological Survey of India.

Karale, R.C. (1992). *Natural Resources Management: A New Perspective.* Department of Space, Government of India, Bangalore.

Lillies and Keifer. *Interpretation of Remote Sensing.*

Navagulad. Scientist ISRO, Published Paper Introduction of Remote Sensing Techniques.

Pathan, S.K. *The Role of Remote Sensing and in Urban Planning: A Case Studies.*

Rao, D.P., N.C. Gautam, R.L. Karale and Baldev Sahai. *Application for Landuse/Landcover Mapping in India.*

Raymonds, *Basaltic and Sandstone Studies*, Text Book.

18

Evaluation of Groundwater Quality in Bharar River Basin, District Chhatarpur, Central India

☆ *Pradeep K. Jain & Anamika Shrivastava*

Introduction

Water is an invaluable commodity available in very limited quantities to man and other living beings. In the present study evaluation of water quality in Bharar river basin, district Chhatarpur have been carried out. The study area lies between 25°0′00″ to 25°10′00″ N latitude and 79°20′00″ to 98°35′00″ E longitude (Figure 18.1) and covers an area of more than 150 sq km. The major litho types of the areas are Bundelkhand granites.

Material and Methods

In the present study 26 groundwater samples (dug-well) were collected on seasonal pattern. Hydrochemical analysis were done for determination of Ca, Mg, Na, K, So_4, Cl, $C^o{}_3$, $HC^o{}_3{}^-$, $P^o{}_4$, $N^o{}_3$, TH, TDS, pH and electrical conductivity (EC). Water samples were

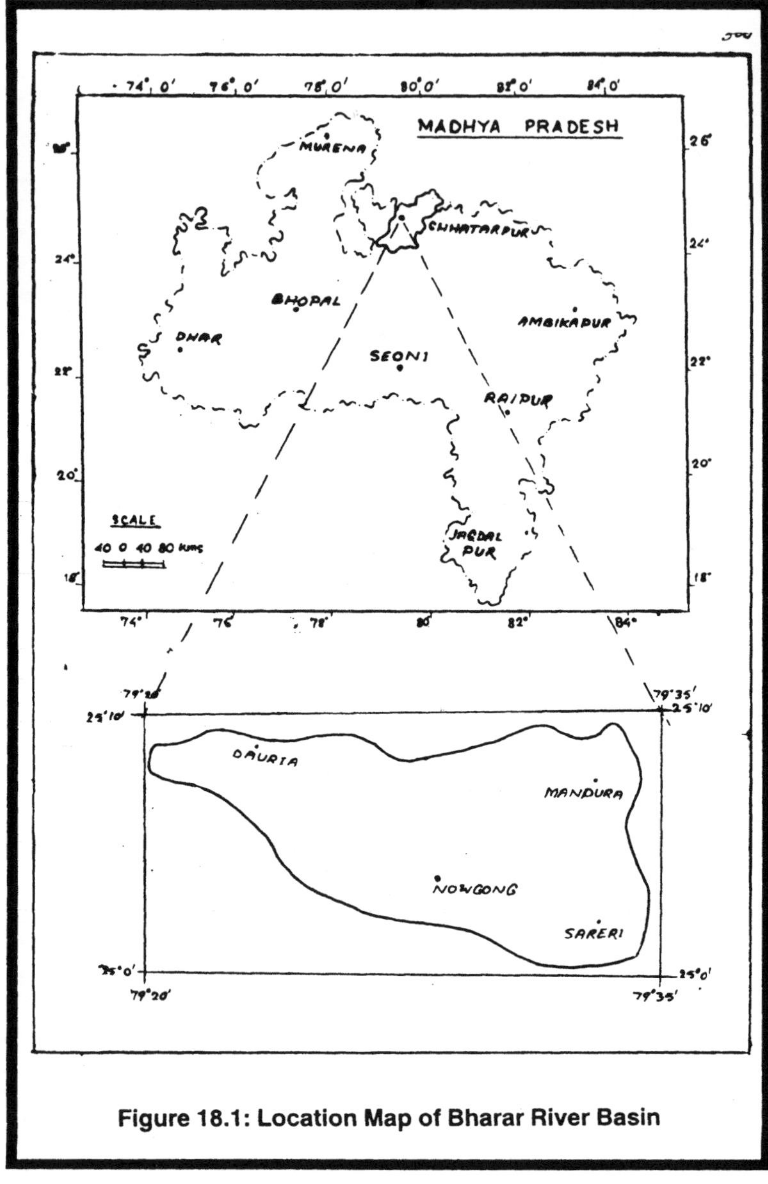

Figure 18.1: Location Map of Bharar River Basin

collected and analysed as per APHA (1985), Trivedi and Goel (1984) and NEERI (1986).

Results and Discussion

The average values of the physico-chemical characteristics of groundwater of the Bharar river basin is presented in Table 18.1. For comparison, this table shows that the principal anion in water is HCO_3^-, followed by Cl^- while Ca^{2+} is dominant cation followed by Mg^{2+} and Nat. The water pH varies from 7.2 to 8.6 with a mean value of 7.73, which is a safe range for drinking as well as for the growth of plants. Somashekar and Ramaswamy (1984) have shown that the alkaline water harbours more plants than acidic waters.

Table 18.1: Physico-chemical Constituents of Bharar River Basin (October–November 1998)

Sl.No.	Constituents	Maximum	Minimum	Mean
1.	Specific Conductance in mohs/cms	1272.00	410.90	733.20
2.	pH	8.60	7.20	7.73
3.	TDS (ppm)	814.08	262.98	470.01
4.	Total hardness	227.00	102.00	158.20
5.	Ca^{++}	94.00	40.00	57.60
6.	Mg^{++}	40.60	12.00	21.17
7.	Na^+	14.00	5.00	8.41
8.	K^+	5.50	1.00	2.90
9.	CO_3^-	88.00	0.00	7.17
10.	HCO_3^-	239.00	124.00	162.70
11.	Cl^-	291.10	14.20	83.80
12.	SO_4^{--}			
13.	NO_3^-	0.60	0.00	0.17
14.	PO_4^{--}	0.40	0.10	0.15

Electrical conductivity (EC) in the Bharar river basin ranges between 410 to 1272.0 micromhs/cm, with a mean value of 733.2 mhos/cm. According to Langenegger (1990) the importance of EC is its measure of salinity, which greatly affects the taste and thus has a significant impact on the users acceptance of the water, as potable (WHO, 1984).

Total hardness (TH) ranges from 102 to 227 ppm with a mean of 158.2 ppm. Hardness is mainly due to the presence of carbonates and bi-carbonated (Calcium and Magnesium) and expressed as an equivalent amount of calcium carbonate. The mean value of hardness in the study area (158.2 ppm) is indicative of moderately hard to hard water. It is temporary types of hardness, which is removed simply by boiling the water. The water of the study area has problem for domestic use in washing, cleaning and laundering. The high concentration of hardness of 150–300 ppm and above may causes heart diseases and kidney problems, Keller (1979). TDS ranges between 262.98 to 814.08 ppm and mean value of 470.01 ppm which is lesser than the maximum permissible limit prescribed by WHO (1984) and ICMR (1975). Carbonate ($C^{o}_{3}{}^{-}$) and bi-carbonate (HC^{o}_{3}) along with hydroxides are responsible for the alkalinity of water. The carbonate and bi-carbonate in the water of the study area is ranges from 0 to 88 ppm and 124 to 239 ppm, respectively.

Sulphate ($SO_4{}^{--}$) and Nitrate ($N^{o}_{3}{}^{-}$) concentration in the study area is low and therefore, poses no problem to the water quality. The Chloride (Cl^-) ranges 14.2 to 291.1 ppm with a mean value of 83.8 ppm. The high concentration of Cl^- is due to sewage disposal near water sources. Which is within permissible limit prescribed by WHO and ICMR.

Sodium (Na^+), Potassium (K^+) in the Bharar river basin occur in small concentrations when compared with WHO and ICMR standards, Calcium and Magnesium is ranges from 40.0 to 94.0 ppm and 12.0 to 40.6 ppm with mean values of 57.06 and 21.17 ppm respectively. High concentration of Ca^{++} and Mg^{--} may occurs from disposal of sewage and domestic wastes of the study area. Sodium and Calcium in the water assigned to weathering of plagioclase feldspars to Kaolinite present in the Bundelkhand granites in the study area (Jain, 1998).

Representation and Interpretation of Chemical Analysis

Water quality maps have been prepared by the authors to represent and interpet the result of the chemical analysis. The interpretation of these maps are described below:

Water Quality Maps

The quality of a region can be studied by platting the result of analysis on maps. Whereas each sampling location a symbol denoting the analysis is entered. This may be small bar graph, a small pie circle, or simply numbers divided by horizontal or vertical lines. The water quality maps for Total Dissolved Solids (TDS), bi-carbonate and total hardness (TH) have been prepared and described below:

TDS Maps

Figure 18.2, illustrates the total dissolved solid map which has been prepared in a horizontal scale of 2 cms equal to 1 km and at a contour interval of 500 ppm and then the different ranges of TDS have been marked in the map.

The TDS varies from 262.97 to 819.08 ppm on paying attention to the total dissolved solid map, it becomes clear that the study area does not show much variation in the TDS and the values are within the permissible limit of drinking purpose. It indicates the presence of good quality of groundwater in the area.

Total Hardness Map

Figure 18.3, illustrates the total hardness map of study area. It has been prepared on a horizontal scale of 2 cms equal to 1 km and at a contour interval of 75 ppm and then the different ranges of total hardness have been marked on the map. The total hardness varies from 102.0 to 227.0 ppm, the figures show alternate pocket of moderatary hard to hard water throughout the study area.

Bi-carbonate Map

Figure 18.4, illustrated the bi-carbonate concentration map of the study area. It has been prepared on a horizontal scale of 2 cms equal to 1 km and at a contour interval of 100 ppm. Bi-carbonate concentration varies from 124.0 to 289.0 ppm. The figure shows that the maximum study area is covered with 100–200 ppm class. There is a cyclic variation in bi-carbonate concentration in the study area.

Evaluation of Water Quality

Water is widely needed for domestic, agriculture and industrial uses, each of which requires water for particular specifications. There are international standards for different purposes given by various organizations.

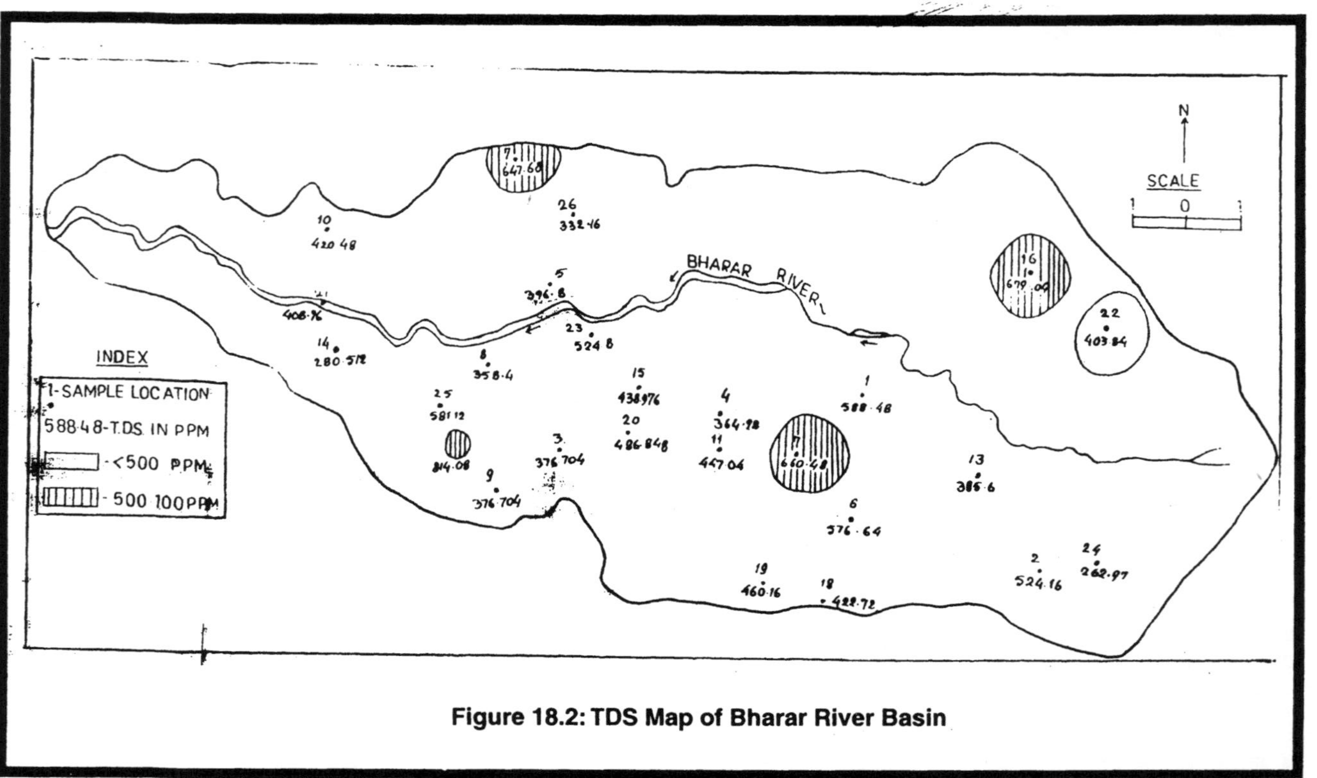

Figure 18.2: TDS Map of Bharar River Basin

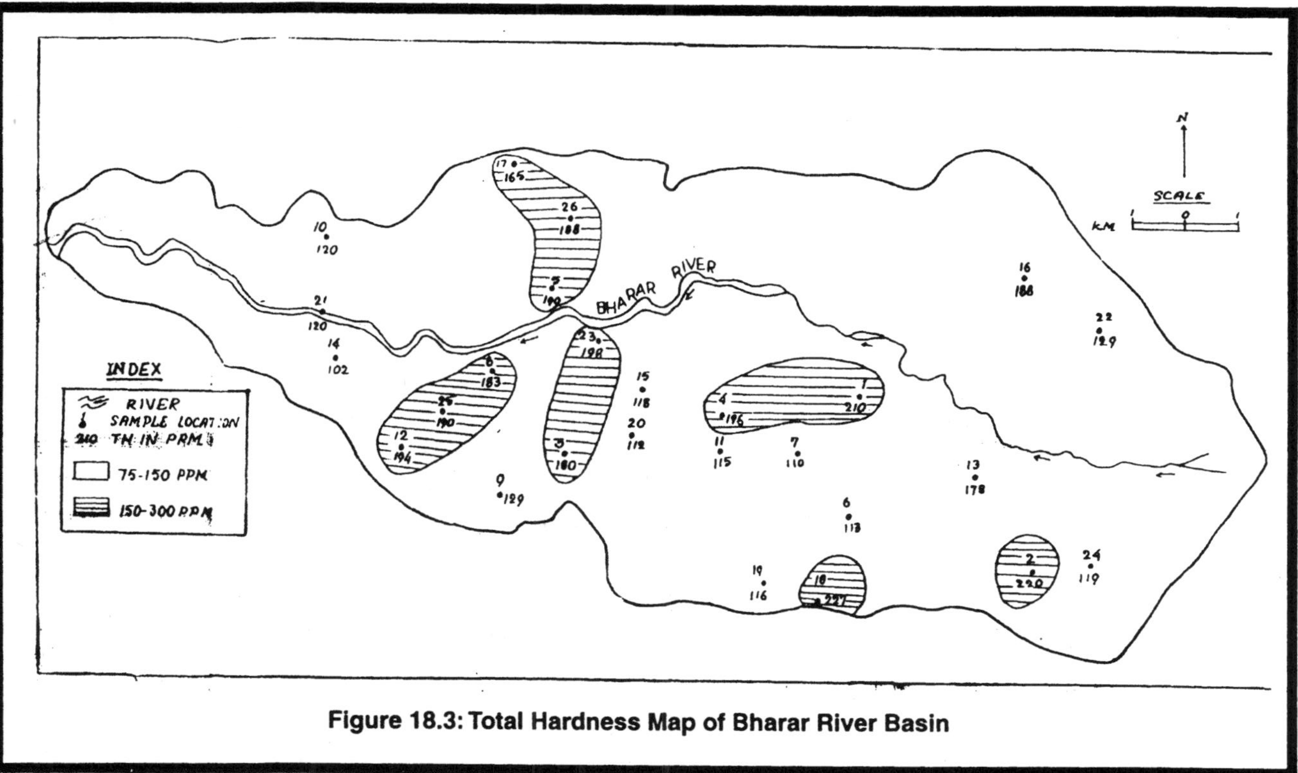

Figure 18.3: Total Hardness Map of Bharar River Basin

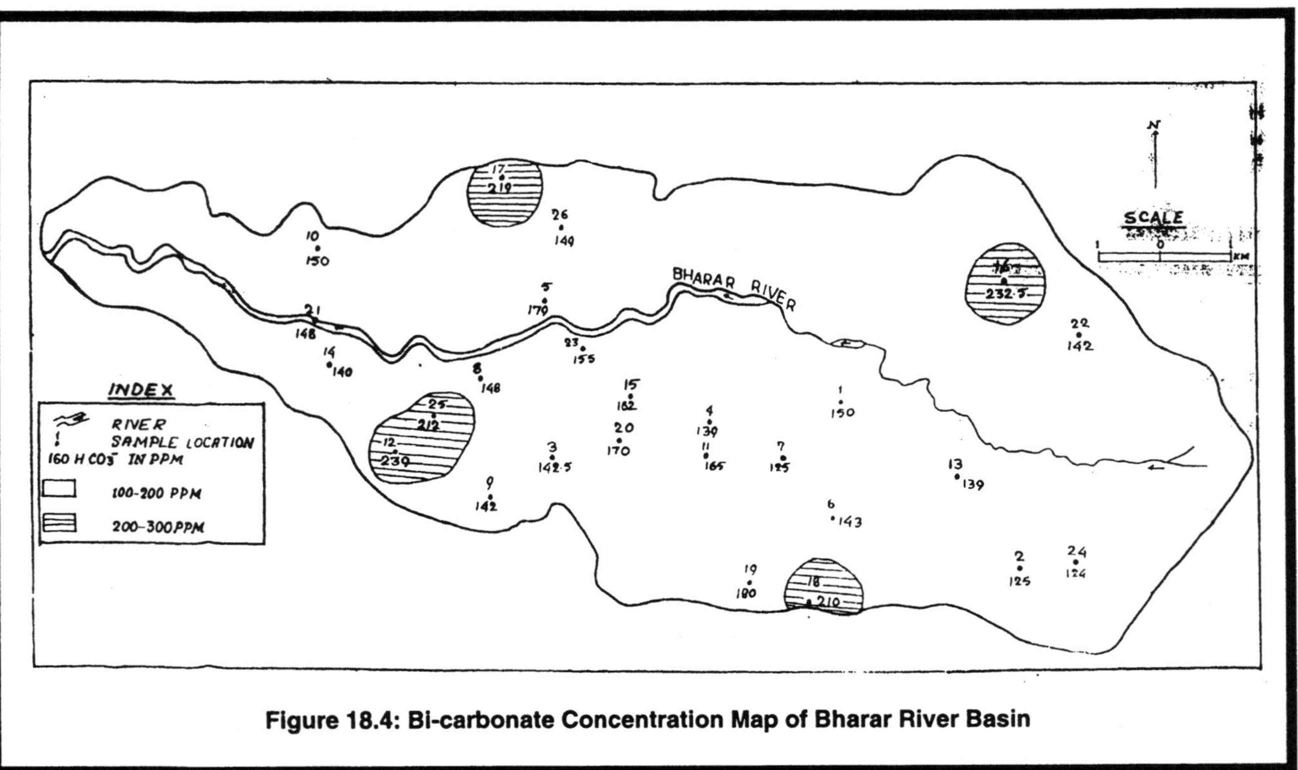

Figure 18.4: Bi-carbonate Concentration Map of Bharar River Basin

Domestic Purposes

Water for domestic use should be colourless and tasteless. The standard by which the suitability of water for domestic use in judged by WHO (1984) and ICMR (1975).

On considering the drinking water standards and the average value of the different chemical constituents presents in surface and groundwater (Table 18.1), it makes clear that the water on the basis of chemical quality is suitable for drinking purposes. At present there is slight effect of pollution on the quality of water due to disposal of sewage and domestic wastes.

Irrigation Purposes

The chemical quality of water is essential factor to be considered in evaluating it suitability for irrigation use. The author has considered the concentration of Total Dissolved Solids (TDS), Electrical Conductivity (EC), per cent Sodium (%Na), Sodium Absorption Ratio (SAR), Residual Sodium Carbonates (RSC) and Potential Salinity (PS) for determination of suitability of water for irrigation purposes (Table 18.2).

Sodium Hazard

Sodium concentration is very important in classifying irrigation water because sodium by the process of base exchange, replace calcium in the soil thereby reduces the permeability of soil which has greater effect on the plant growth, sodium content in chemical analysis is reported by percent sodium. The relative activity of sodium ion in the exchange reaction with soil is expressed in the terms of ratio known as Sodium Absorption Ratio (SAR). There is a significant relationship between SAR values and %NA for irrigation water and the extent to which sodium is absorbed by the soil. The water required for irrigation is classified on the basis of SAR, %NA and Electrical Conductivity is presented in Table 18.2. This table reveals that the water of the study area is suitable for irrigation.

US Salinity Diagram

The US Salinity Laboratory Staff (1954) has constructed a diagram for classification of irrigation water with reference to SAR as an index for Sodium hazard 'S' and Electrical Conductivity (EC) as an Index of Salinity hazard 'C'. In this diagram the values of SAR are plotted on arithmetic scale against specific conductivity (EC) on log scale and different classes of water have been marked.

Table 18.2: Classification of Irrigation Water in Bharar River Basin (1998)

Parameters	*Maximum*	*Minimum*	*Category*	*No. of Sample*	*Water Class*
Na%	11.8	5.63	< 20	26	Excellent
			20–40		Good
			40–60		Permissible
			60–80		Doubtful
			> 80		Unsuitable
EC	1272.0	410.9	< 250	15	Excellent
			250–750	11	Good
			750–2000		Permissible
			2000–3000		Doubtful
			> 3000		Unsuitable
SAR	0.36	0.18	< 10	26	Excellent
			10–18		Good
			18–26		Fair
			> 26		Poor
RSC	– 4.81	– 0.49	< 1.25	26	Good
			1.25-2.50		Medium
			> 2.50		Bad

In the present study the data plots of water samples are shown in Figure 18.5. From this plot water can be classified as good for irrigation because most of the water samples fall in C_2–S_1 and C_3–S_1 categories on the diagram.

Doneen Classification

Doneen (1962) proposed a classification of water based on the salinity, permeability and toxicity of irrigation water. Salinity is caused by high solubility of salts which rapidly accumulate in the soil. The low solubility salts precipitate in the soil as the soil solution increases in salinity and does not play much role in salinisation of the soil. The potential soil salinity (PS) is defined as the concentration

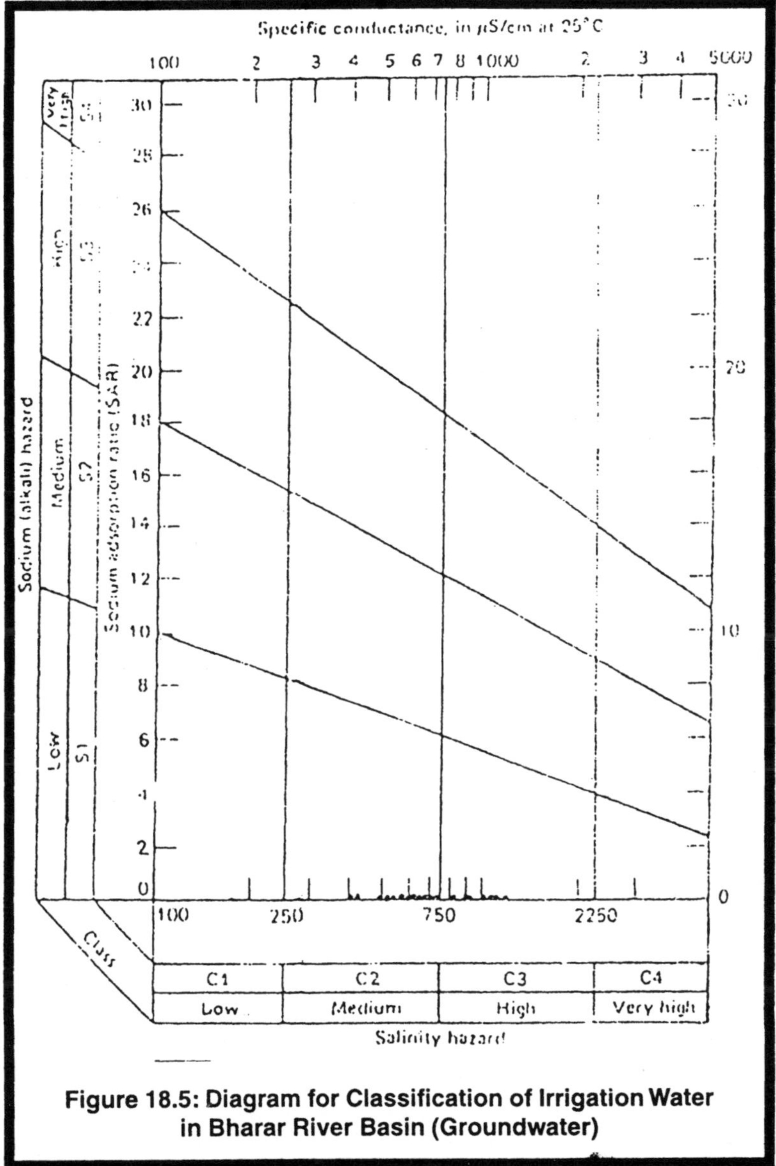

Figure 18.5: Diagram for Classification of Irrigation Water in Bharar River Basin (Groundwater)

of chlorides and half of the sulphate ions. The water of study area have been classified for irrigation purposes (based on PS) and presented in Table 18.2. The table reveals that 56 per cent water samples fall in excellent to good water category and 39 per cent samples come under good to injurious category. Only 5 per cent samples come under injurious to unsatisfactory category. Permeability of the soil is influenced by the sodium content of the irrigation water and it is expressed as permeability index (PI). The Doneen classification for the study area reveals that majority of water samples fall in Class I. It shows clear that water is good for irrigation purpose.

Bi-carbonate Hazard

Bi-carbonate concentration of water has been suggested as an additional criterion for irrigation purpose. If the water contain high concentration of bi-carbonate ion there is a tendency of calcium and magnesium ions to precipitate as carbonates and it is expressed as Residual Sodium Carbonate (RSC). Table 18.2 (for RSC) shows clear that the water of Bharar river basin is free from bi-carbonate hazard and 26 water samples fall in good category.

Conclusion

From the above discussion on the chemical quality data, it is concluded that the surface as well as groundwater of the Bharar river basin is generally suitable for drinking and irrigation purposes.

Though a few groundwater samples have unsuitable for drinking purposes. At present there is slight effect of degradation on the quality of surface water as well as groundwater due to sewage and domestic wastes.

References

APHA (1976). *Standard Methods for the Examination of Water and Wastewater,* 14th edn. American Public Health Association, Washington, DC, 1193 pp.

Doneen, L.D. (1962). The influence of crop and soil on percolating waters. *Proc. Biannual Conferences on Groundwater Recharge.*

Hill, R.A. (1940). Geochemical patterns in Coachella valley, California. *Trans. Amer. Geophyes. Union,* **27**: 46-49.

ICMR (1975). *Manual of Standards of Quality and Drinking Water Supplies,* 2nd edn. Indian Council of Medical Research, New Delhi, Special Report Series, **44**.

Jain, P.K. (1998). Hydrochemical studies of sub-surface of water in upper Urmil river basin, District Chhatarpur, Central India. *Ecol. Env. and Cons.*, **4**: 259-269.

Keller, A.W. (1979). *Environmental Geology*. Charles E. Merrill Publishing Co., Ohio, p. 548.

Langenegger, O. (1990). Groundwater quality in rural areas of western Africa. UNDP Project INT/81/26, p. 10.

NEERI (1986). *Manual on Water and Wastewater Analysis*. National Environmental Engineering Research Institute, Nagpur.

Piper, A.M. (1944). A graphic procedure in the geochemical interpretation of water analysis. *Trans. Amer. Geophysical Union*, **25**: 914-928.

Somashekar, R.K. and S.N. Ramaswamy (1984). Biological assessment of water pollution: A study of river Kapila. *Int. J. Environ. Studies*, pp. 261-267.

Trivedi, R.K. and P.K. Goel (1984). *Chemical and Biological Method for Water Pollution Studies*. Enviromedia Publications, Karad.

US Salinity Laboratory Staff (1954). Diagnosis and improvement of saline and alkaline soils. *US Deptt. of Agri. Handbook*, pp. 160.

Walton, W.C. (1977). *Groundwater Resource Evaluation*. McGraw-Hill Book Co., New York, p. 434.

WHO (1984). *Guidelines for Drinking Water Quality, Vols. 1 and 2*. Geneva, Switzerland.

19

The Use of Histopathological Methods in Bioindication of Ecodegradation

☆ *S. Anitha Kumari & N. Sree Ramkumar*

Introduction

Histopathology has been accepted as an important tool in biomedical pathology for many years. Ever since Virchow (1958) published his historic treatise–Cellular Pathology, Histopathology or more broadly structural pathology has been a corner stone in the larger field of biomedical pathology. In addition to this important role in biomedical pathology, Histopathology also is used extensively in biomedical toxicology. In fact, today; Histopathology is accepted as one of the most important determinants for establishing acceptable concentrations of environmental contaminants, food additives, cosmetic additives and other chemicals etc.

All tissues and organs in the body of an animal may be potential targets for the toxic effects of chemicals and heavy metal pollutants. Histopathological assessments throw light on the nature of tissue alteration and the extent of damage in turn indicates the toxic nature of the compound. Therefore, the study of Histology, can give useful insights on tissue alterations even prior to the external manifestations of the deleterious effects of the toxicants.

Further, pollution of the natural waters causes significant tissue damage in the fish. (Mckim *et al.*, 1970; Reichenbach Klinke, 1972; Horvath and Stammer, 1979; Ferri and Macha, 1980; Rojik *et al.*, 1983; Benedeczky *et al.*, 1984; Galat *et al.*, 1985).

With the rapid industrialization and urbanization in the developing countries like India, the increasing menace due to pollution by Industrial wastes, usually discharged into the freshwater bodies, is of great importance. Besides rendering the water unsuitable for human consumption, the irrational discharge of effluents into the water bodies, frequently causes hazards to aquatic life, especially the fish in which they induce severe pathologic changes at the tissue level.

Hussain Sagar lake, a freshwater body situated in between the twin-cities of Hyderabad and Secunderabad in Andhra Pradesh, India, is a victim of severe pollution which receives industrial effluents and municipal discharges due to rapid urbanization and industrialization in its vicinity. The pollution induced damage to the non-target components of the food chain especially the fish, as evidenced by frequent fish kills in the lake has elicited much concern in the recent years. (Anitha, 1998). The nature and periodicity of fish kills in the lake remains an enigma and several workers have assigned various reasons for fish mortality based on ecological and biochemical parameters (Muley, 1987; Manjula Devi, 1988 and Siddiqui and Rao, 1991).

During the past two decades, water quality biologists have been searching for sensitive indices of sublethal effects of contaminants on fish in an attempt to understand the mode of action of the toxicants and to develop a basis for corrective action in cleaning up the water body before the lake itself is declared dead.

The present investigation is a step in that direction. The report presents the results of an assessment of the possible effects of water pollution on the gill structure in three different species of fish, namely, *Channa punctatus*, *Channa striatus* and *Heteropneustes fossilis* that abound the lake waters with the help of light and scanning Electron microscopy. Five sites were sampled because of their varying degrees of sediment contamination. These include station I, *i.e.* the Begumpet site, is more heavily contaminated followed by station V (Centre of the lake), station II (Khairatabad), Station III (Hyderabad Boats Club) and Station IV (Secunderabad boats club) (Anitha, 1998).

Materials and Methods

The fish collected from the relatively non-polluted water bodies *i.e.* Himayatsagar and Osmansagar were used as the control fish. Both the control fish and the fish from the polluted-waters of Hussainsagar lake, were sacrificed and the gills were isolated and washed in the fish-ringers solution (Ekberg, 1958) and fixed in 10 per cent formalin for 2 days. This was followed by the dehydration in successive grades of alcohol (ethanol), cleared in Xylene and embedded in paraffin wax (58°–60°C). Sections of 6 to 8 microns in thickness were cut using a rotary microtome and were stained with Haematoxylin and eosin stains. The D.P.X. mounted sections were then observed under the light microscope for the Histopathological study. However, for SEM studies, the gills were processed by fixing in modified paraducz-fixative and drying by chemical vacuum desciccators method (Singotamu, 1991). Material was mounted on SEM stubs with double adhesive tape and coated with gold (300°A) in high vacuum evaporator (HUS–5GB). Later the material was scanned by Hitachi SEM S520 operated at 10 KV and pictures were taken at appropriate magnification.

Results and Discussion

The gills in the fish mainly serve for two important functions, respiration and regulation of Osmotic concentration of body fluids. The gills are among the most delicate structures of the teleost body. Their vulnerability is considerable because of their intimate contact with the water in which the fish live. They are liable to damage by any toxicant whether dissolved or suspended in the water. (Rankin *et al.*, 1982; Mallatt, 1985).

In a normal gill tissue, the primary gill lamellae are laterally compressed and are situated alternatively on either side of the inter-branchial septum. A row of secondary gill filaments are present on both the sides of the primary gill lamellae. The primary gill lamellae consists each of a centrally placed rod like supporting axis with blood vessels on either side. The secondary gill lamellae are highly vascularized and are covered with a thin layer of epithelial cells. Blood vessels are extended into each of the secondary gill filaments.

The region between the two adjacent secondary gill lamellae is known as the inter lamellae region (Figures 19.1 and 19.2).

Figure 19.1: Gills Control (*Channa punctatus*) H&E × 320

Figure 19.2: Gills Control (*Heteropneustes fossilis*) H&E × 320
PL: Primary Lamella; SL: Secondary Lamella

When compared to that of the normal gill tissue, light microscopic studies on the gills of all the fishes collected from Hussainsagar lake revealed certain significant changes such as swelling, fusion, atrophy, degeneration of the secondary lamellae and bulging of the tips of the gill filaments Hyperplasia of the inter-lamellar cells and necrosis of the secondary lamellae were also observed in a few cases. In many fishes, complete degeneration of the gill filaments and also distortion of the respiratory epithelium covered with mucus, blood cells and cellular worn off were noted. Few fishes, however, showed sloughing off of the respiratory epithelial layer (Figures 19.3–19.6).

Similar observations like swelling of gill filaments and lamellae, increased mucus production, inflammatory alterations in epithelium and reduction of inter-lamellar spaces were reported in acid-stressed fishes and fishes exposed to various chemicals. (Christie and Battle, 1963; Baker, 1969; Munshi and Singh 1971; Skidmore and Tovell, 1972; Tuurala and Soivio, 1982; Jagoe and Haines 1983).

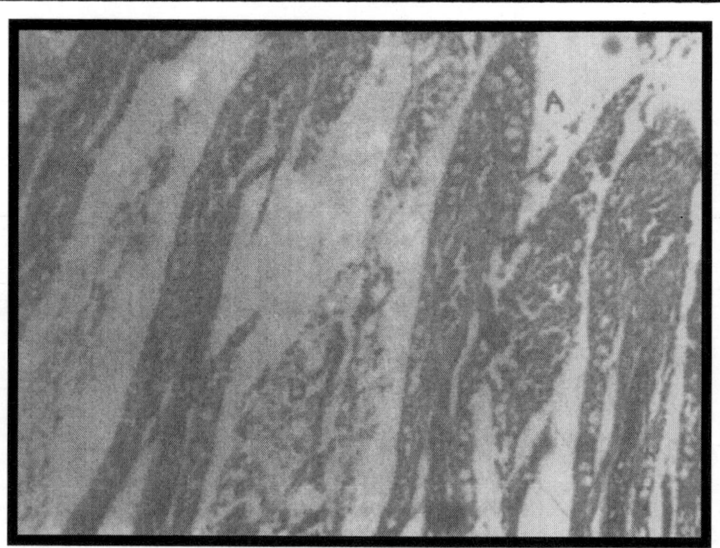

Figure 19.3: Gills Polluted (*Channa punctatus*) H&E × 320 Showing Atrophy and Degeneration of Secondary Lamellae

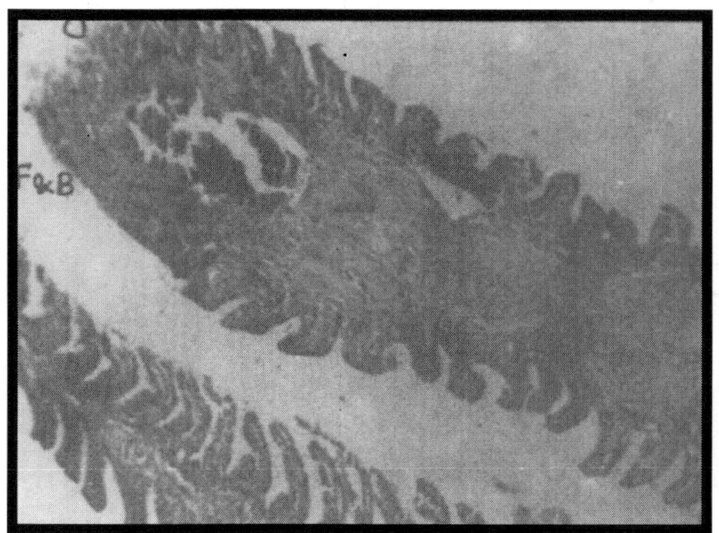

Figure 19.4: Gills Polluted (*Channa punctatus*) H&E × 320 Showing Fusion of Secondary Lamellae and Bulging at the Tips

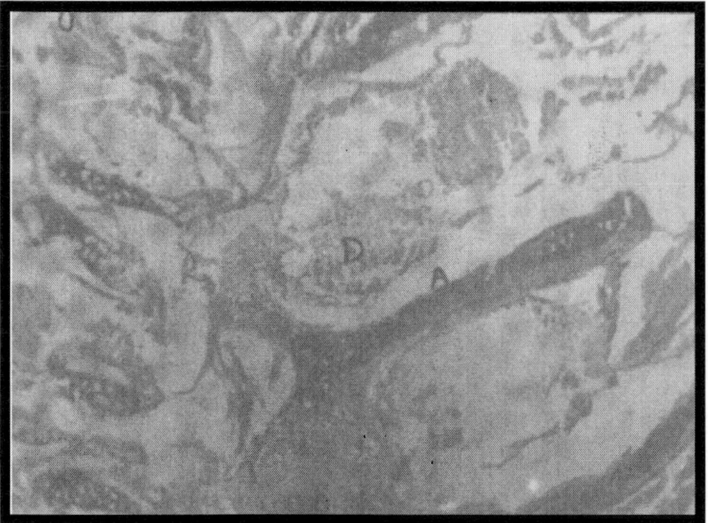

Figure 19.5: Gills Polluted (*Channa striatus*) H&E × 320 Showing Complete Degeneration of Gill Filament Along with Respiratory Epithelium

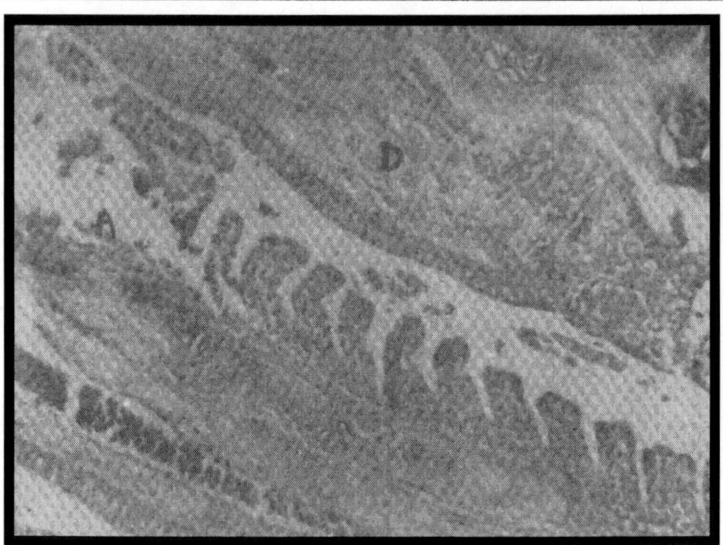

Figure 19.6: Gills Polluted (*Heteropneustes fossilis*) H&E × 320 Showing Atrophy of Secondary Lamellae and Degeneration of Respiratory Epithelium RE: Respiratory Epithelium; A: Atrophy; D: Degeneration; F&B: Fusion and Bulging

Further, similar reports were also made by Srivastava and Srivastava (1984), Asfia Parveen (1988) Anita Evangelin (1988), Vinod Ghanathay (1989) and Anand (1994) in different fishes exposed to various pesticides.

In the present investigation, scanning Electron microscopy has also been done to have a better understanding about the ultratopography of the fish gills. In addition to the normal architecture of the fish gill, as revealed by light microscopy, scanning electron microscopic studies further showed that the branchial or gill arch region is covered with micro-ridged epithelial cells (MREC). The gill filaments and the secondary lamellae are also covered with these cells. Mucus gland openings and chloride cell openings are seen in between these micro-ridged epithelial cells (Figures 19.7–19.10).

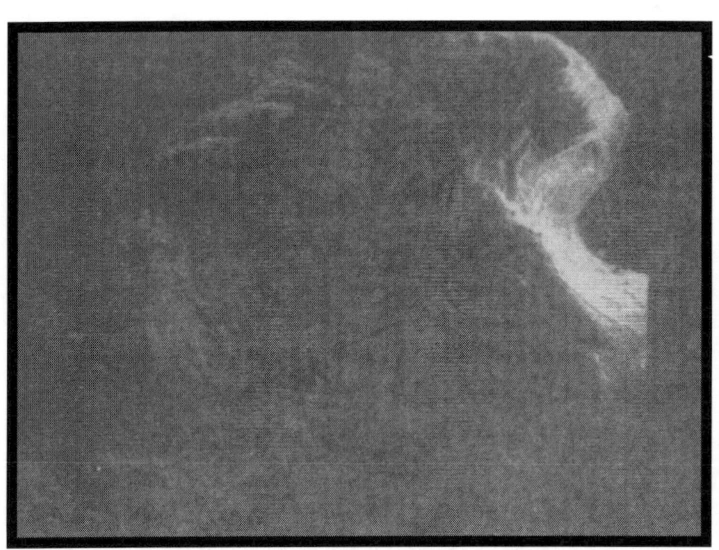

Figure 19.7: Gills Control (*Heteropneustes fossilis*)

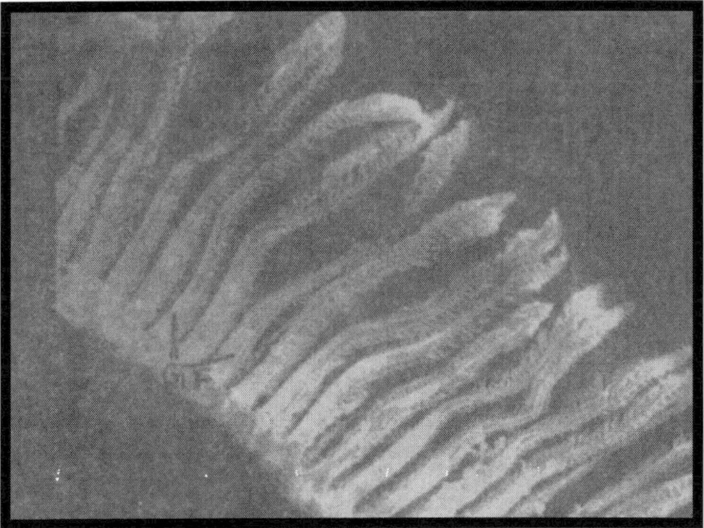

Figure 19.8: Gills Control (*Channa punctatus*)

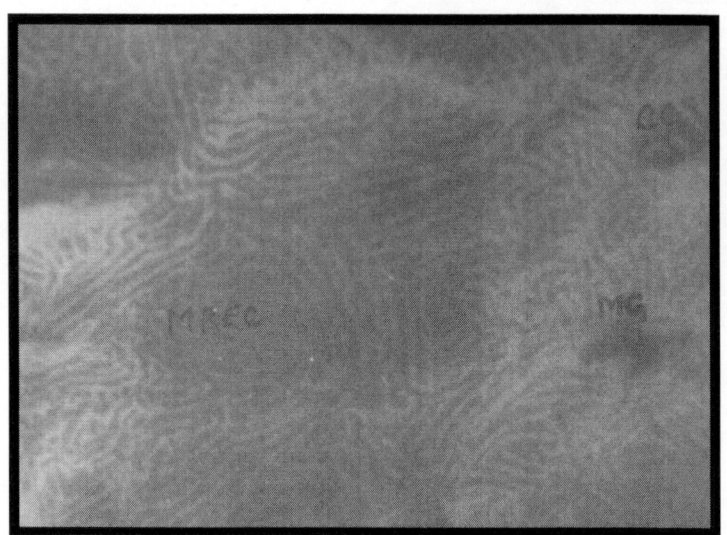

Figure 19.9: Gills Control (*Channa punctatus*)

Figure 19.10: Gills Control (*Channa striatus*)
GH: Gill Head; GR: Grill Raker; GF: Gill Filament;
PL: Primary Lamella; SL: Secondary Lamella; CC: Chloride C
MREC: Microridged Epithelial Cell; MC: Mucus Cell

When compared to that of the normal gill tissue, the surface architecture of the gill from all the three species of fish, investigated from the polluted waters of Hussainsagar Lake, showed similar but prominent changes such as damage, atrophy and fusion of the gill lamellae. The secondary lamellae, where the major part of the gaseous exchange takes places (Hughes and Morgan, 1973) were swollen and fused with each other decreasing the inter-lamellar spaces and thereby decreasing the total respiratory surface. The epithelium got distorted and large quantities of mucus, blood cells and cellular worn-off were seen accumulated on the gill filaments and in the spaces between the secondary lamellae and also in the ridges, channels and micropits on the epithelial surface of the secondary lamellae. The lamellae gave a frizzy appearance under SEM. However, in a few fishes, the epithelial cells of the lamellae did swell up to form puff balls (Figures 19.11–19.17).

Thus, the light microscopic studies corroborated the findings of the SEM study.

Figure 19.11: Gills Polluted (*Heteropneustes fossilis*) Showing Clumping of Gill Filaments

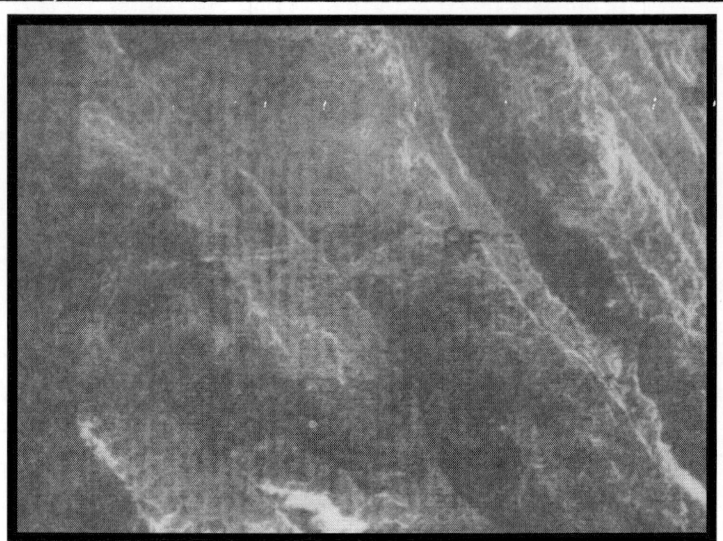

Figure 19.12: Gills Polluted (*Heteropneustes fossilis*) Showing Ruptured and Distorted Respiratory Epithelium

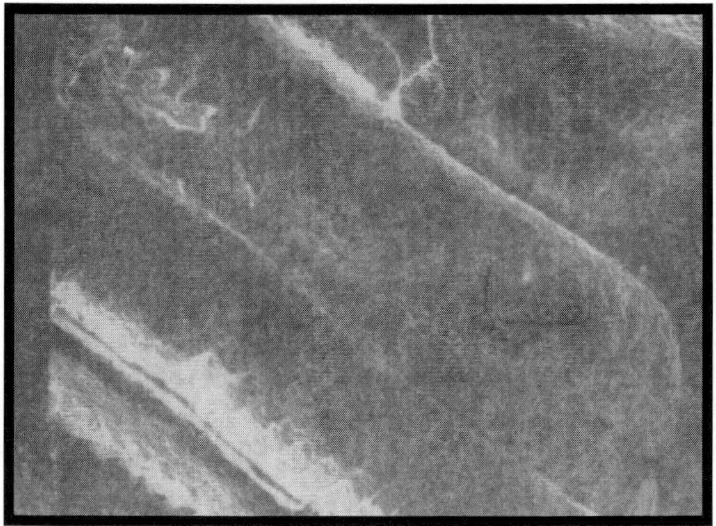

Figure 19.13: Gills Polluted (*Heteropneustes fossilis*) Showing Swelling and Fusion of Secondary Lamellae GF: Gill Filaments; RE: Respiratory Epithelium; SL: Secondary Lamellae

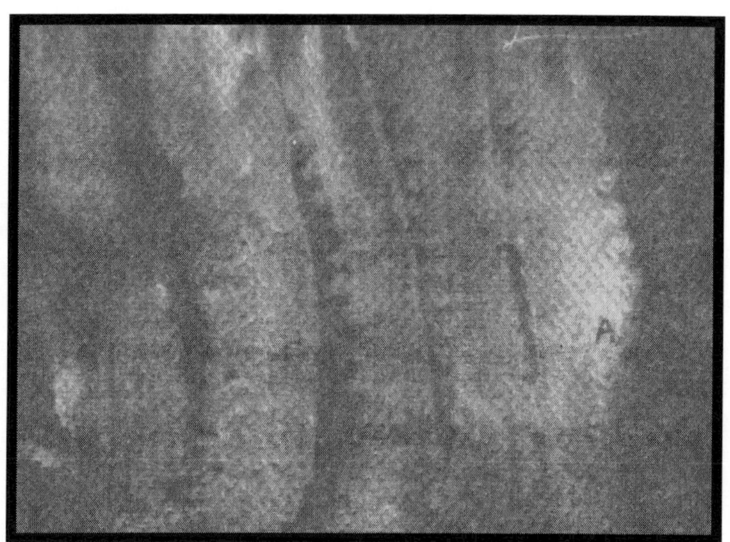

Figure 19.14: Gills Polluted (*Channa punctatus*) Showing Atrophy and Fusion of Secondary Lamellae

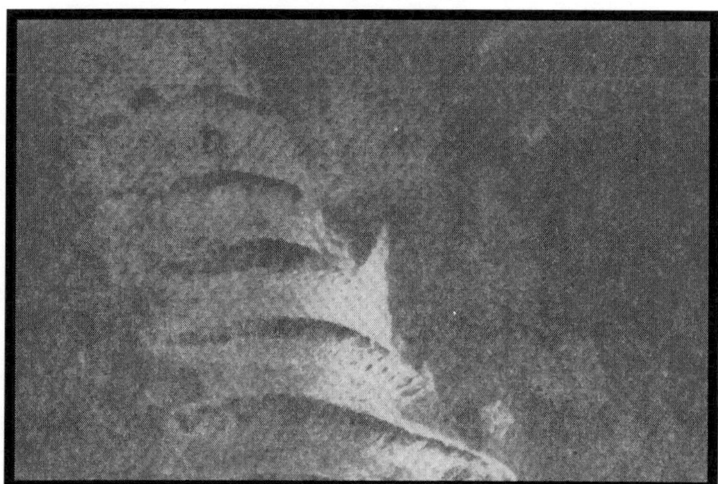

Figure 19.15: Gills Polluted (*Channa striatus*) Showing Rupture and Degeneration of the Gill Filaments

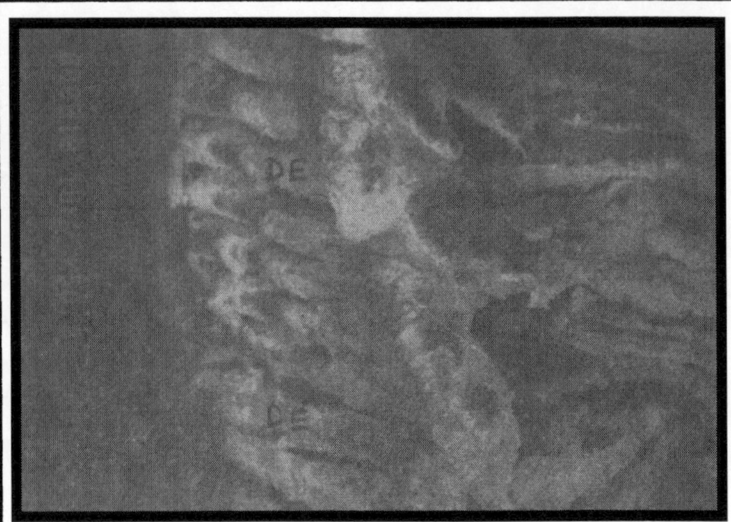

Figure 19.16: Gills Polluted (*Channa striatus*) Showing Complete Rupture of the Gill Filaments with Dissociated Epithelium
A: Atrophy; F: Fusion; D: Degeneration; DE: Dissociated Epithelium

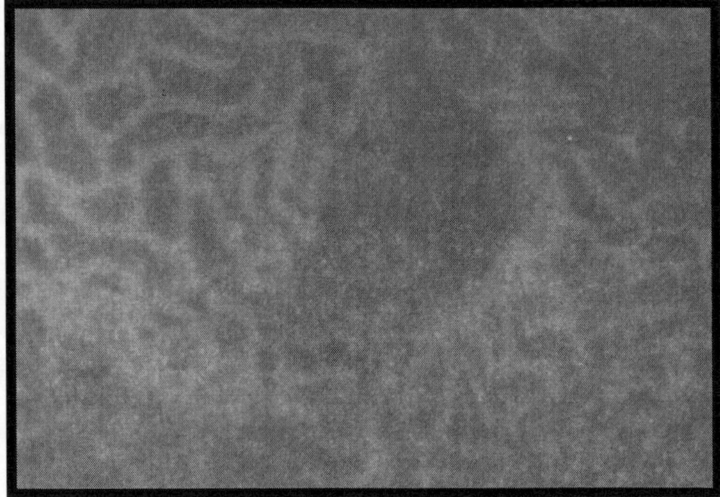

Figure 19.17: Gills Polluted (*Channa punctatus*) Showing Hyperactivity of Mucus Cell

Similar observations were made by several authors on fish gills under SEM, on exposure to different pollutants. For example, Roy *et al.* (1986) observed swelling of gill filaments and lamellae, increased mucus production, inflammatory alterations in epithelium, loss of micridges, shrinkage of blood capillaries and reduction of inter-lamellar spaces in gills of Anabas testudineus (Bloch) exposed to saponin, Leino *et al.* (1987) have observed the reduction in inter-lamellar and inter-filamentary spaces of the gills and also fusion of gill filaments in *semotilus margarita* (cope) in a Canadian lake. Ojha *et al.* (1989) also observed dissociation of filament and lamellar epithelia and fusion of secondary lamellae in the gills of fish, *Garra Lamta*, exposed to biocidal plant sap (*Terminalia tomentosa*). On exposure of the fish, *Channa punctatus*, to acidic water of different pH values, Munshi and Singh (1992) too observed swelling and fusion of lamellae, increase in the number of mucus gland openings and ruptured branchial epithelium. Further, Jadgish *et al.* (1993) also observed distorted epithelium and haemolysed blood in the lamellar channels in the gills of loach, *Noemacheilus rupicola*, exposed to biocidal plant sap.

Thus, all these findings lend ample support to our present observations.

However, one interesting feature that is noticed in the present study is the bifurcation of the gill filaments in the fishes collected from the Begumpet site of the Hussainsagar Lake. The bifurcations were one of the most striking observations of an abnormality not extensively discussed in the literature on pollution effects. Most of the bifurcations occurred at the tips of the filaments consisting of two branches of equal length with few exceptions where one or more isolated filaments rather than groups of filaments only showed the bifurcations. The latter consisted each of more than 2 branches of unequal length and could be seen throughout the gill filament rather than at the tips (Figure 19.18).

A similar observation was also made by other investigators. (Hughes, 1984; Hargis and Zwerner, 1988) who maintained that it is probably due to the physical damage caused by the pollutants to the gill tissue. However, the prevalence of gills with bifurcated filaments in fish collected at the Begumpet site only suggest or point out to the huge accumulation of pollutants at the site and the severity of their effect on fish gills.

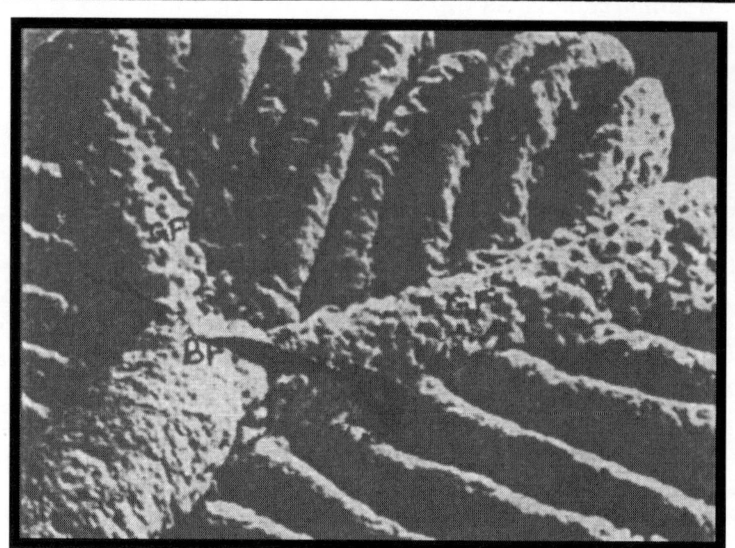

Figure 19.18: Gills Polluted (*Channa punctatus*) Showing Bifurcation of Gill Filaments GF: Gill Filament; BF: Bifurcation

Other workers have also noted an association between morphological abnormalities of the gills and contaminated environments. Hargis and Zwerner (1988) reported a 70 per cent prevalence of gill–growth deformities (including gill bifurcations) in toad fish, *opsamus tau* (Linnaeus) from the Elizabeth river, Virginia, a river whose sediments are heavily contaminated by PAHS (Polycyclic aromatic hydrocarbons) and heavy metals. Bengtsson *et al.* (1985) reported a higher prevalence of the deformities of the gill rakers (including forked gill rakers) in the white fish, *coregomus larvaretus* (linnaeus) from waters contaminated by heavy metals.

Though on one hand, the bifurcations increase the surface area, they also create a problem by making the filaments get entangled and accumulate the water-borne debris. Thus, the respiratory process as well as the osmoregulatory mechanisms are seriously affected. The problem is further aggravated by a reduction in the diffusion-distance between the blood and the polluted lake water, thus facilitating an easy diffusion of the toxic chemicals into the blood.

This situation adversely affects the gaseous and ionic exchanges forcing the fish to spend much energy on maintaining the basal metabolism reducing its capacity to work and survive in nature. (Soivio and Hughes, 1978).

The problem of pollution seems to be more acute at station I, *i.e.* at Begumpet, the entry point of the Kukatpally channel into the Lake due to the discharge of effluents from various industries situated at this station, and station V *i.e.* the centre of the lake, wherein there is always a possibility of accumulation of more pollutants. In the present study, it was noticed that there was huge accumulation of pollutants at site I compared to the other sites and hence, the damage done to the fish is more severe at this site. However, the fish at other sites also showed similar damage done to their gills but on a smaller scale.

Thus, it can be concluded that the present findings will be helpful to check mass killings of the fish in the highly polluted waters of Hussainsagar Lake and conserve its natural resources.

References

Baker, J.T.P. (1969). Histological and electron microscopical observations on copper poisoning in the winter flounder (*Pseudopleuronectes americanus*). *J. Fish Res. Bd. Can.*, **26**: 2785-2793.

Benedeczky, I., P. Biro and Z.S. Schaff (1984). Effect of 2,4-D containing herbicide (Diconirt) on ultrastructure of carp liver cells. *Acta Biol. Szeged.*, **30**: 107-127.

Bengtsson, B., Bengtsson, A. and Himberg, M. (1985). Fish deformities and pollution in some Swedish waters. *Ambio.*, **14**: 32-35.

Christie, R.M. and H.I. Battle (1963). Histological effects of 3-trifluromethyl-4- nitrophenol (TEM) on Larval Lamprey and trout. *Can. J. Zool.*, **41**: 51-61.

Devi, Manjula P. (1988). Ecotoxicological studies on freshwater fish with special reference to pollution. Ph.D. Thesis, Osmania University, Hyderabad.

Ekberg, D.R. (1958). Respiration in tissues of Gold fish adapted to high and low temperatures. *Biol. Bull. V.*, **114**: 308-316.

Evangelin, Anita N. (1988). Studies on some aspects of physiology and Histology of fish *Channa punctatus* (Bloch) during methyl parathion intoxication. Ph.D. Thesis, Osmania University, Hyderabad.

Ferri, S. and N. Macha (1980). Lysosomal enhancement in hepatic cells of a teleost fish induced by cadmium. *Cell. Biol. Int. Rep.*, **4**: 357-363.

Galat, D.L., G. Post, T.J. Keefe and G.R. Boucks (1985). Histopathological changes in the gill, kidney and Liver of Lohonta cut throat trout, Salmo Clarki Henshawi. *J. Fish. Biol.*, **27**: 533-552.

Hargis, W.J. Jr. and Zwerner, D.E. (1988). Some histological gill lesions of several estuarine finfishes related to exposure to contaminated sediments; A preliminary report. In understanding the Estuary: Advances in Chesapeake Bay Research. Proceedings of a Conference, 29-31 March 1988. Chesapeake Research Consortium. pp. 474-487.

Horvath, I. and A. Stammer (1979). Electron microscopical structure of gill lamellas of the ide, Leuciscus idus, with particular regard to the chloride cells and H_2S pollution. *Acta. Biol. Szedged.*, **25**: 133-142.

Hughes, G.M. (1984). General anatomy of the gills. In: *Fish Physiology*, Hoar, W.S. and Randall, D.J. (eds.) Academic Press, New York, **10(A)**: 1-72.

Hughes, G.M. and M. Morgan (1973). The structure of fish gills in relation to their respiratory function. *Biol. Rev.*, **48**: 419-475.

Jagoe, C.H and T.A. Haines (1983). Alterations in gill epithelial morphology of yearling Sunapee trout exposed to acute acid stress. *Trans. Amer. Fish. Soc.*, **112**: 689-695.

Kumar, Anand (1994). Endosulfan induced biochemical and pathophysiological changes in freshwater fish *Clarias batrachus* (Linn). Ph.D. Thesis, Osmania University, Hyderabad.

Kumari, Anitha S. (1998). Effect of water pollution on LDH isozyme in fish from Hussainsagar Lake, Hyderabad, A.P. Ph.D. Thesis, Osmania University, Hyderabad.

Leino, R.L., P. Wilkinson and J.G. Anderson (1987). Histopathological changes in the gills of Pearl Dale, *Semotilus margarita* and fathead minnows, pimephales promelas from experimentally acidified Canadian Lakes. *Canadian Journal of Fisheries and Aquatic Sciences.*, **44**: 126-134.

Mallatt, J. (1985). Fish gill structural changes induced by toxicants and other irritants: A statistical review. *Canadian Journal of Fisheries and Aquatic Resources*, **42**: 630-648.

Mckim, J.M., A.M. Christensen and E.P. Hunt (1970). Changes in the blood of brook trout, *Salvelinus fontinalis* after short term and long term exposure to copper. *J. Fish. Res. Bd. Can.*, pp. 1883-1889.

Muley, E.V. (1987). Preliminary report on the fish kills in the Lake Hussainsagar. Project sponsored by Municipal Corporation of Hyderabad, p. 42.

Munshi, J.S.D. and A. Singh (1992). Scanning electron microscopic evaluation of effects of low pH on gills of *Channa Punctata* (Bloch). *J. Fish. Biol.*, **41**: 83-89.

Munshi, J.S.D. and R.K. Singh (1971). Investigation of the effect of insecticides and other chemical substances on the respiratory epithelium of predatory and weed fishes. *Indian J. Zool.*, **22**: 127-134.

Ojha, J., N.C. Rooj, A.K. Mittal and J.S. Datta Munshi (1989). Light and scanning electron microscopic studies on the effect of biocidal plant sap on the gills of a hill stream fish, *Garra Lamta* (Ham). *J. Fish. Biol.*, **34**: 165-170.

Ojha, Jagdish, Narayan Chandra Rooj and Birendra Kumar (1993). Effect of biocidal plant sap on the gills of a hill stream Loach, *Noemacheilus rupicola*, Mcclelland. Cobitidae, Cypriniformes. *Indian J. Exp. Biol.*, **31**: 650-652.

Parveen, Asfia (1988). Some histological and biochemical variations associated with pesticide treatment in a fish, *clarias batrachus* (Linnaeus). Ph.D. Thesis, Osmania University, Hyderabad.

Rankin, J.C., R.M. Stagg and L. Bolis (1982). Effects of pollutants on gills. In: *Gills*, Houlihan, D.F. (ed.) New York Cambridge University Press, pp. 207-219.

Reïchenbach-Klinke, H.H. (1972). Histologische and Enzymatische Veranderungen nach Schadstoffeinwirkung beim Fish. *Veroff. Inst. Kust-u-Bennenfisch*, **53**: 113-120.

Rojik, I., J. Nemcsok and L. Boross (1983). Morphological and biochemical studies on liver, kidney and gill of fishes, affected by Pesticides. *Acta Biol.*, **34**: 81-92.

Roy, P.K., J.S.D. Munshi and J.D. Munshi (1986). Scanning electron microscopic evaluation of effects of saponin on gills of climbing perch, *Anabas testudineus* (Bloch) Anabantidae; Pisces. *Indian J. Exp. Biol.*, **24**: .511-516.

Siddiqui, S.Z. and Kaza V. Rama Rao (1991). Limnologic investigations on a recent major fish kill (*Notopterus notopterus*) in Hussainsagar, Hyderabad, India. *Poll. Res.*, **10(4)**: 191-198.

Singotamu, L. (1991). An alternative method for preparation of soft Biological samples for scanning electron microscope (SEM) studies. *Proc. Indian Acad. Parasitol.*, **11(1&2)**: P.7-11.

Skidmore, J.F. and P.W.A. Tovell (1972). Toxic effects of Zinc sulphate on the gills of rainbow trout. *Wat. Res.*, **6**: 217-230.

Soivio, A. and G.M. Hughes (1978). Circulatory changes in secondary lamellae of salmo gairdneri gills in hypoxia and anaesthesia. *Ann. Zool. Fenmci.*, **15**: 221-225.

Srivastava, T. and A.K. Srivastava (1984). Histopathology of the gill of *Channa gachua* exposed to sublethal conc. of malathion and chlordane. *Proc. Sem. Eff. Pest. Aq. Fan.*, p. 37.

Tuurala, H. and A. Soivio (1982). Structural and circulatory changes in the secondary lamellae of *salmo gairdneri* gills after sublethal exposures to dehydroabietic acid and Zinc. *Aquatic Toxica.*, **2**: 12-19.

Vinod, V. Ghanathay (1989). *In vivo* effects of BHC on some aspects of physiology and histopathology of fish; *Channa punctatus* (Bloch). Ph.D. Thesis, Osmania University, Hyderabad.

Virchow, R. (1958). Die Celllular pathologie in ihrer Begrundung auf Physsiologische und Pathologische gewelehlre. Berlin. A. Hirschwald.

Index